Integrative Approaches in Ceramic Petrography

Integrative Approaches in Ceramic Petrography

Edited by

Mary F. Ownby, Isabelle C. Druc, *and* Maria A. Masucci

The University of Utah Press
Salt Lake City

The Defiance House Man colophon is a registered trademark of the University of Utah Press. It is based on a four-foot-tall Ancient Puebloan pictograph (late PIII) near Glen Canyon, Utah.

21 20 19 18 17 1 2 3 4 5

LIBRARY OF CONGRESS CATALOGING-IN-PUBLICATION DATA

Names: Ownby, Mary F., editor.
Druc, Isabelle C. (Isabelle Clara), 1962– editor.
Masucci, Maria A., 1958– editor.
Title: Integrative approaches in ceramic petrography / edited by Mary F. Ownby, Isabelle C. Druc, and Maria A. Masucci.
Description: Salt Lake City : The University of Utah Press, [2016] | Includes bibliographical references and index.
Identifiers: LCCN 2016011038 | ISBN 9781607815068 (cloth : alk. paper) | ISBN 9781607815075 (ebook)
Subjects: Petrology in archaeology—Congresses. | Ceramics—Analysis—Congresses. | Pottery—Congresses.
Classification: LCC CC79.P4 I56 2016 | DDC 930.1
LC record available at https://lccn.loc.gov/2016011038

Printed and bound by Edwards Brothers Malloy, Inc., Ann Arbor, Michigan.

Contents

Figures

Tables

Introduction

Mary F. Ownby, Isabelle C. Druc, *and* Maria A. Masucci

Petrographic analysis of ceramics has contributed greatly over many decades to archaeological research addressing topics on ceramic production, technology, and exchange that relate to larger issues of population movement, social change, identity, and cultural interaction. One of the first pioneers in ceramic petrography as its own discipline was Anna O. Shepard (1936, 1942; see also Bishop and Lange 1991). Others made similar significant contributions (Courtois 1976; Cremonte 1988; Hodges 1962; Peacock 1977; Rigby 1953). The early interest in petrographic analysis in the United States waned somewhat, however, when studies of archaeological ceramics turned more toward chemical approaches. Over the past decade petrography has seen a resurgence in the United States and worldwide, particularly in combination with other techniques. In order to explore the current state of ceramic petrographic research, a symposium titled "Petrography's Continued Role in Ceramic Studies: New Advances and Debates" was held at the 2012 Society for American Archaeology meeting. This symposium was organized by Mary Ownby and Sophia Kelly in order to bring petrographers together to present methodological and theoretical issues in the field. The fourteen papers emphasized several developments in methods, explored innovative perspectives on pottery technology, and highlighted archaeological questions that were being addressed through petrographic research, principally those regarding human interaction with the landscape and interaction among groups, technological traditions, and communities of practice. The papers examined ceramics from many places, including North and South America, Greece, and Portugal. This symposium also facilitated discussion of wider issues in the field and future directions.

Based on the results of the symposium, a volume highlighting trends and research in the field was proposed. The importance of petrography in archaeology can be seen by the frequency with which it is being combined with other methods for ceramic analysis and by the recent publications related to petrographic methodology (for example, Goren 2014; Peterson and Betancourt 2009; Quinn 2013; Reedy 2008) and case studies (for example, Falabella and Sanhueza 2013; Kreiter et al. 2013; Pavia et al. 2013; Renson et al. 2014; Sahlén 2013; Stoltman 2015). Another measure of the renewed attention to petrography in archaeology is seen in the sessions focused on this method at several recent Society of American Archaeology meetings. Within the discipline, there is also a need to consider how the use of new instruments, in particular portable XRF (X-ray fluorescence) spectrometers, which obtain chemical data nondestructively and are increasingly used in the field, compares to petrographic results. Presentation of new case studies highlighting the significance of petrographic data for interpreting chemical compositional groups and the limitations of both techniques, particularly in relation to geological homogeneity or heterogeneity, is important.

Further, the current collected papers build on the studies of Dean Arnold and others (Arnold et al. 1991; Arnold et al. 2000) and James Stoltman (2001), while adding more examples on the positive relationship between petrographic and chemical data. This is significant, as some previous studies have shown that the separate data sets can create divergent interpretations (for example, Blomster et al. 2005; Neff et al. 2006; Sharer et al. 2006), whereas their combination can lead to more robust data for a better understanding of pottery's significance in past societies (see Bishop et al. 1982; Day et al. 1999; Eckert and Schleher 2012; Fowles et al. 2007; Minc and Sherman 2011; Porat et al. 1991; Stoltman and Mainfort 2002; Stoner et al. 2008; Wallis 2011).

Other motivations to create this volume included the use of petrographic data to investigate a broader range of archaeological questions than in the past. While some of these research projects focus on ceramic technology and the series of culturally embedded actions and choices made by potters (*chaine opératoire*), others look at the creation of identities through ceramic manufacture and the role of sociocultural settings for pottery production. These investigations have often benefited from combining petrographic data with data from ethnographic research and experimental studies. New developments in image analysis are also highlighting petrography's role in understanding raw material use.

Many of these archaeological topics have been explored in the past, but few books are devoted solely to petrographic application and method. These topics were previously covered in books on ceramic analysis in general (such as Shepard 1956 and Rice 1987). While these books did much to promote the method and outline its utility in archaeological research, they lacked case studies to illustrate its implementation. Such examples were typically published in journals, site reports, or other topical volumes. More recently, entire dissertations have covered the use of petrography for research topics (for example, Sunahara 2009). As noted, new textbooks on the petrographic analysis of archaeological materials have been published (for example, Quinn 2013; Reedy 2008). An entire volume devoted to papers on petrographic research was published by Patrick Quinn (2009). These did much to promote the field and highlight the variety of research conducted globally.

Our goal was to carry on this tradition through the publication of a volume that would highlight the continuing developments in the field, particularly from the perspectives of American petrographers. Researchers in the Western Hemisphere are often differently trained, in particular in North America, due to being based in anthropology departments and studying pottery that is not wheel-made. In fact, in Arizona a group of researchers developed a unique approach to ceramic analysis through the development of petrofacies. These unique sand composition zones are identified through the collection and petrographic analysis of many hundreds of sand samples. This method was developed for southern Arizona (Heidke and Miksa 2000; Miksa and Heidke 2001) but could be utilized in other parts of the world. Sand thin sections are point-counted to provide quantitative data that are assessed in a discriminant model. The statistical results suggest those sands with similar composition that could be grouped together and those that were different, creating geographic boundaries for sand. Point-counting sand temper in sherds and its inclusion in the discriminant model result in the assignments of pottery to production areas in restricted petrofacies zones. This research highlights a unique development in North America that uses geological diversity and pottery practices, including typically acquiring sand raw materials within 3 km of a site (Arnold 1985:51, Heidke 2011:table 4.10). Several chapters in this volume utilize these Arizona petrofacies models. Thus the unique contributions of American researchers are important to highlight. It is also important to continue the tradition of books focused on petrographic research, which are vital for the visibility and progress of the field.

Hence the volume *Integrative Approaches in Ceramic Petrography* was born. The aim was to present papers that integrated ceramic petrographic results with other types of data, information, or approaches. This will both reveal existing trends and, we hope, inspire new directions for the field of ceramic petrography. The chapters highlight a diversity of archaeological questions and approaches with petrography as the central method but offer individual interpretations and perspectives. This reflects the current state of the field, where many different methods are valid and are based on specific research questions, geology, and ceramic material. The intent is to emphasize a more holistic approach by highlighting various ways to do petrography and its many benefits for archaeological research.

The volume is not presenting "new methods" but rather the use of existing techniques combined with other methods (petrography and chemistry, petrography and experimental analysis, petrography and ethnographic research). There is a certain amount of history for the combination of some of these approaches (Stoltman 2001). The current volume builds upon this previous work by presenting recent case studies based on individual research projects and their perspectives. Each researcher selected the techniques that would provide data to address specific research questions. These case studies do not aim to provide an exhaustive review of previous research but present new work utilizing specific materials and methods.

One interesting trend that is common among the case studies is the use of small sample sizes. This seems to reflect the overall development in archaeology for "slow science": the exploration of research questions initially with qualitative data on a few samples to gauge the potential for additional research. In some instances these small data sets are sufficient to address the archaeological questions; in others they indicate useful directions for further research. This new approach is due in part to limits on funding, which necessitate initiating smaller projects first before a larger sum is spent to ensure that the funds will provide the data needed to examine past behaviors.

The fourteen chapters in this volume present research conducted on pottery from Arizona, Arkansas, Belgium, China, Crete, Ecuador, Greece, Guatemala, Peru, Syria, and Turkey and span a time range from 5000 BC to the nineteenth century. The first five present research combining the use of petrography and instrumental techniques, while the subsequent chapters examine ceramic technology, production, and exchange through integrating a more diverse set of data and analytical perspectives.

Chapter 1 by Anabel Ford, Frank Spera, and Connie Christensen presents the results of petrographic and microprobe analysis of volcanic ash temper in Maya ceramics. The difficulties in identifying chemical and petrographic differences with the goal of sourcing the ash reveal how challenging that can be with certain material types.

The geological diversity in the Tucson Basin of the American Southwest facilitated the combination of chemical data (NAA: neutron activation analysis) and petrography in chapter 2, Mary Ownby's study of Tanque Verde Red-on-brown pottery. With only a few thin sections, some of the chemical groups could be assigned to a provenance and the exchange patterns more clearly discerned.

In a similar methodological approach, Maria Masucci, Hector Neff, Michael Glascock, and Jeff Speakman combine petrographic and NAA analysis to examine the technology of fine paste blackwares of the late prehispanic Manteño-Guancavilca society of coastal Ecuador in chapter 3. The petrographic analysis demonstrated continuity in production technology of finewares over time, while the elemental analysis indicated multiple production locales. This means that some vessels likely derived from outside the local area, suggesting that fineware vessels circulated in the region, as in earlier periods, possibly through feasting and gift exchanges.

In chapter 4 Emily Stovel, Beatriz Cremonte, and Ester Echenique analyze Middle Period ceramics from San Pedro de Atacama, Chile, with portable X-ray florescence spectrometry and petrography. Both techniques indicated that similar raw materials had been utilized in producing fine and coarse wares found in both burial and household contexts, suggesting a lack of specialization by ware or use.

In chapter 5 Yukiko Tonoike combines petrography with X-ray diffraction to examine changes in ceramic raw materials in the Halaf to Ubaid period sherds from Tell Ziyadeh, Syria. The results suggest local sources for the pottery, with only a few changes, such as adding some chaff temper, between the two periods in terms of material use.

Chapter 6 by James Heidke provides an excellent example of integrating petrographic data with the information on the unsampled ceramic corpus to look at production and exchange in the Tucson Basin. Several statistical tests revealed that the potters in certain locations were using specific paste recipes to manufacture different wares that had particular functions.

In chapter 7 Clare Burke, Peter Day, and Daniel Pullen examine Early Helladic pottery from the Nemea Valley, Greece, to investigate patterns of consumption and production. The petrographic results indicated that the area received a wide variety of vessel types from producers located at considerable distances during the Early Helladic I. By Early Helladic II, however, pottery was acquired from fewer manufacturers.

In a cross-cultural study examining the use of loess for pottery production, James Stoltman petrographically compares sherds from China and Belgium in chapter 8. Although loess was widely available in both regions, it was predominantly used for fineware/noncooking pottery, suggesting a technological limitation on its employment for cooking vessels.

In chapter 9 Sophia Kelly studies the clays employed to produce Hohokam Red-on-buff pottery from Phoenix, Arizona. She found that clay pellets in the pastes were indicative of changes and an increase in ceramic production through time that may relate to reorganization of pottery production.

While some of these chapters employ petrography and instrumental methods, other chapters combine petrography with ethnography or experimental data. In chapter 10 Isabelle Druc, Kinya Inokuchi, Victor Carlotto, and Pedro Navarro examine the provenance and type of geological setting responsible for the nonlocal intrusive temper observed in many ceramics from Kuntur Wasi, a Formative site in the north Peruvian Andes. To explain the presence of these wares, they consider different scenarios, involving the movement of pots, itinerant potters as documented ethnographically in the region, and questions of on-site or off-site production.

In chapter 11 Kacy Hollenback uses historical data and petrography to investigate pottery production changes after epidemics in North Dakota. She found that gradual changes did take place but that maintenance of ceramic traditions may have been related to confirming identity and coping with societal loss.

In chapter 12 Jerolyn Morrison and Mara Horowitz utilize petrographic data on cookwares from Mochlos, Crete, and Tell Atchana, Turkey, to guide an experimental study of raw material use. Through the production of their own cookware recipes, they found that the ancient potters were working with local materials in particular ways to produce pottery suitable for use in cooking.

An experiment to examine digital image analysis versus petrographic point-counting was carried out by Patrick Livingood and Ann Cordell (chapter 13) on pottery from the Moon Site in Arkansas. Digital image analysis is recommended in cases where the tempering practices for large amounts of pottery need to be understood in a general way, while petrographic point-counting is better for addressing more specific raw material use. This research builds upon previous studies of image analysis (for example, Middleton et al. 1991; Reedy 2006; Velde and Druc 1999), an important methodological issue in the field.

Finally, chapter 14 by Ian Whitbread offers some highly pertinent comments on the development of ceramic petrography as a field and how this volume contributes to that discipline.

The varied approaches presented here elucidate how ceramic petrography as a discipline in its own right has been advancing and is increasingly being used to address many different archaeological questions. As such, this volume should be of interest not just to petrographers but also to ceramists and archaeologists looking for innovative ways to address research questions through petrography and new research questions that petrography can investigate. Avenues for future developments are likely to be found in the continued integration of petrography with other methods. This volume illustrates the utility of this approach and demonstrates that renewed interest in petrography within archaeology is providing significant contributions to the larger discipline. We hope that the contribution of American petrographers to developments in ceramic petrography is also demonstrated. While this was one of the aims of this volume, our overall goal was to show the impact and importance of ceramic petrography to archaeology and the role that petrographers play not just in providing important data but in developing significant interpretations of their own.

As editors and authors, we feel that the field still faces challenges in utilization, such as the preference for chemistry that provides a complementary approach but not identical results and the ready use of in-field nondestructive instruments. But petrography remains the best approach in areas with coarse-tempered pottery and geological variability and a useful first method for all areas and types of ceramics. Although learning the method takes considerable time, the reward is a close examination of potter behavior that is situated in particular sociocultural and economic environments. Petrography can examine the selection and alteration of raw materials, technological considerations of pottery production, and culturally embedded decisions, information that is obscured by chemical methods that homogenize the material. We are delighted that petrography has seen a resurgence after a decline when chemical methods became more popular. We feel that it offers a unique opportunity to understand past behavior. Today ceramic analysts and archaeologists recognize its valuable contribution to information on provenance and technology and its capacity to address larger questions of social interaction, identity, and cultural continuity.

References Cited

Arnold, Dean E.
1985 *Ceramic Theory and Cultural Process*. Cambridge University Press, Cambridge.

Arnold, Dean E., Hector Neff, and Ronald L. Bishop
1991 Compositional Analysis and "Sources" of Pottery: An Ethnoarchaeological Approach. *American Anthropologist* 93:70–90.

Arnold, Dean E., Hector Neff, and Michael D. Glascock
2000 Testing Assumptions of Neutron Activation Analysis: Communities, Workshops, and Paste Preparation in Yucatán, Mexico. *Archaeometry* 42:301–316.

Bishop, Ronald L., and Frederick W. Lange (eds.)
1991 *The Ceramic Legacy of Anna O. Shepard*. University Press of Colorado, Niwot.

Bishop, Ronald L., Robert L. Rands, and George R. Holley
1982 Ceramic Compositional Analysis in Archaeological Perspective. In *Advances in Archaeological Method and Theory* 5, edited by Michael B. Schiffer, pp. 275–330. Academic Press, New York.

Blomster, Jeffrey P., Hector Neff, and Michael D. Glascock
2005 Olmec Pottery Production and Export in Ancient Mexico Determined through Elemental Analysis. *Science* 307:1068–1072.

Courtois, Liliane
1976 *Examen au microscope pétrographique des céramiques archéologiques*. Notes et Monographies Techniques, no. 8. CRA, CNRS, Paris.

Cremonte, Beatriz
1988 Técnicas alfareras tradicionales en la Puna: Inti-Cancha. *Arqueología Contemporánea* 2:5–29.

Day, Peter M., Evangelia Kiriatzi, Alexandra Tsolakidou, and Vassilis Kilikoglou
1999 Group Therapy in Crete: A Comparison between Analyses by NAA and Thin Section Petrography of Early Minoan Pottery. *Journal of Archaeological Science* 26:1025–1036.

Eckert, Suzanne L., and Kari L. Schleher
2012 Integrating Petrographic and Chemical Compositional Analyses: Santa Fe Black-on-White Ceramic Production and Distribution in the Española Basin. Paper presented at the Society for American Archaeology Annual Meeting, Memphis.

Falabella, Fernanda, and Lorena Sanhueza
2013 Studying Technological Practices at a Local

Level: Neutron Activation and Petrographic Analyses of Early Ceramic Period Pottery in Central Chile. *Archaeometry* 55:33–53.

Fowles, Severin M., Leah Minc, Samuel Duwe, and David V. Hill
2007 Clay, Conflict, and Village Aggregation: Compositional Analyses of Pre-Classic Pottery from Taos, New Mexico. *American Antiquity* 72:125–152.

Goren, Yuval
2014 The Operation of a Portable Petrographic Thin-Section Laboratory for Field Studies. *New York Microscopical Society Newsletter* (September 2014):1–17.

Heidke, James M.
2011 Prehistoric Pottery Containers from the Julian Wash Site, AZ BB:13:17 (ASM). In *Craft Specialization in the Southern Tucson Basin: Archaeological Excavations at the Julian Wash Site, AZ BB:13:17 (ASM), Part 1: Introduction, Excavation Results, and Artifact Investigations*, edited by Henry D. Wallace, pp. 263–294. Anthropological Papers No. 40. Center for Desert Archaeology, Tucson.

Heidke, James M., and Elizabeth J. Miksa
2000 Correspondence and Discriminant Analyses of Sand and Sand Temper Compositions, Tonto Basin, Arizona. *Archaeometry* 42: 273–299.

Hodges, Henry W. M.
1962 Thin Sections of Prehistoric Pottery: An Empirical Study. *Bulletin of the Institute of Archaeology* 3:58–68.

Kreiter, Attila, Szabolcs Czifra, Éva Széles, Mária Tóth, and Orsolya Viktorik
2013 Petrographic, LA-ICP-MS and XRD Analyses of Hallstatt Ceramics from a Scythian Age Settlement in North Hungary. In *Counterpoint: Essays in Archaeology and Heritage Studies in Honour of Professor Kristian Kristiansen*, edited by Sophie Bergerbrant and Serena Sabatini, pp. 477–490. British Archaeological Reports International Series 2508. Archaeopress, Oxford.

Middleton, Andrew P., Morven N. Leese, and Michael R. Cowell
1991 Computer-Assisted Approaches to the Grouping of Ceramic Fabrics. In *Recent Developments in Ceramic Petrology*, edited by Andrew P. Middleton and Ian C. Freestone, pp. 265–267. Occasional Paper No. 81. British Museum, London.

Miksa, Elizabeth J., and James M. Heidke
2001 It All Comes Out in the Wash: Actualistic Petrofacies Modeling of Temper Provenance, Tonto Basin, Arizona, USA. *Geoarchaeology* 16:177–222.

Minc, Leah, and R. Jason Sherman
2011 Assessing Natural Clay Composition in the Valley of Oaxaca as a Basis for Ceramic Provenance Studies. *Archaeometry* 53: 285–328.

Neff, Hector, Jeffrey Blomster, Michael D. Glascock, Ronald L. Bishop, M. James Blackman, Michael D. Coe, George L. Cowgill, Ann Cyphers, Richard A. Diehl, Stephen Houston, Arthur A. Joyce, Carl P. Lipo, and Marcus Winter
2006 Smokescreens in the Provenance Investigation of Early Formative Mesoamerican Ceramics. *Latin American Antiquity* 17:104–118.

Pavia, Jennifer A., Kathleen M. Marsaglia, and Scott M. Fitzpatrick
2013 Petrography and Provenance of Sand Temper within Ceramic Sherds from Carriacou, Southern Grenadines, West Indies. *Geoarchaeology* 28:450–477.

Peacock, David P. S. (ed.)
1977 *Pottery and Early Commerce: Characterization and Trade in Roman and Later Ceramics.* Academic Press, London.

Peterson, Sarah E., and Philip P. Betancourt
2009 *Thin-Section Petrography of Ceramic Materials.* INSTAP Archaeological Excavation Manual 2. INSTAP Academic Press, Philadelphia.

Porat, Naomi, Joseph Yellin, Lisa Heller-Kallai, and Ludwik Halicz
1991 Correlation between Petrography, NAA, and ICP Analyses: Application to Early Bronze Age Egyptian Pottery from Canaan. *Geoarchaeology* 6:133–149.

Quinn, Patrick Sean
2013 *Ceramic Petrography: The Interpretation of Archaeological Pottery and Related Artefacts in Thin Section.* Archaeopress, Oxford.

Quinn, Patrick Sean (ed.)
2009 *Interpreting Silent Artefacts: Petrographic Approaches to Archaeological Ceramics.* Archaeopress, Oxford.

Reedy, Chandra L.
2006 Review of Digital Image Analysis of Petrographic Thin Sections in Conservation Research. *Journal of the American Institute for Conservation* 45(2):127–146.

2008 *Thin-Section Petrography of Stone and Ceramic Cultural Materials*. Archetype Publications, London.

Renson, Virginie, David Ben-Shlomo, Jan Coenaerts, Katia Charbit-Nataf, Melissa Samaes, Nadine Mattielli, Karin Nys, and Philippe Claeys

2014 Coupling Lead Isotope Analysis and Petrography to Characterize Fabrics of Storage and Trade Containers from Hala Sultan Tekke (Cyprus). *Archaeometry* 56:261–278.

Rice, Prudence M.

1987 *Pottery Analysis: A Sourcebook*. University of Chicago Press, Chicago.

Rigby, George R.

1953 *The Thin-Section Mineralogy of Ceramic Materials*. 2nd ed. British Ceramic Research Association, Stoke-on-Trent, England.

Sahlén, Daniel

2013 Selected with Care?—The Technology of Crucibles in Late Prehistoric Scotland: A Petrographic and Chemical Assessment. *Journal of Archaeological Science* 40:4207–4221.

Sharer, Robert J., Andrew K. Balkansky, James H. Burton, Gary M. Feinman, Kent V. Flannery, David C. Grove, Joyce Marcus, Robert G. Moyle, T. Douglas Price, Elsa M. Redmond, Robert G. Reynolds, Prudence M. Rice, Charles S. Spencer, James B. Stoltman, and Jason Yaeger

2006 On the Logic of Archaeological Inference: Early Formative Pottery and the Evolution of Mesoamerican Societies. *Latin American Antiquity* 17:90–103.

Shepard, Anna O.

1936 The Technology of Pecos Pottery. In *Pottery of Pecos*, vol. 2, edited by Alfred V. Kidder and Anna O. Shepard, pp. 389–587. Papers of the Phillips Academy Southwestern Expedition 7. Yale University Press, New Haven.

1942 Rio Grande Glaze Paint Ware: A Study Illustrating the Place of Ceramic Technological Analyses in Archaeological Research. In *Contributions to American Anthropology and History No. 39*, pp. 129–262. Carnegie Institution, Washington, D.C.

1956 *Ceramics for the Archaeologist*. Volume 609, Carnegie Institution, Washington, D.C.

Stoltman, James

2001 The Role of Petrography in the Study of Archaeological Ceramics. In *Earth Sciences and Archaeology*, edited by Paul Goldberg, Vance T. Holliday, and C. Reid Ferring, pp. 297–326. Kluwer Academic/Plenum Publishers, New York/Boston.

2015 *Ceramic Petrography and Hopewell Interaction*. University of Alabama Press, Tuscaloosa.

Stoltman, James, and Robert C. Mainfort

2002 Minerals and Elements: Using Petrography to Reconsider the Findings of Neutron Activation in the Compositional Analysis of Ceramics from Pinson Mounds, Tennessee. *Midcontinental Journal of Archaeology* 27:1–33.

Stoner, Wesley, Christopher A. Pool, Hector Neff, and Michael D. Glascock

2008 Exchange of Coarse Orange Pottery in the Middle Classic Tuxtla Mountains, Southern Veracruz, Mexico. *Journal of Archaeological Science* 35:1412–1426.

Sunahara, Kay Sachiko

2009 *Ancient Maya Ceramic Economy in the Belize River Valley Region: Petrographic Analyses*. British Archaeological Reports International Series 2018. Archaeopress, Oxford.

Velde, Bruce, and Isabelle Druc

1999 *Archaeological Ceramic Materials: Origin and Utilization*. Springer, New York.

Wallis, Neill J.

2011 *The Swift Creek Gift: Vessel Exchange on the Atlantic Coast*. University of Alabama Press, Tuscaloosa.

CHAPTER 1

Petrography First

Issues of Identification and Sourcing Volcanic Ash Temper in Maya Pottery

Anabel Ford, Frank J. Spera, *and* Connie Christensen

Archaeological study of ceramics in the Maya area (fig. 1.1) has traditionally focused on surface characteristics and general aspects of shape that provide the basis for chronological assessments. The assumptions have been that style and manufacture are local and change can be traced over time, not unlike the back fenders of Cadillacs over the decades of the twentieth century. Chronological dimensions are indeed critical to understanding social changes, yet knowledge of the ceramics remains undeveloped. An appreciation of pottery manufacture and distribution provides a window into social interaction with the environment and the use of resources. Without analyses of the paste characteristics, the links to the landscape are obscure. Chronology is fundamental to research in any area, yet the geochemical characterization of ceramics is connected to the local geology and changes in the makeup of pottery manufacture can reflect changes in the social and physical environment. To understand the basis of ceramic manufacture, a more detailed examination of both the composition and form of the geomaterials is required. Before any more detailed analyses of ceramics, we call for petrography first. On this petrographic foundation, detailed geochemistry of the paste with such refined techniques as the electron microprobe, laser ablation mass spectrometry, and instrumental neutron activation analysis (INAA) takes on greater meaning.

Our investigation is focused on the volcanic ash tempered ceramics of the Late Classic (AD 600–900) Maya area, first identified by Shepard (1937). Over the past eight decades there have been debates as to the nature of the volcanic ash in the Maya area, specifically its presence, origin, and significance. While interest in the ash has waxed and waned, it has predominantly been used as a chronological marker. The descriptions of the presence and proportions of ash-tempered ceramics have not been systematic (compare Gifford 1976; Jones 1986; Sunahara 2003). Consequently any efforts to synthesize these data on volcanic ash and understand the variability in its use and composition have been frustrated by the lack of consistent data and reporting.

In this chapter we look at a collection of samples from the El Pilar area for which we have developed a standard for describing the ceramic sherds that accounts for the general macroscopic characteristics as well as the petrographic qualities that unite them. Our petrographic analyses reveal great similarities among the samples, in accord with other reports from the region (Jones 1986; Sunahara 2003). While the source of the volcanic ash had been conjectured on the basis of associated minerals and the variable inclusion of calcite, these are not uniquely distinguishing characteristics. We discuss the importance of petrography as the first step in the descriptions of the Late Classic Maya volcanic ash temper

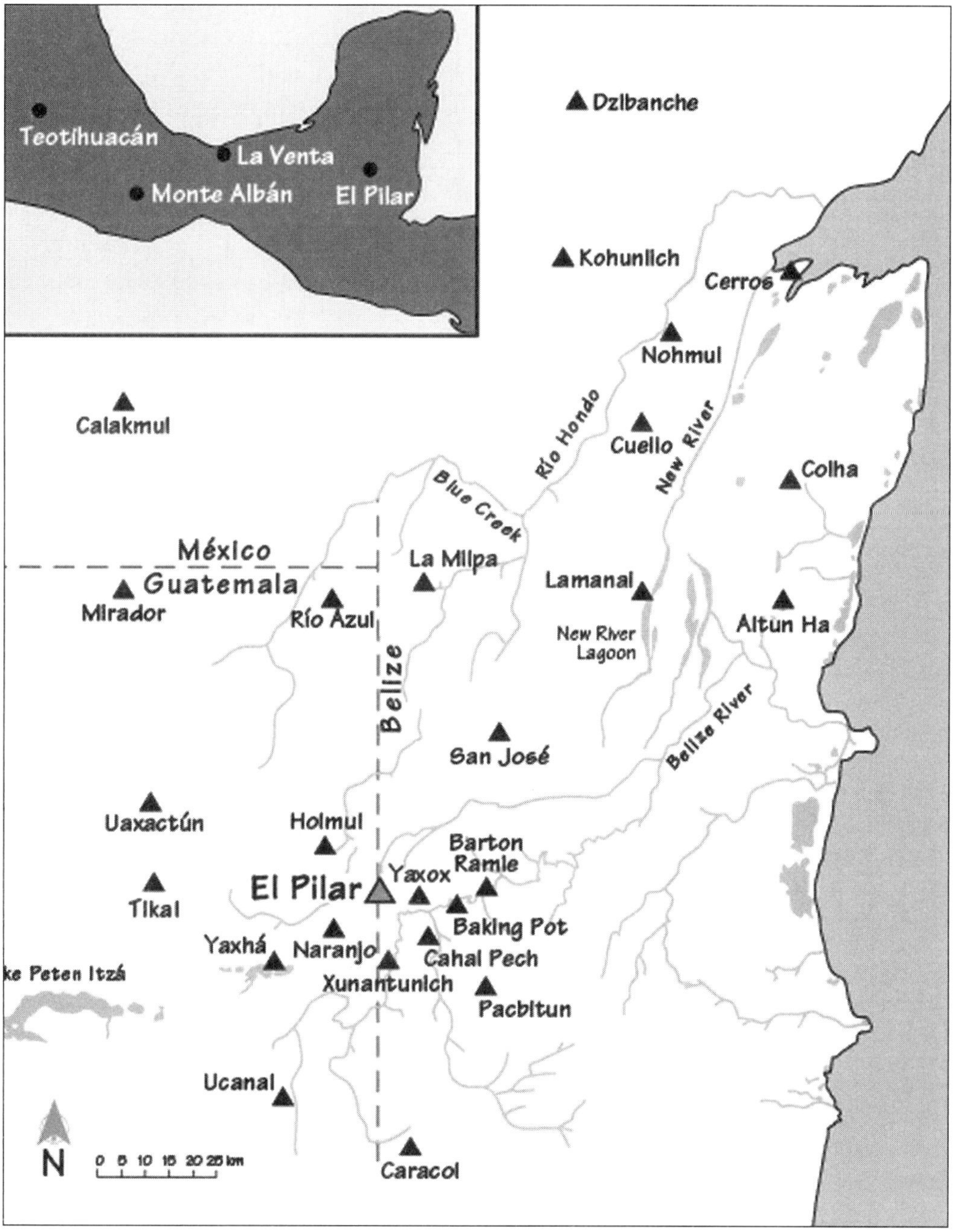

FIGURE 1.1. Maya area of southern Mesoamerica including Belize, Guatemala, and Mexico, with El Pilar and other Maya sites indicated.

ceramics. The abundance of ash revealed in the petrographic analyses provides evidence that the ash is a local resource and the glass shard size distribution is consistent with an airfall character. Yet mineralogy is insufficient to isolate a specific volcanic event.

The origin of the volcanic materials needs to be assessed with geochemical analyses to narrow down potential volcanic events. Hence we performed electron microprobe analyses in order to determine the major element composition of the glass shards that make up the bulk of the temper. Our evaluation of the major elements from the microprobe analyses is the basis for isolating potential sources by comparison with glass from known volcanoes in the region (Simkin and Siebert 2002). This exercise revealed new ambiguities. We draw the conclusion that the ash is from a rhyolitic eruption based on the uniformly high silica (SiO_2) content of the glass incorporated as temper. The high-silica rhyolitic composition of the glass shards, however, is not in itself diagnostic of a specific volcanic source. In addition, the possibility that volcanic glass has reacted with the clay matrix implies that in order to identify a unique source we must understand the nature of the chemical exchange that takes place during reactions of pristine glass with clay matrix at firing temperatures. This consideration calls for an experimental study of the impacts of the firing environment on glass shard composition, a subject of future research. This chapter presents the details of the analyses, beginning with a petrographic study of volcanic ash samples of Late Classic Maya pottery and then moving into our more detailed geochemical examinations.

Background: Archaeological and Geological Contexts

Maya ceramics have been well described in the literature, beginning with the Carnegie studies that involved Shepard (1939), through the extensive Barton Ramie project (Willey et al. 1965; Gifford 1976), and up to the active projects in the region today (for example, Jones 1986; LeCount 1996; Pagliaro 2011; Sunahara 2003). Ceramic paste composition as analyzed petrographically from the earliest periods of the Preclassic (ca. 1000–800 BC through AD 250) are dominated by local calcite (limestone) tempering, as are those of the Early Classic (ca. AD 250–600). The Maya mastered the unstable additive and produced finely made ceramics as well as the necessary common cooking and storage vessels. In the Late Classic (AD 600–900), however, there is a distinction of ceramic pastes, with the abrupt presence of serving bowls, plates, and water jars as well as vases tempered with volcanic ash, leaving the cooking and storage vessels consistently tempered with the local calcite. What was the motivation driving this rather major pottery production change in the Maya area?

It is important to know the geological context of the local and regional setting of Maya ceramic production. Clearly, basic materials of ceramic manufacture are voluminous and would be procured in close proximity to the production area (Arnold 1985:32–56; Shepard 1956:165). In fact, Arnold (1985:57) states that the threshold for procurement of clay is 7 km and 6–9 km for temper. Furthermore, common ceramic vessels designed for daily household use would require the most reliable sources for production, concentrating on readily available local materials. Trade items typically travel long distances. Ceramics of this nature are recognized as unique in form, paste, and decoration and are restricted in their distribution, generally to the elite cultural fraction of the population (Brumfiel and Earle 1987; LeCount 1996; Rice 1987). Therefore, when we consider the production of ceramics in the Maya area, it is important to know what raw materials are locally available and what ceramic wares are unique.

Local and Regional Geological Landscape

The Maya lowlands, the area in which the volcanic ash tempered ceramics are found, are characterized by low-lying karst topography formed by the dissolution of limestone bedrock, sometimes with dolomite and often interbedded with other evaporites, including gypsum. Elevation in this karst terrain fluctuates between approximately 100 and 300 m above sea level (Deevey et al. 1979; Rosenmeier et al. 2002). Groundwater is well below the surface, as is common in karst

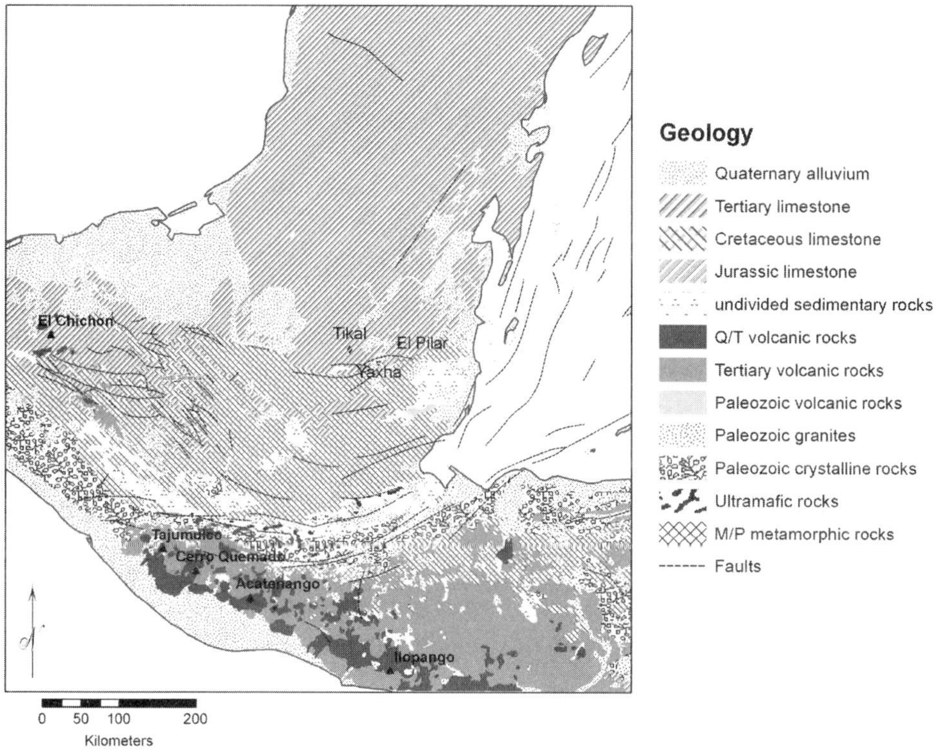

FIGURE 1.2. Geology of the Maya region.

regions the world over. The water table is locally perched to form the numerous lakes and seasonally inundated depressions (Rosenmeier et al. 2002) where locally important impermeable clay rich sediments are concentrated, generally as a result of calcite dissolution. Mollisol is the common soil type on the well-drained ridges that developed as a thin layer over friable, calcium carbonate-rich material (Brenner et al. 2002). Montmorillonite is the common clay material found in the depressions throughout the lowlands, while kaolinite is characteristic of the narrow and restricted Belize River valley alluvium (our XRF analyses in 2010).

The limestone forming the karst topography of the Maya area (fig. 1.2) is Cretaceous to Tertiary in age and overlies older limestone and crystalline basement rock (Alvarado et al. 2007: 351–352; Vinson 1962:432). The two large limestone units in the area are separated by faults that essentially bisect the southern lowlands. The younger Tertiary strata are located to the north of the lakes and the older Cretaceous strata are to the south (Instituto Geográfico Nacional 1970). The Maya mountains, rising 1,100 m in the south east of the Maya area, date to the Paleozoic, more than 250 million years ago.

The local rock types found in the lowland limestone strata and the Maya mountains provide a variable source of ceramic production materials for the Maya potters. The locally available strata of the lowlands are composed of marls and limestone with detrital and weathered volcaniclastic sediments associated with marine, nearshore, and terrestrial environments of the late Cretaceous (ca. 65 Ma: Alvarado et al. 2007:374–376; Vinson 1962:442–454). This thick Cretaceous unit is composed of a limestone representing a reef environment with interlayers of silt, bands of flint/chert, calcareous conglomerates, and dolomitic and marly limestone that formed over 140 million years ago (Alvarado et al. 2007:366–368; Vinson 1962:432–442). The Maya Mountains are composed of Paleozoic sedimentary strata intruded by Paleozoic (ca. 405–420 Ma) granodiorite to granite plutons (between 65–78 weight percent SiO_2) and associated metavolcanic and metasedimentary rocks. Within the Maya Mountains, the Bladen Formation is composed of volcaniclastic material dated at ca. 406 Ma (Martens et al. 2010:815). These are the local geological materials that would have been available to the ancient Maya.

More than 400 km to the south-southwest of the Maya lowlands lies a mountainous and volcanic region known as the Central American Volcanic Belt. This belt runs from southern Mexico through Costa Rica. The majority of the active volcanoes in this region are a product of the subduction of the oceanic Cocos Plate beneath the continental to transitional crust of the North American and Caribbean plates. The area contains a few large stratovolcanoes and silicic calderas along with many basaltic volcanoes. West of the Maya lowlands is El Chichón (Luhr et al. 1984:70; Tilling 2009:4), a small andesitic tuff cone and lava dome complex associated with the Chiapanecan Volcanic Belt.

Volcanoes known to be active around the Late Classic Period that we use as comparison include El Chichón, Ilopango, Acatenago, Fuego, Pacaya, and San Miguel (Simkin and Siebert 2002). The most promising candidates for the volcanoes that might have dispersed air fall ash (glass shards plus phenocrysts) into the lowlands are El Chichón and Ilopango. The andesitic El Chichón, like the majority of active Central American volcanoes (Ford and Rose 1995:156), is characterized by plagioclase and hornblende phenocrysts with lesser amounts of augite, biotite, and other minor accessory minerals (Espíndola et al. 2000:97; Luhr et al. 1984:69; Varekamp et al. 1984). Significantly, the composition of the devitrified groundmass and mesostatic glass present in the proximate El Chichón eruptive products is not a high silica rhyolitic glass. In contrast, Ilopango eruptive products are considerably more silicic in bulk composition and include some silicic mesostatic glass. The phenocrysts assemblage consists mainly of plagioclase and hornblende with minor amounts of hypersthene (Hart 1981:49–57).

In sum, the Maya lowlands are made up of Mesozoic marine limestones consistently lacking in fresh volcanic ash materials. Records of volcanic ash associated with the local Maya Mountains are in fact Paleozoic well-lithified metavolcanics and not considered a source of ash for temper material. No young fresh ash deposits are known in the lowlands today, despite rather intensive efforts to locate such sources (Glicken and Ford 1983; Siebe et al. 2007). The major active volcanoes of the region with the potential to produce volcanic airfall ash are all no less than 400 km away from the study area, and evidence of their impact in the lowlands is scant. The long list of active volcanoes in the region and the potential of some volcanoes to have been active in the Late Classic from AD 600 to 900 provide a guide to the distant origin of volcanic ash.

Characterization of Volcanic Ash in the Maya Area

Research on Maya ceramics has attempted to characterize the volcanic ash tempering and account for its sudden dominance in the archaeological collections dating to the Late Classic Period. Calcite tempering, commonly found in the local Maya lowland ceramics, is consistently used throughout prehistory from the earliest records of ceramics (ca. 800–1000 BC) through the conquest and into the present (Reina and Hill 1978). It is clear that the Maya were skilled masters of the use of calcite, an unstable tempering agent that undergoes decarbonation to lime and carbon dioxide gas (CaO and CO_2, respectively) at about 800°C. They produced a variety of Preclassic utilitarian vessels as well as finewares tempered with calcite. Given the accomplished manufacturing strategy of the ancient Maya, and the careful ability to restrict firing temperature to below the breakdown limit of calcite (Ford and Lucero 2000), the presence of volcanic ash is all the more mysterious. The Maya clearly had a proven ability to manage the unstable limestone as a temper. The use of volcanic ash evidently was not a necessity; it must have represented an opportunity.

Shepard's early recognition of diagnostic pristine volcanic glass shards was unexpected, given the Cretaceous/Eocene carbonates of the Maya lowlands. Consequently, volcanic ash has been interpreted as an exotic tempering agent (Jones 1986; Shepard 1937, 1951, 1956, 1962; Sunahara 2003). As noted by Shepard and confirmed by all subsequent research (Ford and Glicken 1987; Ford and Rose 1995; Ford and Spera 2007; Jones 1986; Sunahara 2003), ceramics tempered with volcanic ash occur in association with the Late Classic Period (AD 600–900). The consistency of the pastes points to a reliable and steady procurement of volcanic ash over the long period of use (Shepard 1939, 1942; Smith 1955). Yet even with thorough work and exhaustive study, Shepard (1942:24) was unable to account for the large quantities of ash-tempered pottery, presuming that a source existed in the lowlands.

Best characterized petrographically (as recommended by Jones 1986:38), the volcanic ash temper is notably homogeneous and dominated by glass shards making up from 35 to 50 percent of the vessel (Sunahara 2003:94, 98, 126). Paste colors are equally uniform light values, with the majority falling into shades of pale yellow-browns based on the Munsell Soil Color Chart (compare Sunahara 2003:93). The ash is associated with minerals typically found as phenocrysts in volcanic rocks of intermediate to silicic composition, such as biotite, plagioclase feldspar, and hornblende (Jones 1986:42, 45), together with minor amounts of detrital quartz and calcite (Sunahara 2003:94, 97). Based on the petrography, Kay Sachiko Sunahara (2003: 94–100) separates the types of volcanic ash tempering by the relative abundance of calcites, from trace amounts of micritic calcite (made up of particles ranging in diameter up to 4 μm) formed by the recrystallization (solution/reprecipitation) of lime mud to 3–10 percent of primary crystalline calcite.

While there is agreement on the general quality and character of the volcanic ash tempered ceramics from the Late Classic Period, there is little agreement on the origin of the ash. Lea Jones (1986:44) favors the importation of ash, though she recognizes the difficulty of accepting this hypothesis. Sunahara (2003:133) suggests importation of finished vessels without addressing implications of the wide distribution of the ash-tempered products in all contexts from ritual ceremonial to households, large and small.

Volcanologists working with Anabel Ford (Ford and Glicken 1987; Ford and Rose 1995; Ford and Spera 2007) see the ash as fresh and unworked by any mechanical degradation process and hence as evidence of an airfall context. The preservation of delicate tricuspate glass shards representing fragments of vesicle walls is classic evidence for explosive, volatile-driven eruptions and subsequent airfall deposition. Yet no significant deposits accumulated.

The distribution and abundance of the volcanic ash in the Maya ceramics and the uniform nature of its size and petrographic composition point to a local source, as suggested by Shepard (1939) and Arnold (1985:57–58). Ford and her colleagues agree but recognize the mystery of the source volcano. We know that volcanic eruptions are capable of distributing volcanic ash in great amounts and over long distances. The eruption of Mount St. Helens in 1980 is an example. Millimeters of ash were distributed more than 600 km away and at an estimated surface density of 159 kg per acre (Arnold 1985:59). The ash temper of the Maya ceramics is consistent with such wind-borne deposits brought into distant regions and available in the Late Classic Period. Regrettably, such deposits are not long lived and are subject to weathering and erosion over the centuries. This may account for our difficulty in finding the local deposits today.

Petrography is a critical step in clearly identifying the presence of volcanic ash. We cannot use color or texture to confirm the presence of ash (Jones 1986:38). Nor can the use of hydrochloric acid securely isolate the ceramics tempered with volcanics, as calcite can be included in the pastes of the vessels and the archaeological context is embedded in carbonates. An experienced geologist can use a hand lens to confirm the presence of volcanic ash (Harry Glicken, personal communication, 1983) but not its intricate descriptions. A petrographic microscope is the only secure means of identifying characteristics of the ash and is used in the lab to identify the mineralogy of the ash, the size of the glass, amount of inclusions, and specific characteristics of the glass shards. Thus the most accurate means to identify discrete qualities of volcanic ash is petrographically. In this way the quantity and qualities of the ash can be described and mineral associations enumerated. Documenting the pristine shapes of the ash fragments, cuspate and thin, and the associated minerals (typical phenocrysts in volcanic rocks) provided the basis for the inquiry.

The descriptions of the petrographic samples are unanimous among the several analysts. Given the abundance of volcanic ash tempered pottery of the Late Classic Maya, making up from 27 to 43 percent of the collections (Sunahara 2003:126), coupled with its presence in common bowls, plates, and narrow-orifice jars as well as specialized vases and wide social distribution in ceremonial and residential contexts, an immediate and locally accessible source is the best fit (see Arnold 1985:32–60). Proximity of the main components of ceramics, the clay and the temper, is paramount. Importation of the tempering materials, as Jones (1986) suggests, would be prohibitive (compare Ford and Glicken 1987:492), with closest known distances exceeding 200 km. Importation of the vessels, as Sunahara (2003) proposes, is even more unlikely given breakability and the ubiquitous distribution. None of these suggestions considers the simple explanation of airfall into the local area.

The logical source is an active volcano, and petrographic analyses provide essential data on geological description and composition, size fraction, associated minerals, and mixing. With petrography first, the stage is set to investigate the more specific geochemical options.

Volcanic Ash in the El Pilar Area

The analytic focus of this study is on Late Classic Maya ceramic samples from the El Pilar collections. More than 10,000 Late Classic Maya ceramic sherd samples are archived in the Meso-American Research Center Laboratory at the University of California–Santa Barbara (UCSB), along with all related data analyses, including those from the current study.

We followed basic steps in the assessment of the volcanic ash in ceramic sherds. To isolate the potential volcanic ash tempered samples, all the ceramic collections from the El Pilar area, including test excavations and full-scale excavations from all contexts, were sorted by vessel shape and assigned chronological periods.

The diagnostic vessels were subjected to a detailed visual analysis that classified the collections by period, coding for Munsell soil color and reaction to hydrochloric acid and metric measurements of thickness and diameter, as well as sorting for a series of paste characteristics. Once the macroscopic analyses were complete, the collections were organized by shape and paste within chronological periods. The Late Classic Period fine-tempered ceramics with light Munsell values tend to be ash tempered. Our volcanic ash-tempered samples were drawn from these organized collections for petrographic analysis.

We extracted a purposeful sample clearly representing the macroscopic qualities of ash-tempered ceramics of the Late Classic Period. The selected samples were derived from middens and stratigraphic construction fill of residential contexts.

Macroscopic Analyses

We selected a total of 18 samples for our petrographic study: 4 samples by shape from midden collections of four residential units and 14 samples from a stratigraphic building sequence at one residential unit (table 1.1). Each sample was subjected to 10× objective macroscopic analyses to verify the presence of volcanic ash. The final selection was subjected to detailed petrographic analyses to determine the specific characteristics of the glass shards only visible with microscopy.

In line with previous research, we sought to gather comparative data. Munsell colors, all of light hue and varying in the brown and yellow shades, were comparable to what Sunahara (2003) had noted. Reaction to hydrochloric acid was negative to slight and not necessarily associated with the calcite inclusion. Other features such as the presence of slip or painting were recorded.

TABLE 1.1. Results of macroscopic analyses of El Pilar volcanic ash samples

Sample	Site	Phase	Strata	Form	Slip	Munsell	HCL	Provenience
1902	281-036		0	bowl	monochrome		1	surface
1478	278-082		2	plate	monochrome		1	midden test
4513	272-005		2	plate	monochrome		1	const. test
9055B	282-010		2	jar	monochrome	10YR52	2	const. test
16579CE	272-145	1	413	plate	polychrome	10YR66	1	cobble fill
16579CF	272-145	1	413	bowl	monochrome	10YR64	2	cobble fill
16578F	272-145	1	411	bowl	polychrome	10YR64	1	plaster fill
16270B	272-145	3	306	bowl	polychrome	10YR64	1	collapse
16386T	272-145	3	315	bowl	monochrome	10YR66	1	collapse
16478	272-145	3	315	bowl	monochrome	10YR66	1	collapse
16547B	272-145	3	390	plate	ash	10YR66	1	cobble fill
16548P	272-145	3	395	plate	polychrome	10YR54	2	cobble fill
16580C	272-145	5	414	vase base	polychrome	10YR66	1	cobble fill
16589E	272-145	5	423	bowl	ash	75YR66	1	cobble fill
16552B	272-145	6	396	jar	monochrome	10YR66	2	cobble fill
16556A	272-145	6	401	bowl base	monochrome	10YR53	1	cobble fill
16535J	272-145	7	314	plate	monochrome	10YR56	1	cobble fill
16535I	272-145	7	314	bowl	monochrome	10YR64	1	cobble fill

All 18 samples showed evidence of red slip, of which 5 showed evidence of polychrome (see table 1.1). We included samples from the basic vessel shapes recognized for Late Classic volcanic ash tempering: bowls (9), jars (2), plates (6), and vases (1). The 18 volcanic ash ceramic samples from the El Pilar area were professionally prepared for petrography as standard uncovered thin sections. These thin sections were analyzed and described using a petrographic microscope with special attention to the volcanic ash characteristics, associated minerals, matrix characteristics, and the presence of calcite.

Petrographic Analyses

The analyses of the volcanic ash samples used a standard petrographic microscope with 10× and 40× magnification using reflected and cross-polarized light. Our study focused exclusively on the glass, which shows remarkable similarity among the 18 volcanic ash ceramic samples (table 1.2), consistent with the studies of Shepard (1937, 1939, 1942, 1962) as well as Jones (1986:45) and Sunahara (2003:96, 98). The glass shards, which range in size from 50 to 375 microns with the majority ca. 150 microns, were cuspate and easy to identify (fig. 1.3).

As with Sunahara's estimates, we found the ash made up a major part of the paste, from 30 to 50 percent. Biotite, plagioclase, and quartz were also noted, with quartz being detrital and not related to the ash (with rounded and sometimes frosted grains), whereas the feldspar and biotite phenocrysts were fresh and associated with the ash, based on their composition and forms. There was no evidence of hornblende phenocrysts (table 1.2). Uncommon euhedral zircon grains were also noted. These results verify that volcanic ash is the tempering material and that the composition, size, sorting, and shape are consistent with airfall distribution; the ash was not quarried or transported by humans or water.

This means that the ash literally fell onto the lands of the producers and the material accumulated during the Late Classic period. Yet the exact volcanic source or sources cannot be distinguished with petrographic analyses. We propose that petrography is the first step to review the geochemical options available to isolate potential sources. In this study we use the analyses of petrographic thin sections as the foundation for the electron microprobe analyses that can target the ash shards and provide major elemental data on their composition. Volcanic

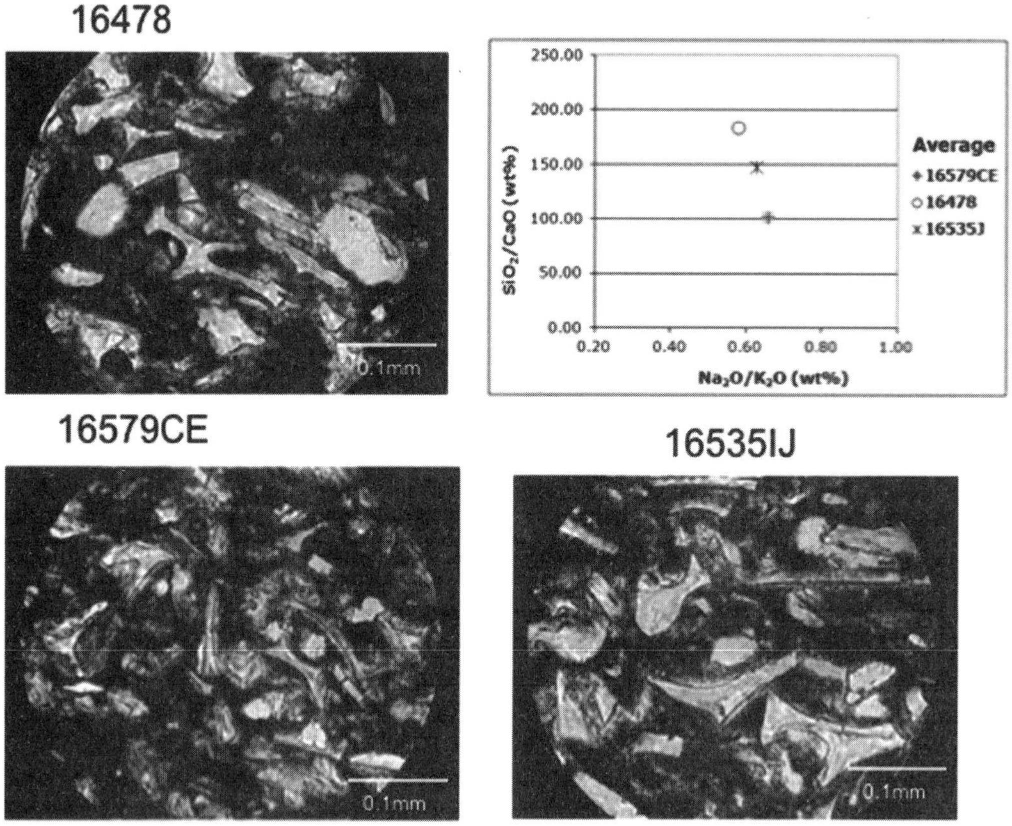

FIGURE 1.3. Photomicrographs and geochemistry of the El Pilar area samples. Note the cuspate shape of the ash shards as well as their density and size. While the petrographic view shows commonalities, the geochemistry is variable. Magnification is 40×.

eruptions are distinguished by their elemental compositions, so we have an opportunity to look for matches in the major elemental suite. This can only be accomplished after petrographic examinations.

Microprobe Analyses

Based on the petrographic identification of the volcanic ash shards in the Maya pottery sherds, we conducted microprobe analyses focused on 15 individual glass shards that were petrographically visible in every ceramic sherd thin-section sample (see fig. 1.3). This data set includes only glass analyses for each of the approximately 15 shards of the ceramic sherd samples. The thin sections were analyzed on the electron microprobe at the University of California–Santa Barbara for major element concentrations. A beam current of 15 nanoamperes and 15kV of accelerating voltage were used for analyses. Analyses were conducted with a 30-second count time consisting of 15 lots of 2-second intervals. In order to mitigate the occurrence of Na loss during the analyses, the program CalcZAF™ was utilized. This program uses an algorithm to determine the initial Na content of the glass. While ten elements were measured (Si, Al, Fe, Na, K, Ca, Mg, Mn, Cr, and Ti), we focused on four major oxides (SiO_2, CaO, Na_2O, K_2O) and their ratios, a standard used by volcanologists to identify potential volcanic sources (table 1.3). Calcium (CaO) is an omnipresent component of the Maya lowlands and has the potential to influence the ceramic composition. Silica (SiO_2) is the most significant attribute of the glass and with soda (Na_2O) and potash (K_2O) is used to classify basic volcanic

TABLE 1.2. Results of petrographic analyses of El Pilar volcanic ash samples

Sample	% Ash	Matrix	Associated minerals: Quartz	Plagioclase	Biotite	Carbonate	Ash grain size microns
1902	50	yellow red-brown		angular twinned plagioclase grains		small rounded carbonate fragments	120–220
1478	40	yellow red-brown	semirounded quartz grains	small twinned plagioclase		small round-ed carbonate fragments	~150
4513	50	yellow red-brown	rounded quartz grains			small round-ed carbonate fragments	130–170
9055B	45	gray-brown	subangular quartz grains			no	70–110
16579CE	50	red-brown	angular quartz	twinned plagioclase grains		yes	~100
16579CF	40	red-brown	angular quartz	twinned plagioclase grains		no	125–250
16578F	50	gray-brown	angular quartz	twinned plagioclase grains	a few biotite grains	rounded carbonate inclusions	250–375
16270B	30	red-brown	rounded quartz grains (averaging 50 mm)	twinned plagioclase grains		carbonate grains	75–125
16386T	50	red-brown	rounded quartz grains (averaging 50 mm)	twinned plagioclase grains (averaging 50 mm)		carbonate grains	75–125
16478	50	red-brown	angular quartz grains (averaging 100 mm)		biotite (200 mm)	no	175–250
16547B	50	red-brown	angular quartz	twinned plagioclase grains		rounded carbonate inclusions	125–250
16548P	40	red-brown	angular quartz (averaging 60 mm)	twinned plagioclase grains		rounded carbonate inclusions	~150
16580C	30	red-brown	angular quartz grains (averaging 50 mm)			rounded carbonate inclusions	50–100
16589E	50	red-brown	angular quartz (200–50 mm)	twinned plagioclase grains (about 100 mm)		rounded carbonate in-clusions mostly associated with the outer slip	125–250
16552B	30	red-brown	angular quartz (100 mm)	twinned plagioclase grains (200 mm)	abundant biotite (200 mm length)	rounded carbonate clasts	100–125
16556A	40	red-brown	angular quartz grains (200 mm)			yes	175–250
16535J	50	red-brown	angular quartz	twinned plagioclase grains	few biotite (120 mm)	yes (1 grain)	~250
16535I	40	red-brown	angular quartz grains		biotite	no	125–250

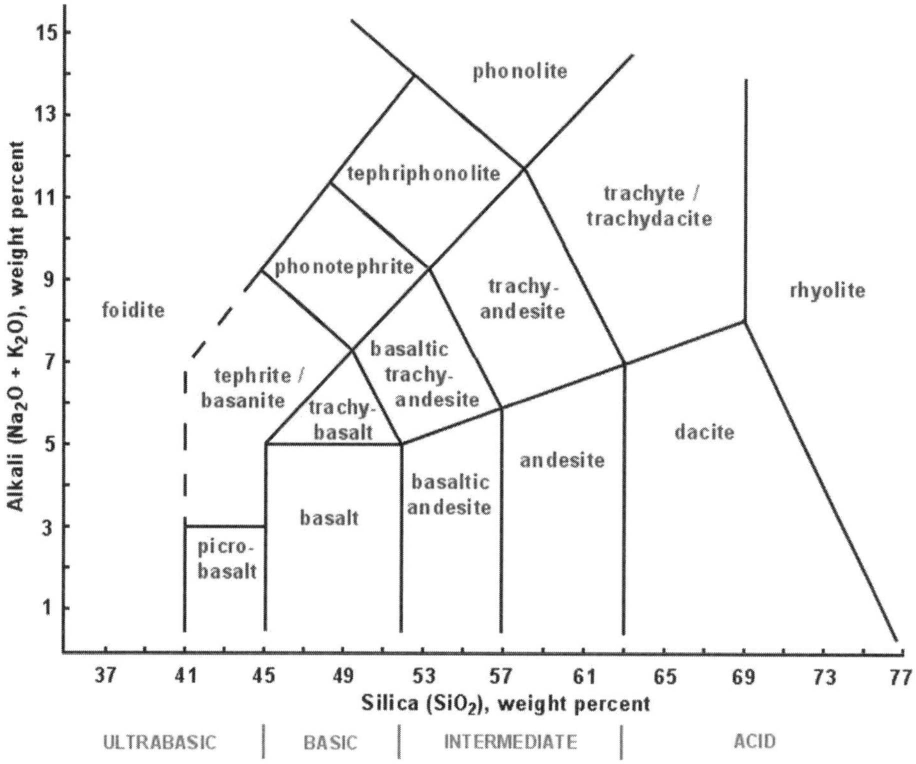

FIGURE 1.4. Alkali composition for different volcanic rock types (after Le Maitre 2002).

rock types (fig. 1.4). Together these four major elements, the most abundant oxides, and their ratios were used to distinguish the potential volcanic sources.

Working with the ratios of silica and lime (SiO_2/CaO) to soda and potash (Na_2O/K_2O), typically a means of determining composition and number of volcanic sources involved, we found considerable variation (fig. 1.5a and 1.5b). The variation at face value suggested multiple volcanoes, more than expected for the time frame. The considerable intrasample homogeneity (fig. 1.5a) even with great intersample heterogeneity (fig. 1.5b) is noteworthy. These collections all represent Late Classic products and as such most likely represent comparable use of raw materials. Yet the intersample heterogeneity is significant, suggesting the possibility of several sources or events over the course of the 300-year period.

Interestingly, while the ranges for some oxides varied widely, the SiO_2 averages were focused tightly around 76 weight percent SiO_2 with little

TABLE 1.3. Microprobe results for the El Pilar samples

Sample	Mean SiO_2	Mean SiO_2/CaO	Mean Na_2O/K_2O
1902	79.2	170	0.68 ± 0.07
1478	78.7	83.9 ± 9.7	0.70 ± 0.06
4513	79.3	156 ± 30	1.01 ± 11.00
9055B	78.5	73 ± 7.1	0.28 ± 0.02
16579CE	79.3	97.7 ± 24	0.66 ± 0.06
16579CF	78.3	125.2 ± 31	0.60 ± 0.05
16578F	78.1	124.77 ± 29	0.59 ± 0.05
16270B	78.6	89.6 ± 8.1	0.61 ± 0.06
16386T	79.2	80.57 ± 7.4	0.59 ± 0.04
16478	79.3	180 ± 12	0.58 ± 0.05
16547B	79.1	75.35 ± 11	0.64 ± 0.04
16548P	76.9	84.4 ± 2.8	0.51 ± 0.03
16580C	77.7	91.4 ± 7.5	0.63 ± 0.04
16589E	77.5	73.0 ± 11	0.61 ± 0.05
16552B	79.0	92.4 ± 21	0.50 ± 0.05
16556A	79.3	77.8 ± 6.8	0.63 ± 0.04
16535J	78.7	139.2 ± 35	0.63 ± 0.05
16535I	78.5	93.9 ± 5.5	0.61 ± 0.04

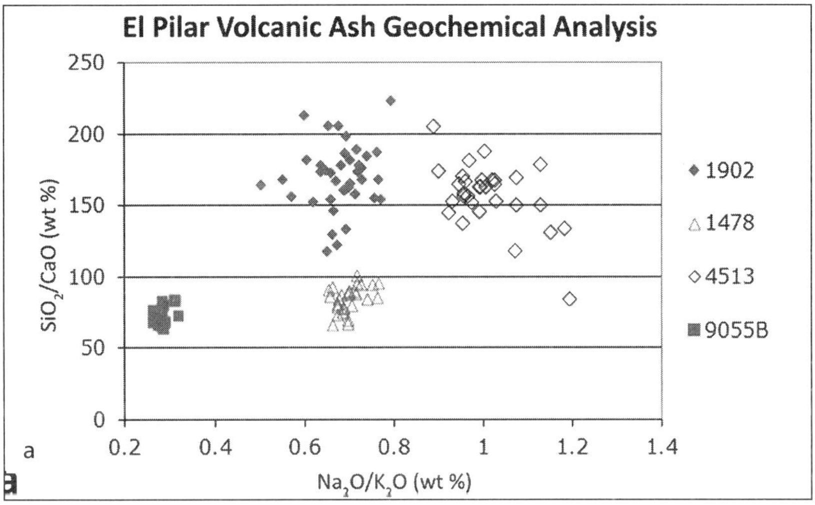

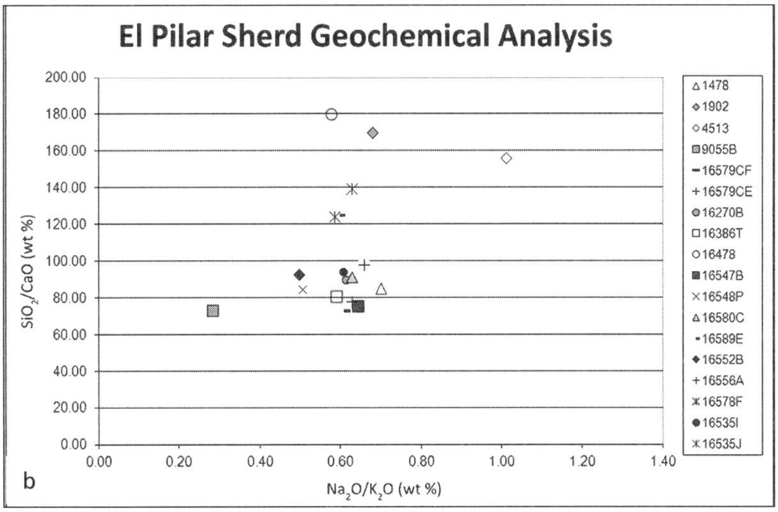

FIGURE 1.5. (*a*) results of microprobe ratios for El Pilar samples, examining homogeneity within sample and heterogeneity among samples; (*b*) results of microprobe ratios averages for El Pilar samples, showing heterogeneity among samples.

variation, reflecting a rhyolitic bulk composition (see fig. 1.4). This is important because the majority of the regional volcanoes are of intermediate composition, in the basaltic andesite to andesite range, with considerably lower silica content than in the Maya samples (roughly 55–65 weight percent silica versus 77 weight percent silica). This observation would, at face value, exclude El Chichón, a basaltic eruption (fig. 1.6).

Based on the SiO_2 results, we should consider the probability of a rhyolitic volcano as a candidate. This could point to Ilopango, but the other major oxides do not match Ilopango. These data raise the question of how the firing environment impacts the geochemistry of the volcanic ash. The untested assumption has been that the volcanic context provided the ultimate in temperature exposure and, along with the evidence

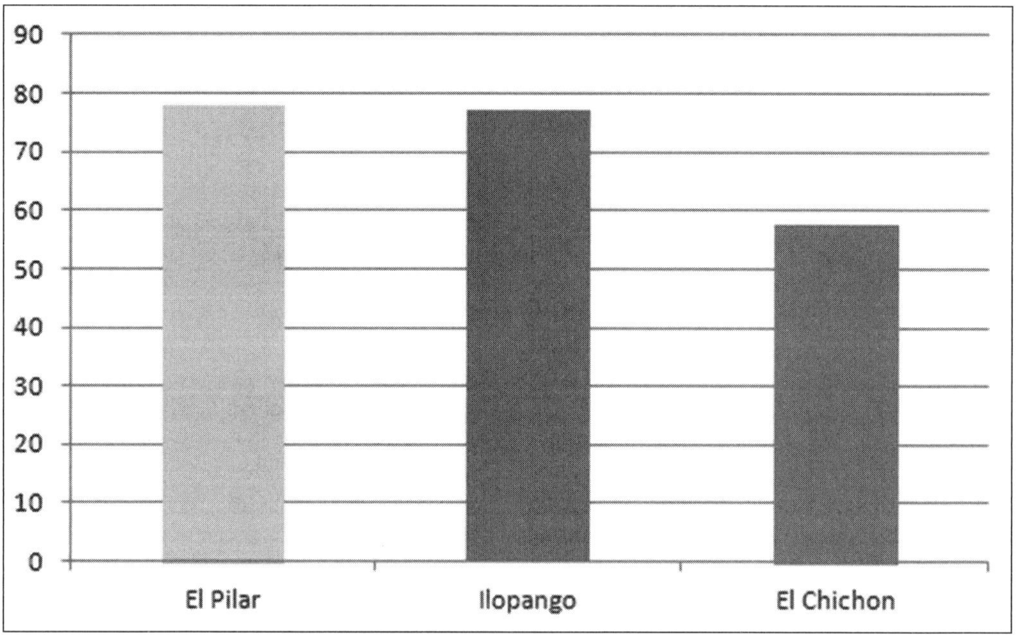

FIGURE 1.6. Percentage of SiO_2 for the volcanic ash temper of the El Pilar samples compared to those from candidate volcanoes.

of the fresh and unweathered condition of the glass, that no significant changes would occur with firing and use over time. Further reflection, however, suggests the possibility that there were chemical reactions between pristine glass and clay matrix at the time of firing. This possibility can be tested directly by experiment.

The geochemical ratios vary, so the actual source of the volcanic ash cannot be determined with petrography and microprobe analyses. It is true that we have moved closer to the nature of the source of volcanic ash in Maya ceramics. The descriptions based on the petrographic analyses support an airfall origin, and the microprobe data for the SiO_2 point to a rhyolitic volcanic source.

Questions still remain. The homogeneity of major elements for the glass shards in each pottery sherd, but the heterogeneity evident across sampled sherds presents a problem. Other factors beyond simply source may affect the results. No association can be determined by vessel form or shape and all are from the same chronological period. It may well be that a major factor that differentiates each vessel sample is the way in which it was fired.

Can the firing environment change the composition of the glass shards by reaction with the clay-calcite matrix? What is the relationship of the clay to the temper? If there is chemical exchange and reaction in the firing environment, is it a closed system, as assumed for NAA (neutron activation analysis), or are there reaction products that leave the solid system and enter the atmosphere? These questions cannot be answered by either petrographic or microprobe analyses of the archaeological samples. To consider the issue of firing environment and the metamorphosis that would be encountered in the process, we need to develop firing experiments and look at the petrography first and the microprobe second to determine how firing might change the geochemistry of the glass. Only with a greater understanding of the firing environment can we draw conclusions from these complex data.

Conclusion

The appreciation of the presence of volcanic ash and the recognition of its importance in the procurement, production, and distribution of Late Classic Maya pottery came with petrographic analyses. Basic petrographic descriptions help to confirm the presence of volcanic ash, clarify descriptions of the glass shards, and identify paste composition that varies little from vessel to vessel among the samples.

El Pilar, only 50 km from Tikal, evinces commonalities with its neighbors in the basic petrographic characteristics of the ceramics as described by Shepard, Jones, and Sunahara. The ash bears the hallmarks of airborne ashfall, arriving from volcanic eruptions far south of the Maya lowlands. The lack of lowland deposits today is not entirely unexpected, as the glass will weather, decompose, and move into the surrounding soil matrix, perhaps accounting for the silica-rich soil found in the clays of the area.

Where does the volcanic ash come from? It must be from a high-silica rhyolitic volcano that erupted in the Late Classic. But is it Ilopango? There was a major eruption of 18 cubic kilometers of ejecta that spread throughout the region and indeed the world (Dull et al. 2001). While there was sufficient volume to provide for the Maya ceramic production needs, could ash have fallen in sufficient quantities to account for its use in the Late Classic? And more significantly, can we identify elemental characteristics to fingerprint it? New data suggest otherwise (Coffee et al. 2014).

Our research demonstrates the necessity of petrography before any other analytical steps are taken. Basic petrographic descriptions are indispensable to the interpretation of elemental data. Yet, even with these fundamental data, we have discovered that we are still unable to pinpoint the source of the volcanic ash used by the ancient Maya. The major elements, gathered from the microprobe analyses, do not match any known local volcano. Are essential geochemical changes produced in the firing environment? The next step is to pursue firing experiments and follow the same protocol with first petrography then detailed geochemical analyses. In that way we can begin to unravel the mystery of the volcanic ash shards in the Maya pottery sherds.

Acknowledgments

We thank a number of students who have worked with us over the years: Johanna Kienowitz, Christy Till, Lindsey Dandridge-Perry, Brianne Caitlin, Marina Bravo-Foster, and Olivia Anderson. We also acknowledge the National Science Foundation for its support of the technical laboratory analyses.

References Cited

Alvarado, Guillermo E., Carlos Dengo, Uwe Martens, Jochen Bundschuh, Teresita Aguilar, and Samuel B. Bonis

2007 Stratigraphy and Geologic History. In *Central America: Geology, Resources and Hazards*, edited by Jochen Bundschuh and Guillermo E. Alvarado, pp. 345–394. Taylor and Francis Group, London.

Arnold, Dean E.

1985 *Ceramic Theory and Cultural Process*. Cambridge University Press, Cambridge.

Brenner, Mark, Michael F. Rosenmeier, David A. Hodell, and Jason H. Curtis

2002 Paleolimnology of the Maya Lowlands: Long-Term Perspectives on Interactions among Climate, Environment, and Humans. *Ancient Mesoamerica* 13(1): 141–157.

Brumfiel, Elizabeth M., and Timothy K. Earle

1987 Specialization, Exchange, and Complex Societies: An Introduction. In *Specialization, Exchange, and Complex Societies*, edited by Elizabeth M. Brumfiel and Timothy K. Earle, pp. 1–9. Cambridge University Press, Cambridge.

Coffey, K. T., A. K. Schmitt, A. Ford, F. J. Spera, C. Christensen, and J. Garrison

2014 Volcanic Ash Provenance from Zircon Dust with an Application to Maya Pottery. *Geology* 42:595–598.

Deevey, Edward S., Don S. Rice, Prudence M. Rice, Hague H. Vaughan, Mark Brenner, and Michael S. Flannery

1979 Mayan Urbanism: Impact on a Tropical Karst Environment. *Science* 206(4416):298–306.

Dull, Robert A., John R. Southon, and Payson Sheets

2001 Volcanism, Ecology and Culture: A Reassessment of the Volcán Ilopango TBJ Eruption in the Southern Maya Realm. *Latin American Antiquity* 12(1):25–44.

Espíndola, Juan Manuel, José Luis Macías, Robert I. Tilling, and Michael F. Sheridan

2000 Volcanic History of El Chichón Volcano (Chiapas, Mexico) during the Holocene, and

Its Impact on Human Activity. *Bulletin of Volcanology* 62:90–104.
Ford, Anabel, and Harry Glicken
1987 The Significance of Volcanic Ash Tempering in the Ceramics of the Central Maya Lowlands. Paper presented at the 1985 Maya Ceramics Conference, Washington, D.C.
Ford, Anabel, and Lisa Lucero
2000 The Malevolent Demons of Ceramic Production: Where Have All the Failures Gone? *Estudios de Cultural Maya* 21:57–74.
Ford, Anabel, and William I. Rose
1995 Volcanic Ash in Ancient Maya Ceramics of the Limestone Lowlands: Implications for Prehistoric Volcanic Activity in the Guatemala Highlands. *Journal of Volcanology and Geothermal Research* 66(1–4):149–162.
Ford, Anabel, and Frank Spera
2007 Fresh Volcanic Glass Shards in the Pottery Sherds of the Maya Lowlands. In *Archaeological Investigations in the Eastern Maya Lowlands: Papers of the 2006 Belize Archaeology Symposium*, edited by Jaime J. Awe, John M. Morris, Sherilyne Jones, and Christophe Helmke, pp. 111–118. Research Reports in Belizean Archaeology, Vol. 4. Institute of Archaeology, National Institute of Culture and History, Belmopan.
Giffórd, James C.
1976 *Prehistoric Pottery Analysis and the Ceramics of Barton Ramie in the Belize Valley.* Memoirs of the Peabody Museum of Archaeology and Ethnology, Harvard University, Vol. 18. Peabody Museum of Archaeology and Ethnology, Harvard University, Cambridge, Massachusetts.
Glicken, Harry
1983 Use of Hand Lens to Identify Volcanic Ash. Personal communication to A. Ford, Cayo, Belize.
Glicken, Harry, and Anabel Ford
1983 *Field Survey for Volcanic Ash in the Belize River Area after the 1982 El Chichón Eruption.* On file, ISBER/MesoAmerican Research Center.
Hart, William J. E.
1981 The Panchimalco Tephra, El Salvador, Central America. Ph.D. dissertation, Rutgers University, New Brunswick, New Jersey.
Instituto Geográfico Nacional
1970 *Mapa Geológico de la República de Guatemala, scale 1:500,000.* Instituto Geográfico Nacional, Madrid.
Jones, Lea D.
1986 *Lowland Maya Pottery: The Place of Petrological Analysis.* BAR International Series 288. British Archaeological Reports, Oxford.
LeCount, Lisa Jeanne
1996 Pottery and Power: Feasting, Gifting, and Displaying Wealth among the Late and Terminal Classic Lowland Maya. Ph.D. dissertation, University of California, Los Angeles.
Le Maitre, R. W.
2002 *Igneous Rocks: A Classification and Glossary of Terms: Recommendations of the International Union of Geological Sciences Subcommission on the Systematics of Igneous Rocks.* Cambridge University Press, Cambridge.
Luhr, James F., Ian S. E. Carmichael, and Johan C. Varekamp
1984 The 1982 Eruptions of El Chichón Volcano, Chiapas, Mexico: Mineralogy and Petrology of the Anhydrite Bearing Pumices. *Journal of Volcanology and Geothermal Research* 23:69–108.
Martens, Uwe, Bodo Weber, and Victor A. Valencia
2010 U/Pb Geochronology of Devonian and Older Paleozoic Beds in the Southeastern Maya Block, Central America: Its Affinity with Peri-Gondwanan Terranes. *Geological Society of America Bulletin* 122(5/6):815–829.
Pagliaro, Jonathan
2011 Rethinking Thinking: An Ecologically Based Consideration of Contextual Diversity in Late Classic Ceramic Assemblages from the Upper Belize River Area, Belize. Ph.D. dissertation, Southern Methodist University, Dallas, Texas.
Reina, Ruben E., and Robert M. Hill
1978 *The Traditional Pottery of Guatemala.* Texas Pan American Series. University of Texas Press, Austin.
Rice, Prudence
1987 Economic Change in the Lowland Maya Late Classic Period. In *Specialization, Exchange, and Complex Societies*, edited by Elizabeth M. Brumfiel and Timothy K. Earle, pp. 76–85. Cambridge University Press, Cambridge.
Rosenmeier, Michael F., David A. Hodell, Mark Brenner, Jason H. Curtis, and Thomas P. Guilderson
2002 A 4,000-Year Lacustrine Record of Environmental Change in the Southern Maya Lowlands, Petén, Guatemala. *Quaternary Research* 57(2):183–190.

Shepard, Anna O.
1937 Ceramic Technology. *Carnegie Institution of Washington Year Book 1936–1937*:144–145.
1939 Technological Notes on the Pottery of San Jose. In *Excavations at San Jose, British Honduras*, edited by J. Eric S. Thompson, pp. 251–277. Carnegie Institution of Washington Publications No. 506. Carnegie Institution of Washington, Washington, D.C.
1942 Classification of Painted Wares. In *Late Ceramic Horizons at Benque Viejo, British Honduras*, edited by J. Eric S. Thompson, pp. 11–17. Carnegie Institution of Washington Publications, No. 528. Carnegie Institution of Washington, Washington, D.C.
1951 Ceramic Technology. *Carnegie Institution of Washington Year Book 1950-1951*:241–244.
1956 *Ceramics for the Archaeologist*. Volume 609, Carnegie Institution, Washington, D.C.
1962 Ceramic Development of the Lowland and Highland Maya. Paper presented at the 35th International Congress of Americanists.

Siebe, Christina, Claus Seibe, Sergio Rodríguez, Juan Manuel Espíndola, and Anabel Ford
2007 *Field Survey for Volcanic Ash in the El Pilar Area*. On file, ISBER/MesoAmerican Research Center.

Simkin, Tom, and Lee Siebert
2002 Global Volcanism FAQs: Smithsonian Institution. In *Global Volcanism Program Digital Information Series, GVP-5*, edited by Tom Simkin and Lee Siebert (January 2013). Smithsonian Institution, Washington, D.C. http://www.volcano.si.edu/reports/usgs/.

Smith, Robert Eliot
1955 *Ceramic Sequence at Uaxactun, Guatemala* I. Middle American Research Institute, Tulane University, New Orleans.

Sunahara, Kay Sachiko
2003 Ancient Maya Ceramic Economy in the Belize River Valley Region: Petrographic Analyses. Ph.D. dissertation, McMaster University, Ontario.

Tilling, Robert I.
2009 El Chichón's "Surprise" Eruption in 1982: Lessons for Reducing Volcano Risk. *Geofísica Internacional* 48(1):3–19.

Varekamp, Johan C., James F. Luhr, and Karen L. Prestegaard
1984 The 1982 Eruptions of El Chichón Volcano (Chiapas, Mexico): Character of the Eruptions, Ash-Fall Deposits, and Gasphase. *Journal of Volcanology and Geothermal Research* 23:39–68.

Vinson, George L.
1962 Upper Cretaceous and Tertiary Stratigraphy of Guatemala. *Bulletin of the American Association of Petroleum Geologists* 46(4): 425–456.

Willey, Gordon R., Jr. William R. Bullard, John B. Glass, and James C. Gifford
1965 *Prehistoric Maya Settlements in the Belize Valley*. Peabody Museum, Cambridge, Massachusetts.

CHAPTER 2

The Importance of Petrography for Interpreting Compositional Data

A Case Study of Tanque Verde Red-on-Brown

MARY F. OWNBY

Petrography has contributed to archaeological research since the 1930s, but its popularity waned in the 1970s with the development of chemical techniques such as neutron activation analysis (NAA), inductively coupled plasma mass spectrometry (ICP-MS) and laser-ablation inductively coupled plasma mass spectrometry (LA-ICP-MS), scanning electron microscopy with energy dispersive spectrometry (SEM-EDS), and X-ray fluorescence analysis (XRF) (Shepard 1936; Stoltman 2001). For some projects, such techniques often have difficulties in determining a provenance for the analyzed samples due to behavioral changes to the raw clay made by the potter (Arnold et al. 1991). In geologically diverse areas petrography is still an excellent method for this type of research. In fact, using both chemical and petrographic techniques is probably the ideal methodology for exploring provenance questions and producing data on production and distribution of pottery (see the introduction for a list of papers successfully using a combined data set). This information can in turn be related to specific economic and social structures in place for moving pottery.

The current study illustrates the utility of combining petrographic and chemical data to clarify production and consumption patterns for Tanque Verde Red-on-brown pottery, a Classic Period (AD 1150–1300) ware found throughout southern Arizona. This ceramic type occurs in both bowl and jar forms and features an orange to buff background with exterior red painted designs. While the pottery has been found at numerous sites in the region, Classic period ceramic manufacture is still poorly understood. This is in contrast to the production of earlier decorated wares in the pre-Classic Sedentary period (AD 950–1150), in which specific places of manufacture have been petrographically identified (Heidke et al. 2002; Heidke 2009). Significant cultural changes are evident between these two periods, including the abandonment of ballcourts, the occupation of new site locations, and the construction of platform mounds, indicating increasing status differentiation and the initiation of new social and economic institutions. Pottery production and exchange were also likely to have changed, but information needed to assess this is lacking. Toward this goal, petrographic analysis of Tanque Verde Red-on-brown samples has been carried out to augment and clarify the results from compositional analyses of almost 700 samples of this pottery type (Harry et al. 2002). This provided provenance information to compare to the published results from the study of the pre-Classic Sedentary period decorated wares.

Previous Research

The first research on Tanque Verde Red-on-brown pottery utilized petrography. Robert Wallace (1957) examined 11 sherds from University Indian Ruin in the eastern Tucson Basin. While 10 of the ceramic samples were locally produced based on the concordance between the prevalent geology and sand temper inclusions, 1 sherd had a sand composition that could not be local. These results were compared to vessels with similar decoration recovered at the sites of Jackrabbit and Ash Hill near Sells, southwest of Tucson in the Papagueria area (Wallace 1957:209, 219). The results indicated that these ceramics had not been produced in the Tucson Basin and suggested that several production locations must have existed for Tanque Verde Red-on-brown.

Further petrographic analysis of Tanque Verde Red-on-brown from several sites in the Tucson Basin supported these initial results, indicating several production areas (Lombard 1987a). For the current study, the analysis of samples from the Marana Mound site is significant and revealed that the sherds came from at least four different production areas, none of which could be local. Sherds from the San Xavier Bridge site, however, did appear to have been locally produced, indicating that some locations produced and consumed their own Tanque Verde Red-on-brown (Lombard 1987b).

Petrographic work in Tucson by Diana Kamilli (1996) indicated that 2 of 3 Tanque Verde Red-on-brown sherds from the Gibbon Springs site in the northeast Tucson Basin were not locally produced. The sand temper of these nonlocal sherds suggested that they may have been made at the northern end of the Tucson Mountains, some 25 km to the west. Petrographic analysis of 35 samples from TaDai and Bid Kih in the Avra Valley (Brawley Wash) west of Tucson revealed that 17 were likely to have been made in the area of these sites (Kamilli 1994). The other 17 were made with sand temper compositions located farther from the sites but still within the Avra Valley, while 1 had a sand temper probably found outside this area. The results revealed both local production of this ware and its acquisition from several other producers within the Avra Valley and beyond (Kamilli 1994:650).

More recently, petrographic examination of 25 Tanque Verde Red-on-brown samples from the Zanardelli site in the southern Tucson Basin also indicated exchange of this pottery (Heidke and Miksa 2011). Of these samples, 14 were not made locally but were produced near Black Mountain, to the northwest of the site: 4 were made near the Tucson Mountains, 1 contained sand from the Catalina Mountains to the northeast of Zanardelli, and only a single sample was produced with local sand temper. These results confirmed that Tanque Verde Red-on-brown was being exchanged in the Tucson Basin, and there were multiple production locations.

Chemical analyses have also been used to examine the production and exchange of Tanque Verde Red-on-brown. Samples from three Phoenix area sites (Las Colinas, Pueblo Grande, and Fortified Hill), and two Tucson area sites (University Indian Ruin and Martinez Hill) were analyzed by X-ray fluorescence spectrometry (Crown et al. 1988:47–54). The results, while qualitative and only utilizing six elements, indicated that samples from each of the sites had been produced with distinct materials, with some overlap for the Tucson area sherds. Comparison of sherds and clay samples from Las Colinas suggested that the vessels were likely to have been made locally (Crown et al. 1988: 64–67).

Within Tucson, a major study of Tanque Verde Red-on-brown using NAA was carried out by Paul Fish et al. (1992) analyzing 347 sherds mostly from sites in the Marana Mound Community. The results identified three compositional groups related to the community, Marana A, Marana BC, and Marana D, two groups with samples from other sites in the Tucson Basin, one group with samples from Phoenix Basin sites, and a final group with samples from the Papagueria. Attempts to match pottery samples chemically to clay samples were not successful. The results did illustrate the presence of several production locations that distributed pottery mostly to surrounding areas. Some vessels were

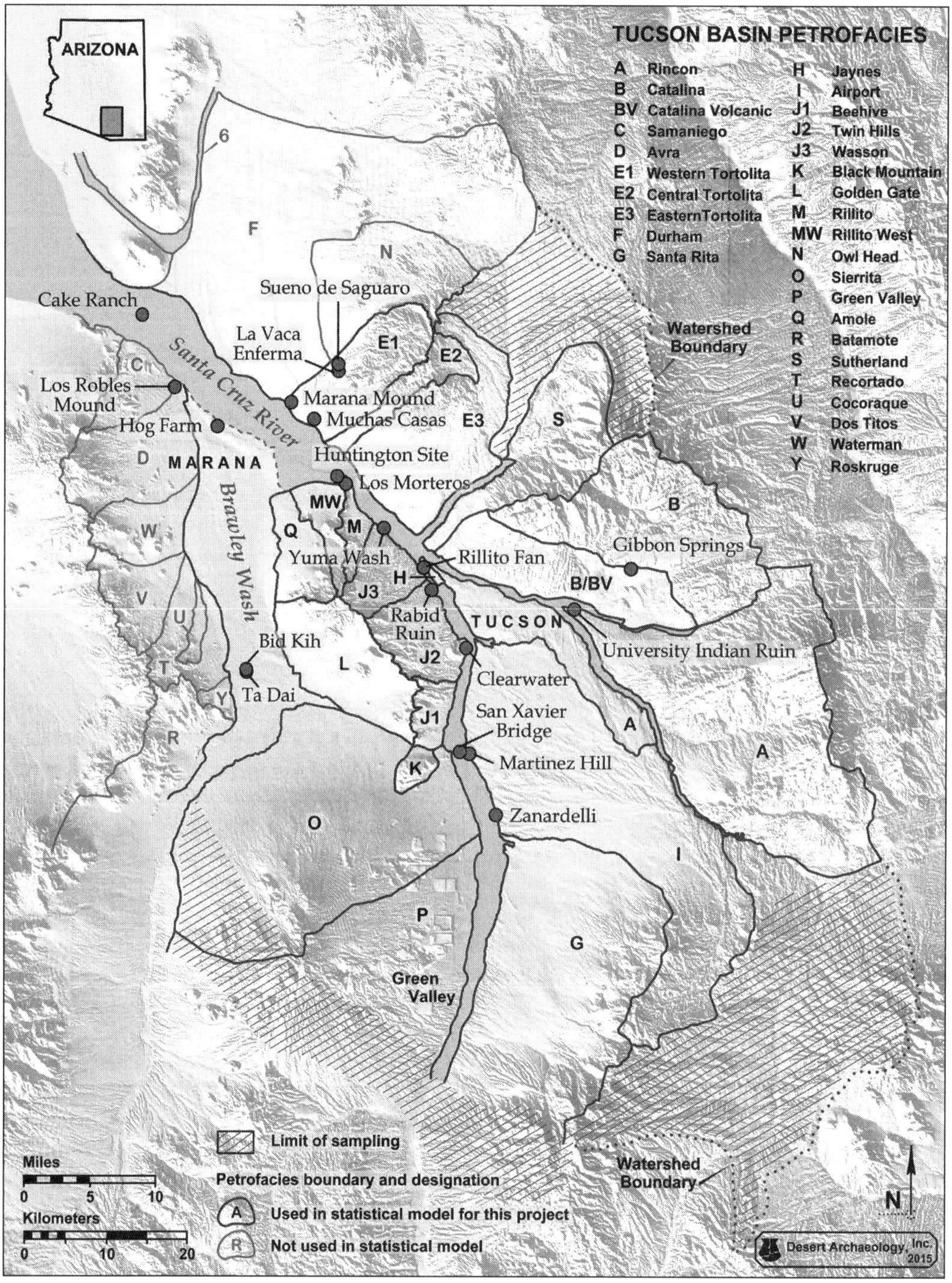

FIGURE 2.1. Map of Tucson Basin Petrofacies, showing sites mentioned in text (prepared by Catherine Gilman).

exchanged over longer distances between the two major basins. (Similar results were attained through acid-extraction ICP analysis of the same samples of Tanque Verde Red-on-brown by Burton and Simon [1996], but this method was critiqued by Hector Neff et al. [1996].)

The NAA study by Fish et al. (1992) was expanded by Karen Harry (1997, 2003), who acquired samples from sites in the Los Robles Mound Community across the Santa Cruz River from the Marana Mound Community. Additional samples from other sites in the Tucson Basin and several sites in the Phoenix Basin increased the total number of samples to 721. The statistical analysis of the data confirmed the previously identified compositional groups, Marana A and BC, Phoenix, and Papagueria, but several new groups (E, F, and G) also appeared, while a south Tucson group was defined. The identification of production sites remained elusive despite some hints provided by a comparison of the sherd and clay data. Harry's study did include the petrographic analysis of 25 samples. Difficulties arose in identifying a provenance due to the mixed sand compositions that were not comparable to existing sand samples. This hindrance has now been overcome due to additional sand samples and a greater understanding of geomorphic processes that create mixed sand compositions (Ownby and Miksa 2012).

Samples and Methodology

Tanque Verde Red-on-brown samples from the sites of Yuma Wash (70 samples) and Clearwater (10 samples) were examined petrographically (Ownby et al. 2011; Ownby and Miksa 2012). The thin-section samples were selected by Elizabeth Miksa and James Heidke based on binocular temper examination. The analysis included quantitative point-counting (see the discussion in the introduction), so the section was taken parallel to the vessel wall rather than perpendicular as is standard in order to provide more area for analysis. Staining for feldspars was applied. This facilitated the qualitative analysis and point-counting using the multiple intercept method and approximately 400 counts (Orton 2000:186).

This approach was selected to facilitate the comparison of the sand temper to sand samples previously collected from around the Tucson Basin (Miksa 2011). These samples have been used to create a map of discrete sand composition zones, known as petrofacies (fig. 2.1). This is based on point-counting the sand samples using the Gazzi-Dickinson method that factors in whether the sand was collected close to the outcrops or farther away (Lombard 1987a; Miksa and Heidke 2001). Typical counted parameters include monomineralic grains and rock fragments classified into generic types, such as mafic volcanic or tectonic metamorphic (see Miksa 2011:table 15.3). The resulting counts are subject to a log-ration transformation using the amount of quartz, a grain that is present in almost all sand samples, to eliminate zeros and make the data more amenable to statistical analysis. The point-count data are entered into a discriminant model that groups them compositionally into distinct spatially defined composition zones that are plotted on a map (Heidke and Miksa 2000). Currently 20 petrofacies are used in the Tucson Basin model (see Heidke in this volume for further use of these petrofacies). The point-count data from the sand temper was acquired in an identical way and entered into the discriminant model as unknowns. The model's classification of the sherd samples into the defined sand compositional groups typically confirmed their assignment to specific Tucson Basin petrofacies.

In order to relate the petrographic results to the large NAA database, a subset of samples was sent to MURR (University of Missouri Research Reactor) for chemical analysis (see Glascock 1992 for the methodology). I chose the 31 samples from Yuma Wash and Clearwater based on the number of sherds assigned to each petrofacies and sherd size. The resulting compositional data were statistically analyzed, but without the elements nickel and arsenic (many values below the detection limit) and sodium (a mobile element). Next the data were transformed into base 10 logarithms to create similar variances among the elements (Baxter 2003:75). Statistical analysis of the compositional data utilized principal components analysis (PCA), hierarchical cluster analysis with Ward's method (HCA),

Table 2.1. Samples from the NAA database used in this analysis and the number of sites included in each area

Region/Community group and number of sites	Number of samples
Marana Mound Community (13)	388
Los Robles Mound Community (16)	140
South and East Tucson Basin (10)	104
Phoenix Basin & northern Santa Cruz River (5)	49
Papagueria (1)	10
Total	691

discriminant analysis (DA), and K-means cluster analysis (KCA) (Baxter 2003:73–83, 92–99, 105–111, 688). Each statistical test was run several times with different elements to identify those that helped in clustering the samples. The number of elements needed to be reduced, as there were only 31 samples compared with 33 elements. The petrographic data were invaluable for this procedure, because they provided an independent means for assessing the statistical results.

Reexamination of the NAA database compiled by Fish et al. (1992) and Harry (1997, 2003) followed a similar procedure. The data were transformed into base 10 logarithms and the same statistical analyses used. Clay samples and sherds with zero values for elements were eliminated, resulting in 691 samples (table 2.1). This large data set was initially analyzed alone before adding the Yuma Wash and Clearwater samples, because the current approach to the data is not analogous to the original statistical treatment. Fish et al. (1992) and Harry (1997, 2003) utilized PCA, cluster analysis using squared mean Euclidean distances and the average-linkage method, and Mahalanobis distances for group refinement. The current method is similar, but the tests used different distances and statistics. These differences in approach meant that it was important to reanalyze the data first to determine the comparability of the results from the two approaches. Once the original data set had been examined statistically, the data from the Yuma Wash and Clearwater sherds were added and the analyses were run again.

Results

Yuma Wash and Clearwater Samples

The petrographic analysis of the 80 samples revealed that they derived from a number of different petrofacies. Some of the Yuma Wash samples had a unique sand composition, however, that was identified as local to the site. This sand contained fragments of volcanic rocks derived from the nearby Tucson Mountains but also granite and gneiss grains that were likely from flood deposits near the site (fig. 2.2). This composition was termed "Wasson, mix," as it is a mix of the Wasson (J3) Petrofacies sand and flood deposits. It was identified in 29 of the 70 samples, while the typical Wasson Petrofacies sand was found in 3 samples. Together the 32 samples are considered locally produced.

Other samples are likely to be nonlocal, as their sand temper was acquired farther than 3 km from the site, a distance that ethnographic data suggest is often the limit for potters to walk to acquire sand temper (Heidke 2009:table 4.10). Of these samples, 19 had sand temper from the Twin Hills (J2), Beehive (J1), and Black Mountain (K) petrofacies to the south of Yuma Wash (see fig. 2.1), while 10 had a sand composition derived from an area in the northern Tucson Mountains. This sand composition is newly defined and appears to be a mix of Tucson Mountain volcanic rock fragments and granodiorite from the Amole pluton. Further, 2 samples had sand temper from the Catalina (B) Petrofacies and 4 had Western Tortolita (E1) Petrofacies sand (see fig. 2.2). Both petrofacies are located across the Santa Cruz River from Yuma Wash. Sand that resembled that found in the Tortolita Petrofacies was found in 3 samples, but with small volcanic rock fragments that may indicate that it was collected near the Santa Cruz River. The petrographic results revealed that 46 percent (32 of 70 samples) of the analyzed Tanque Verde Red-on-brown samples were locally made at Yuma Wash, while the remaining 54 percent (38 of 70 samples) came from surrounding areas.

Similarly, 3 of the 10 Clearwater samples were made with the local Twin Hills Petrofacies sand, while 6 samples contained sand from the Beehive and Black Mountain petrofacies to the

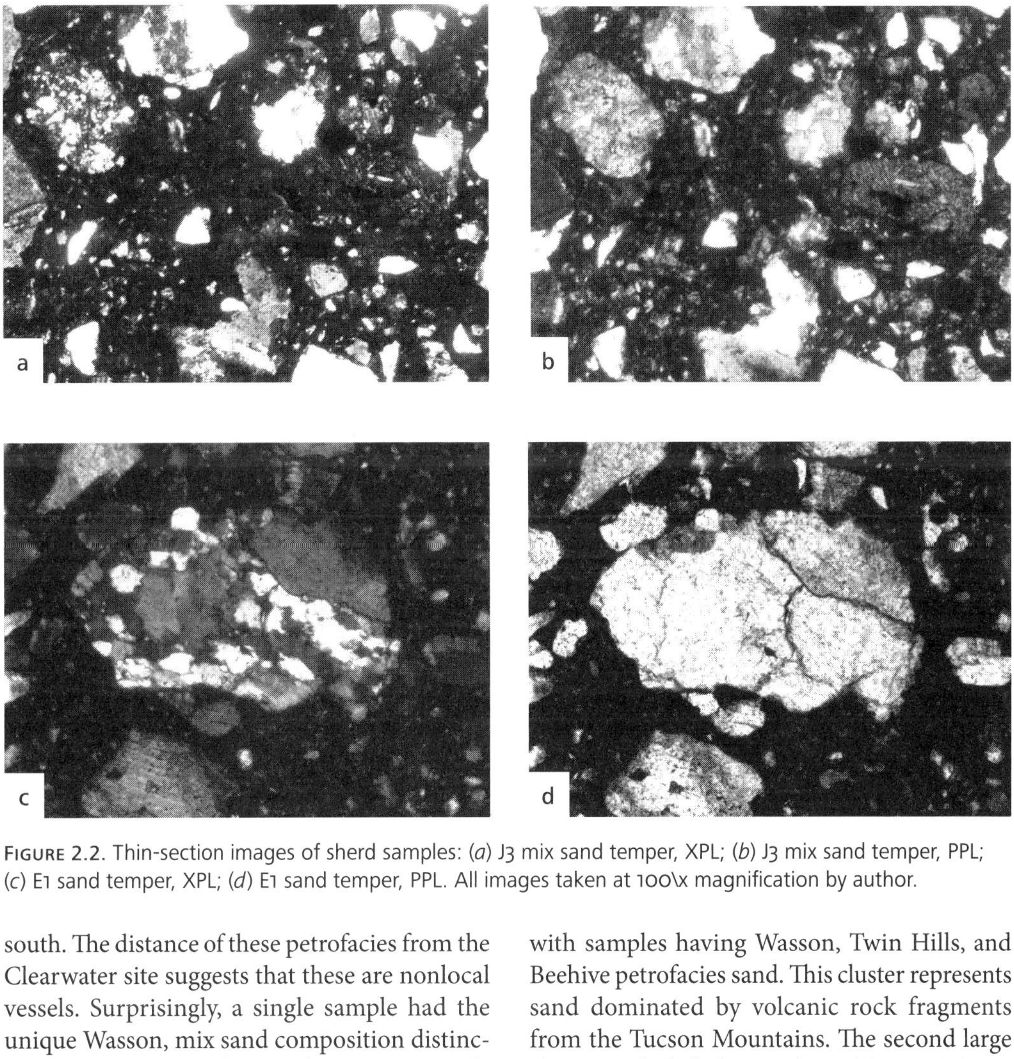

FIGURE 2.2. Thin-section images of sherd samples: (*a*) J3 mix sand temper, XPL; (*b*) J3 mix sand temper, PPL; (*c*) E1 sand temper, XPL; (*d*) E1 sand temper, PPL. All images taken at 100\x magnification by author.

south. The distance of these petrofacies from the Clearwater site suggests that these are nonlocal vessels. Surprisingly, a single sample had the unique Wasson, mix sand composition distinctive to the Yuma Wash site. Thus Tanque Verde Red-on-brown pottery found at Clearwater was both locally produced (30 percent) and acquired from sites to the north and south (70 percent).

From these petrographically analyzed samples, a subset of 31 samples was sent for NAA. Principal components analysis of the resulting NAA data revealed that most of the samples could be put into groups based on their petrofacies assignments. The results of the HCA also supported the petrographic assignments with clusters for samples with the northern Tucson Mountains sand composition and those with the Wasson, mix sand (fig. 2.3). Samples with the latter composition were placed in a large cluster with samples having Wasson, Twin Hills, and Beehive petrofacies sand. This cluster represents sand dominated by volcanic rock fragments from the Tucson Mountains. The second large cluster included the northern Tucson Mountains samples in addition to sand tempers with frequent inclusions of granite and gneiss. Thus there are petrographic reasons for the clustering. Samples placed in a cluster with other samples having different sand tempers may reflect an influence from the clay chemistry. Previous work has suggested alluvial clays from the Santa Cruz River were probably used for Tanque Verde Red-on-brown, which may cause some chemical overlap in samples with various sand tempers (Harry 2003:75–77).

While these chemical similarities and differences are important, they did not prevent the DA from successfully grouping all samples by

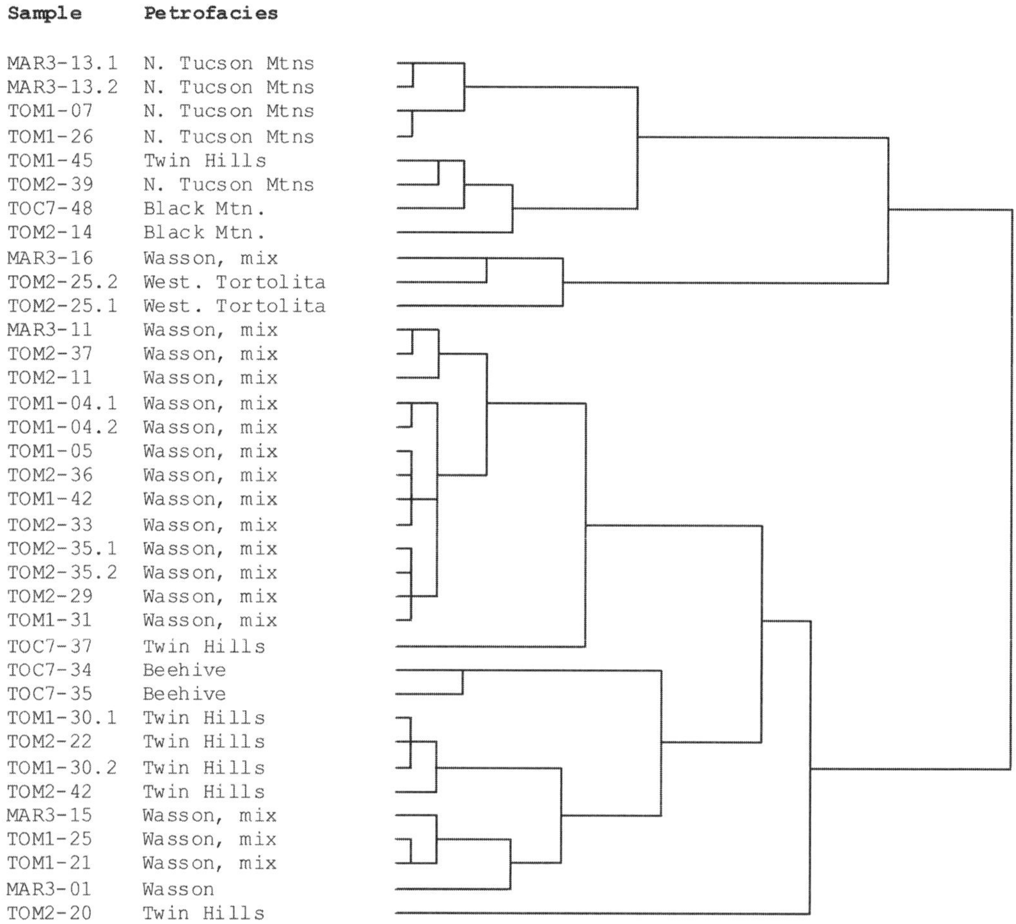

FIGURE 2.3. Results of hierarchical cluster analysis of NAA data from Yuma Wash and Clearwater samples.

their petrofacies (fig. 2.4). The central cluster includes most of the volcanic sand tempered sherds, though, surprisingly, the sample with Wasson Petrofacies sand is separated from this group. Also unexpected is the separation of the samples with Beehive Petrofacies sand from the central cluster. Both of these distinct groups also have sand with volcanic rock fragments. Once again the influence of clay chemistry may play a role, as the discriminant analysis utilizes the combined elemental differences.

The results of the KCA also showed that some samples were grouped with those having a different sand temper. However, the sample with Western Tortolita Petrofacies sand and the sample with Beehive Petrofacies sand were clearly separated. The placement of several samples with Twin Hills Petrofacies sand in their own group highlights that, while the sand can be analogous, chemical differences due to amount of sand temper, size of inclusions, and/or clay can still be present. Overall, the KCA supports the results seen for the other statistical analyses and suggests that the petrographic results were correctly grouping the majority of the samples.

Comparative NAA Data Set

The statistical reanalysis of the large NAA data set compiled by Paul Fish et al. (1992) and Karen Harry (1997, 2003) produced similar results, although the number of unassigned samples was

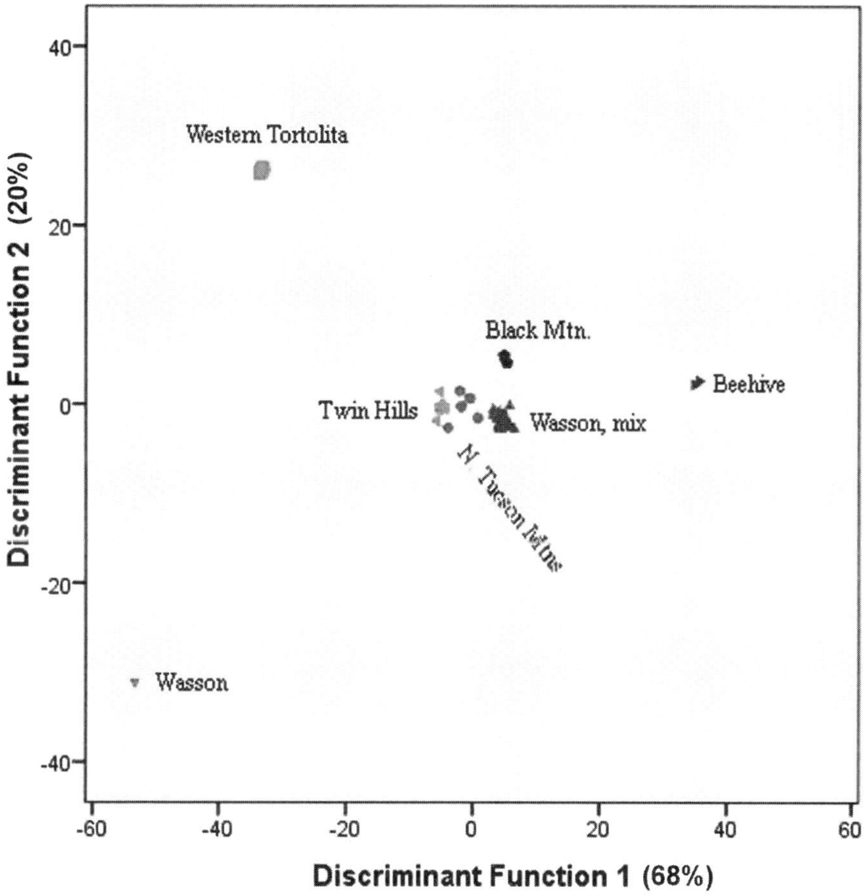

FIGURE 2.4. Results of discriminant analysis of NAA data from Yuma Wash and Clearwater samples. Groups are labeled by their petrofacies assignment.

reduced from 46 percent to 12 percent. This reflects the use of different tests and different criteria for group inclusion. One notable difference is the creation of a NAA Group J, identified in the HCA as a distinct cluster of unassigned samples connected to NAA Group BC. When the Yuma Wash and Clearwater samples were added, this group was associated with samples having Twin Hills and Beehive petrofacies sand. Therefore this set of samples was given its own designation and confirmed as a distinct group by the DA (fig. 2.5).

Other groups could also be linked to the Yuma Wash and Clearwater samples. NAA Group A consistently contained samples with the northern Tucson Mountains sand. NAA Group BC had samples with the distinct Wasson, mix sand. But NAA Groups E, F, and G did not include any of the petrographic samples, suggesting that these groups had sherds made from different raw materials. This is not surprising, as Yuma Wash and Clearwater undoubtedly did not receive Tanque Verde Red-on-brown pottery from every production location.

Along this same line, the South Tucson Group also did not have any petrographic samples. Surprisingly, a group of 3 samples was consistently placed outside of the other compositional groups for most of the tests. These samples are designated "outliers" and derived from two sites in the eastern Tucson Basin. Tentatively, it would appear likely that they were produced at a site in

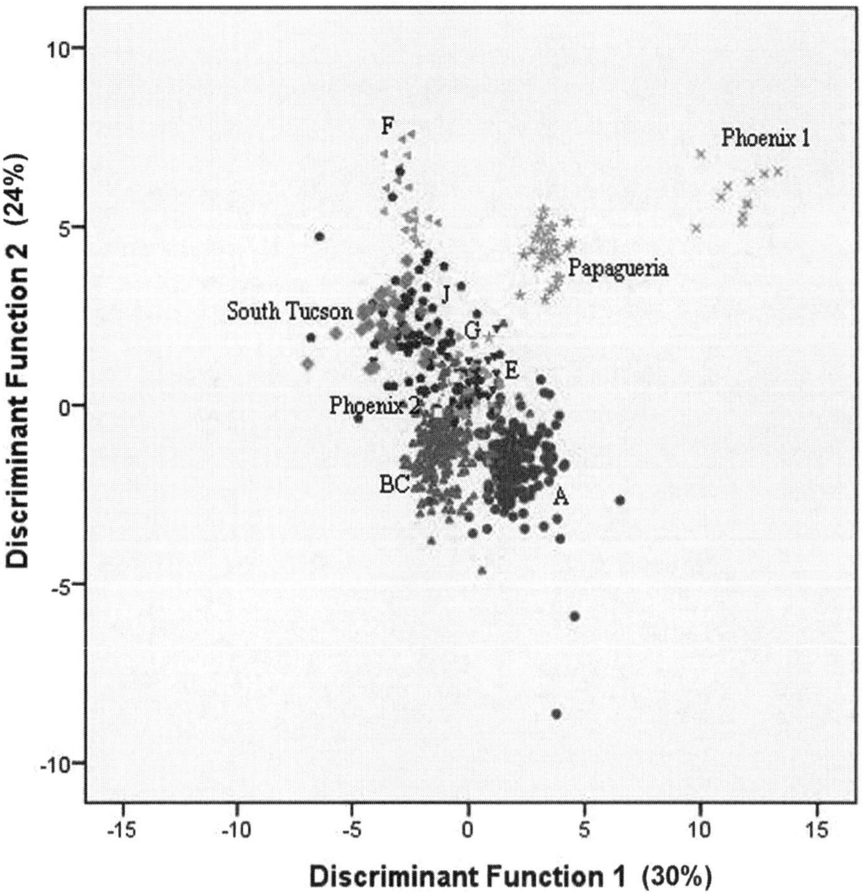

FIGURE 2.5. Results of discriminant analysis of NAA data from all samples. Groups are labeled by NAA compositional group.

this area, possibly University Indian Ruin, which has been suggested as a manufacturing location (Wallace 1957).

NAA groups with samples derived from outside the Tucson Basin also lacked petrographic samples, resulting in no suggested provenance. The reanalysis indicated that the original NAA Phoenix group could be split into two groups, but the Phoenix 2 NAA group includes only 4 samples. The NAA Papagueria group remained unchanged. The unassigned samples are likely related to either additional areas of production represented at very low percentages in the sherds selected for NAA analysis or the utilization of atypical raw materials for these samples.

Discussion

The ability to combine petrographic and chemical data in this study for examining Tanque Verde Red-on-brown has provided an improved understanding of its production and exchange. Three compositional groups identified in the NAA data can now be assigned a provenance. Thus NAA Group A samples were probably produced in the northern Tucson Mountains, possibly at the Los Morteros and/or Huntington site (fig. 2.6). NAA Group BC samples were likely made either at Yuma Wash or in its vicinity. NAA Group J samples were probably manufactured near the Tucson Mountains south of Yuma Wash, almost certainly at several known

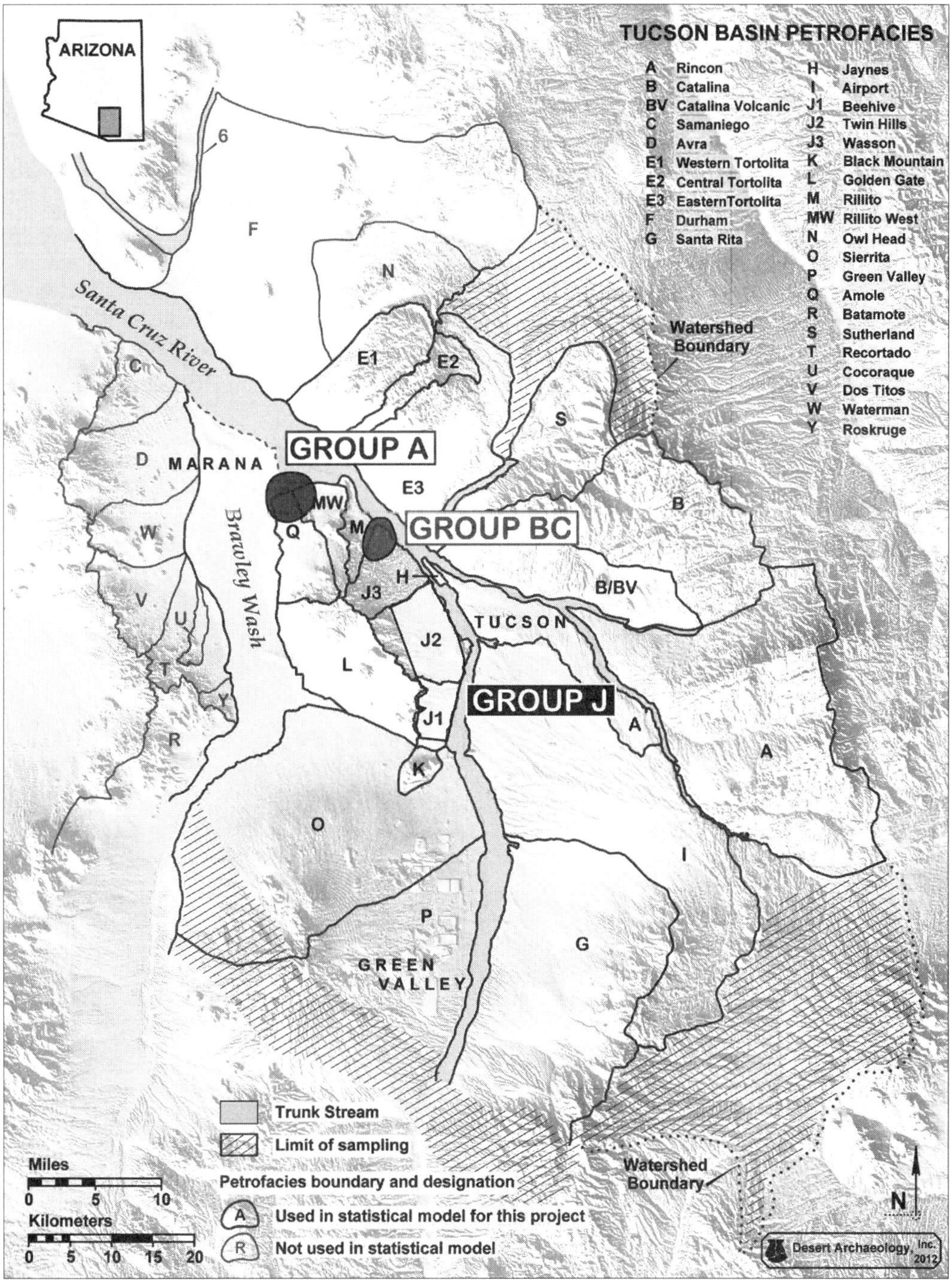

FIGURE 2.6. Production locations for three of the compositional groups (prepared by Catherine Gilman).

large Classic Period sites. Therefore there were a number of specialized producers along the eastern side of the Tucson Mountains, whose products were distributed throughout the basin and beyond.

The distribution of the pottery from the NAA compositional groups shows remarkable variability, keeping in mind the influence of sample sizes. Pottery from NAA Group A is widely distributed, possibly due to the proposed location of production on the Santa Cruz River. Similarly, sites with NAA Group BC pottery are also widely present in southern Arizona. NAA Group E was found in significant numbers only at two sites in the Los Robles Mound Community, Hog Farm and Cake Ranch. This may suggest that one of the sites is the production location. This should be verified petrographically, as prevalence of pottery at a site does not necessarily indicate its source. Similarly, NAA Group G pottery was found predominantly at La Vaca Enferma and Sueno de Saguaro in the Marana Mound Community. This may also indicate where the pottery was produced.

Most of the NAA Group F pottery was found at the site of Muchas Casas and has a limited distribution. Petrographic analysis may indicate whether this site is where potters produced the NAA Group F samples. NAA Group J pottery was found mostly at several sites in the central Tucson Basin but also at a few sites in the northern Tucson Basin, Phoenix, and the Papagueria. The South Tucson NAA Group vessels were mostly present at sites in south and east Tucson. Sherds from the Phoenix 1 NAA Group were only found at Las Colinas, their possible provenance. Phoenix 2 NAA Group sherds were found at Fortified Hill south of Phoenix, perhaps their location of production, and Marana Mound in Tucson. The Papagueria sherds were found in this area, at five sites outside the Tucson Basin, and at Marana Mound. Additional petrography is needed to associate these compositional groups with a provenance.

In examining patterns of consumption, the influence of sample size by site is a major factor, especially as the majority of the samples are from the Marana Mound Community. Thus the percentage of pottery from the compositional groups within each major area was examined (table 2.2 and fig. 2.7). This reveals that the Marana Mound Community acquired pottery from most of the identified groups, except the Phoenix 1 NAA Group. Los Robles Mound Community had a similar pattern of consumption but did not acquire pottery assigned to the South Tucson, Phoenix, and Papagueria NAA groups. This may be the result of the smaller size of the sample from this community or reflect patterns of exchange. Sites in the southern and eastern Tucson Basin acquired most of their Tanque Verde Red-on-brown from producers within this area or along the Tucson Mountains. Pottery assigned to NAA groups E, F, and G and from outside the Tucson Basin was not found in the samples from these sites. The Phoenix Basin and northern Santa Cruz sites appear to have acquired a range of pottery from all of the groups except NAA groups E, G, and South Tucson. Finally, the Papagueria seems to have only consumed pottery assigned to this area, perhaps because of its distance from the two major Hohokam areas.

It is clear that consumption of Tanque Verde Red-on-brown was different at various sites. Sites with potters making these vessels could also acquire them from other producers. Furthermore, sites from which a large sample was analyzed typically show evidence for consumption of pottery from producers located at some distance. This implies that low levels of exchange from certain producers will only be identified if numerous samples are analyzed. In general it appears that the acquisition of Tanque Verde Red-on-brown occurred within a 20- km radius of the site. Thus most of the pottery produced in the northern Tucson Basin is found at sites in this area, while pottery from manufacturers in the southern Tucson Basin remained predominantly in that area. Less common was the movement of pottery over greater distances.

Overall, a number of specialized producers of Tanque Verde Red-on-brown pottery seem to have existed. The inhabitants of Tucson Basin sites appear to have obtained vessels from many sources rather than from a particular manufac-

TABLE 2.2. Samples from each region placed into compositional groups

Region/ Community groups	A	BC	E	F	G	J	S. Tucson	Phoenix 1	Phoenix 2	Papagueria	Total percentage
Marana Mound Community	31.5	41.7	4.7	5.2	5.2	10.5	0.3	0.0	0.3	0.6	100
Los Robles Mound Community	28.6	35.7	13.4	1.8	8.9	10.7	0.9	0.0	0.0	0.0	100
South and East Tucson Basin	22.1	36.8	0.0	0.0	0.0	12.6	28.4	0.0	0.0	0.0	100
Phoenix Basin & northern Santa Cruz River	6.8	11.4	0.0	2.3	0.0	2.3	0.0	25.0	6.8	45.5	100
Papagueria	0.0	0.0	0.0	0.0	0.0	0.0	0.0	0.0	0.0	100.0	100

Note: Based on percentage assigned to each group from the total number of samples.
[a] See table 2.1 for number of samples in each group.

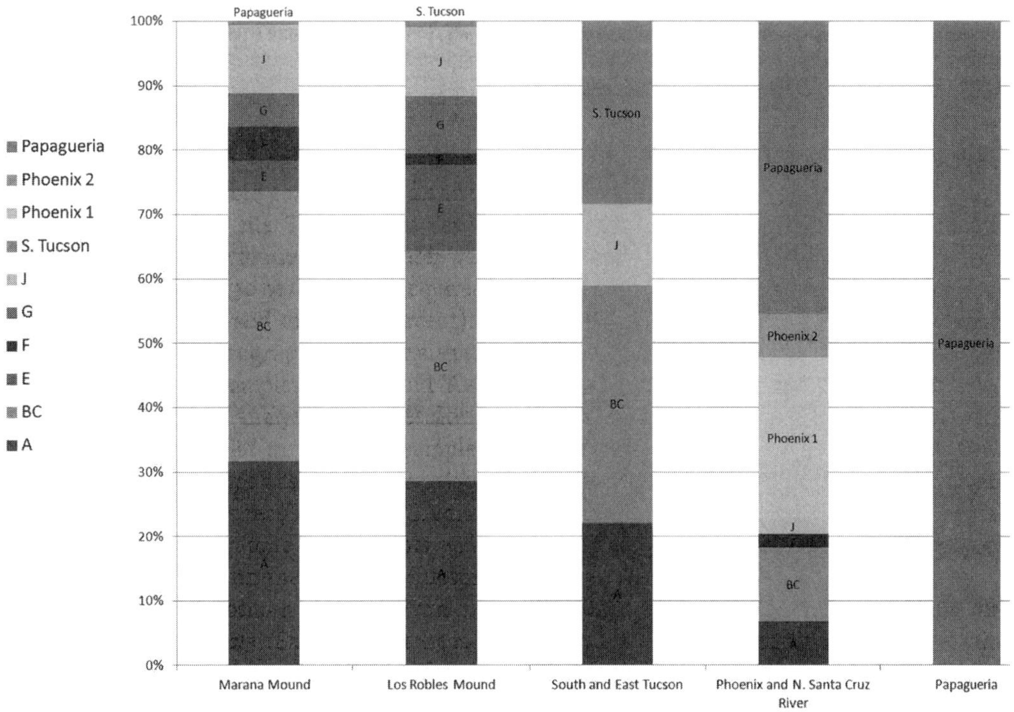

FIGURE 2.7. Consumption by area, showing percentage of each compositional group.

turer. The significance of having a range of such pottery from many production locations is difficult to establish but may relate to economic or social ties with particular groups. This may suggest a system of redistributive exchange, where pottery from many producers arrived at a central area and could then be acquired by many consumers. Particularly for pottery from producers at some distance, this may have facilitated the distribution of their wares. But the nature of this exchange and whether the mound sites played a primary role is still unclear. The results of the current study suggest that Tanque Verde Red-on-brown vessels were not produced by potters at either the Marana or Los Robles Mound sites themselves, although manufacture did occur within what has been defined as their respective community systems. The overall pattern appears to continue the pre-Classic exchange of decorated wares but with additional producers and expanded exchange networks (Heidke et al. 2002; Heidke 2009). These patterns may reflect other cultural changes occurring in the Classic period.

Conclusions

Although combining chemical compositional data with petrographic analysis is not new, the popularity of the latter technique had waned due to the large sample sizes and quantitative results that could be obtained through chemical analyses such as NAA. The increased use of petrographic data to assist in the interpretation of chemical compositional data is encouraging, as it is clear that each technique alone has its limitations. This study has provided further evidence for the utility of combining methods to reach a better understanding of patterns of production, consumption, and distribution. Furthermore, this study is unique in utilizing an existing compositional database and including additional samples for petrography afterward. Through the chemical analysis of only 31 of the samples examined through petrography, much more information on the provenance of the compositional groups was acquired. This revealed that potters at a number of sites in the Tucson Basin were producing Tanque Verde Red-on-brown for local consumption and exchange. Such movement of pottery likely had economic and social impacts that were significant in a period of cultural change.

Acknowledgments

This work would not have been possible without the permission and support of Drs. Suzanne Fish, Paul Fish, and Karen Harry. The NAA data were generated at the University of Missouri Research Reactor with funding support, in part, by grants from the National Science Foundation's Archaeometry Program (9102016 and 9503035). The petrographic samples were from excavations funded by the Town of Marana and City of Tucson and under the direction of Deborah Swartz from Desert Archaeology, Inc. (DAI). Petrographic analysis was facilitated by Dr. Elizabeth Miksa and Carlos Lavayen at DAI with great encouragement from Dr. William Doelle and valuable input by James Heidke. NAA data acquisition was facilitated by Archaeology Southwest and Dr. Deborah Huntley. Catherine Gilman of DAI made the wonderful maps. All of the analyzed samples are currently housed at the Arizona State Museum, while the point-count and NAA data are held by the author.

References Cited

Arnold, Dean E., Hector Neff, and Ronald L. Bishop
1991 Compositional Analysis and "Sources" of Pottery: An Ethnoarchaeological Approach. *American Anthropologist* 93:70–90.

Baxter, Michael
2003 *Statistics in Archaeology*. Hodder Arnold, London.

Burton, James H., and Arleyn W. Simon
1996 A Pot Is Not a Rock: A Reply to Neff, Glascock, Bishop, and Blackman. *American Antiquity* 61:405–413.

Crown, Patricia L., Larry A. Schwalbe, and J. Ronald London
1988 X-Ray Fluorescence Analysis of Materials Variability in Las Colinas Ceramics. In *The 1982–1984 Excavations at Las Colinas: Vol. 4. Material Culture*, edited by David R. Abbott, Kim E. Beckwith, Patricia L. Crown, R. Thomas Euler, David A. Gregory, J. Ronald London, Marilyn B. Saul, Larry A. Schwalbe, Mary Bernard-Shaw, Christine. R. Szuter, and Arthur W. Vokes, pp. 29–72. Archaeological Series No. 162. Arizona State Museum, University of Arizona, Tucson.

Fish, Paul R., Suzanne K. Fish, Stephanie Whittlesey, Hector Neff, Michael Glascock, and J. Michael Elam
1992 An Evaluation of the Production and Exchange of Tanque Verde Red-on-Brown Ceramics of Southern Arizona. In *Chemical Characterization of Ceramic Pastes in Archaeology*, edited by Hector Neff, pp. 233–254. Monographs in World Archaeology No. 7. Prehistory Press, Madison.

Glascock, Michael D.
1992 Neutron Activation Analysis. In *Chemical Characterization of Ceramic Pastes in Archaeology*, edited by Hector Neff, pp. 11–26. Monographs in World Archaeology No. 7. Prehistory Press, Madison.

Harry, Karen G.
1997 Ceramic Production, Distribution, and Consumption in Two Classic Period Hohokam Communities. Ph.D. dissertation, Department of Anthropology, University of Arizona, Tucson.
2003 *Economic Organization and Settlement Hierarchies. Ceramic Production and Exchange among the Hohokam*. Praeger Publishers, Westport.

Harry, Karen G., Paul R. Fish, and Suzanne K. Fish
2002 Ceramic Production and Distribution in Two Classic Period Hohokam Communities. In *Ceramic Production and Circulation in the Greater Southwest: Source Determination by INAA and Complementary Mineralogical Investigations*, edited by Donna M. Glowacki and Hector Neff, pp. 99–109. Monograph No. 44. Cotsen Institute of Archaeology, University of California, Los Angeles.

Heidke, James M.
2009 Multi-Village Specialized Craft Production and the Distribution of Hohokam Sedentary Period Pottery, Tucson, Arizona. In *Interpreting Silent Artefacts: Petrographic Approaches to Archaeological Ceramics*, edited by Patrick Sean Quinn, pp. 227–244. Archaeopress, Oxford.

Heidke, James M., and Elizabeth J. Miksa
2000 Correspondence and Discriminant Analyses of Sand and Sand Temper Compositions, Tonto Basin, Arizona. *Archaeometry* 42: 273–299.
2011 Sedentary and Late Classic Period Ceramic Production and Distribution in the Southern Tucson Basin: Evidence from the Punta de Agua Site Complex, AZ BB:13:16 (ASM), and the Zanardelli Site, AZ BB:13:1 (ASM). In *Archaeological Excavations at the Zanardelli Site, AZ BB:13:1 (ASM)*, edited by Ellen C. Ruble, pp. 133–159. Technical Report No. 2004-01. Desert Archaeology, Tucson.

Heidke, James M., Elizabeth J. Miksa, and Henry D. Wallace
2002 A Petrographic Approach to Sand-Tempered Pottery Provenance Studies: Examples from Two Hohokam Local Systems. In *Ceramic Production and Circulation in the Greater Southwest: Source Determination by INAA and Complementary Mineralogical Investigations*, edited by Donna M. Glowacki and Hector Neff, pp. 152–178. Monograph No. 44. Cotsen Institute of Archaeology, University of California, Los Angeles.

Kamilli, Diana C.
1994 A Petrographic Study of Pottery from the Schuk Toak District. In *Archaeological Studies of the Avra Valley, Arizona: Excavations in the Schuk Toak District: Vol. 2. Scientific Studies and Interpretations*, Allen Dart, pp. 627–652. Anthropological Papers No. 16, Center for Desert Archaeology, Tucson.
1996 Petrographic Report on Ceramics from the Gibbon Springs Project. In *Excavation of the Gibbon Springs Site: A Classic Period Village in the Northeastern Tucson Basin*, edited by Mark C. Slaughter and Heidi Roberts, pp. B.1–B.12. Archaeological Report No. 94-87. SWCA, Tucson.

Lombard, James P.
1987a Provenance of Sand Temper in Hohokam Ceramics, Arizona. *Geoarchaeology* 2:91–119.
1987b Supplementary Report on Temper Sand Characterization and Supporting Data for the Ceramic Petrography Study. In *The Archaeology of the San Xavier Bridge Site (AZ BB:13:14), Tucson Basin, Southern Arizona*, edited by John C. Ravesloot, pp. 409–416. Archaeological Series No. 171. Arizona State Museum, University of Arizona, Tucson.

Miksa, Elizabeth J.
2011 Half Million Points and Counting: Two Decades of Petrofacies Modeling in the Greater Tucson Basin and Avra Valley. In *Craft Specialization in the Southern Tucson Basin: Archaeological Excavations at the Julian Wash Site, AZ BB:13:17 (ASM): Part 2, Synthetic Studies*, edited by Henry D. Wallace, pp. 553–617. Anthropological Papers No. 40. Center for Desert Archaeology, Tucson.

Miksa, Elizabeth J., and James M. Heidke
2001 It All Comes Out in the Wash: Actualistic Petrofacies Modeling of Temper Provenance, Tonto Basin, Arizona, USA. *Geoarchaeology* 16:177–222.

Neff, Hector, Michael D. Glascock, Ronald L. Bishop, and M. James Blackman
1996 An Assessment of the Acid-Extraction Approach to Compositional Characterization of Archaeological Ceramics. *American Antiquity* 61:389–404.
Orton, Clive
2000 *Sampling in Archaeology.* Cambridge Manuals in Archaeology. Cambridge University Press, Cambridge.
Ownby, Mary F., Carlos P. Lavayen, and Elizabeth J. Miksa
2011 Binocular and Petrographic Analysis of Pottery from the Yuma Wash Site. In Excavations at the Yuma Wash Site and Outlying Settlements (draft), edited by Deborah L. Swartz. Anthropological Papers No. 49. Center for Desert Archaeology, Tucson.
Ownby, Mary F., and Elizabeth J. Miksa
2012 Overview of Petrofacies Modeling with Examples from Classic Period Hohokam Sites. *Journal of Arizona Archaeology* 2:23–36.
Shepard, Anna O.
1936 The Technology of Pecos Pottery. In *The Pottery of Pecos*, vol. 2, edited by Alfred V. Kidder and Anna O. Shepard, pp. 389–587. Papers of the Phillips Academy Southwestern Expedition No. 7. Yale University Press, New Haven.
Stoltman, James B.
2001 The Role of Petrography in the Study of Archaeological Ceramics. In *Earth Sciences and Archaeology*, edited by Paul Goldberg, Vance T. Holliday, and C. Reid Ferring, pp. 297–326. Kluwer Academic/Plenum Publishers, New York/Boston.
Wallace, Robert M.
1957 Petrographic Analysis of Pottery from University Indian Ruin. In *Excavations, 1940, at the University Indian Ruin, Tucson, Arizona*, edited by Julian D. Hayden, pp. 209–219. Technical Series No. 5. Southwestern Monuments Association, Globe, Arizona.

CHAPTER 3

The Organization of Ceramic Production and the Origins of Complexity in the Late Prehispanic Coastal Societies of Ecuador

Maria A. Masucci, Hector Neff, Michael D. Glascock, *and* Jeff Speakman

Ethnohistoric accounts place the coastal plain regions of Ecuador as part of the territories of a set of related groups that included the Manteños or Paches along the central coast and the Guancavilcas to the south, although ethnohistorians and colonial accounts disagree on the boundaries and relationship of these territories and their peoples (fig. 3.1; Benzoni 1967; Cieza de León 1864 [1532–1550]; Jijón y Caamaño 1941–1946; Saville 1910:7–8; Stothert 2013). The Guancavilcas may have been part of a larger confederation with the Manteños or a separate identity (Velasco 1842). The research presented here on provenance of early Manteño-Guancavilca ceramics at sites in the southern coastal plain may help clarify this relationship between the southern and central coast societies.

Archaeologists have traditionally identified all sites and the entire central and southern coastal plain region as a form of a Manteño or more recently a Manteño-Guancavilca (hereafter referred to as M-G) cultural phase (AD 800–1530; fig. 3.2) based on the presence of a common set of ceramic types. The key M-G diagnostic type consists of black burnished fineware ceramics. These wares occur in a range of forms and styles including bell-rimmed jars, bowls, and pedestal plates, often in elaborate human or animal forms or decorated with modeled animal, human, or anthropomorphic faces or human forms and pattern burnishing (fig. 3.3; compare Bushnell 1951; Marcos 2013; Stothert 2001). The reliance on decorated fineware ceramics as the diagnostic to define regional cultural phases might be misleading. Looking beyond a focus on the decorated finewares, distinctions among local sequences are distinguishable and the black M-G pottery appears more as an imposition on local traditions than as part of common cultural developments. Recent discussions of M-G finewares have stressed the possible ritual and symbolic role of the ceramics and their potential use by local elites in the transformation and reorganization of regional populations along the Ecuadorian littoral (Stothert 2001, 2006):

> Elites may have manufactured and manipulated symbolically charged goods for religious purposes and for legitimizing their sociopolitical roles. The fancy pottery of the later phases is evidence that local elites from all over Manteño territory shared a common cult or were integrated in some other fashion. (Stothert 2001:308)

The existence of enclaves or colonies (Muse 1989:189) has been proposed as an alternative that could explain the presence of black M-G pottery (see Zeidler and Pearsall 1994:205). To complicate the situation, most archaeologists have begun to refer to the southern coastal peoples as Guancavilca in order to emphasize a

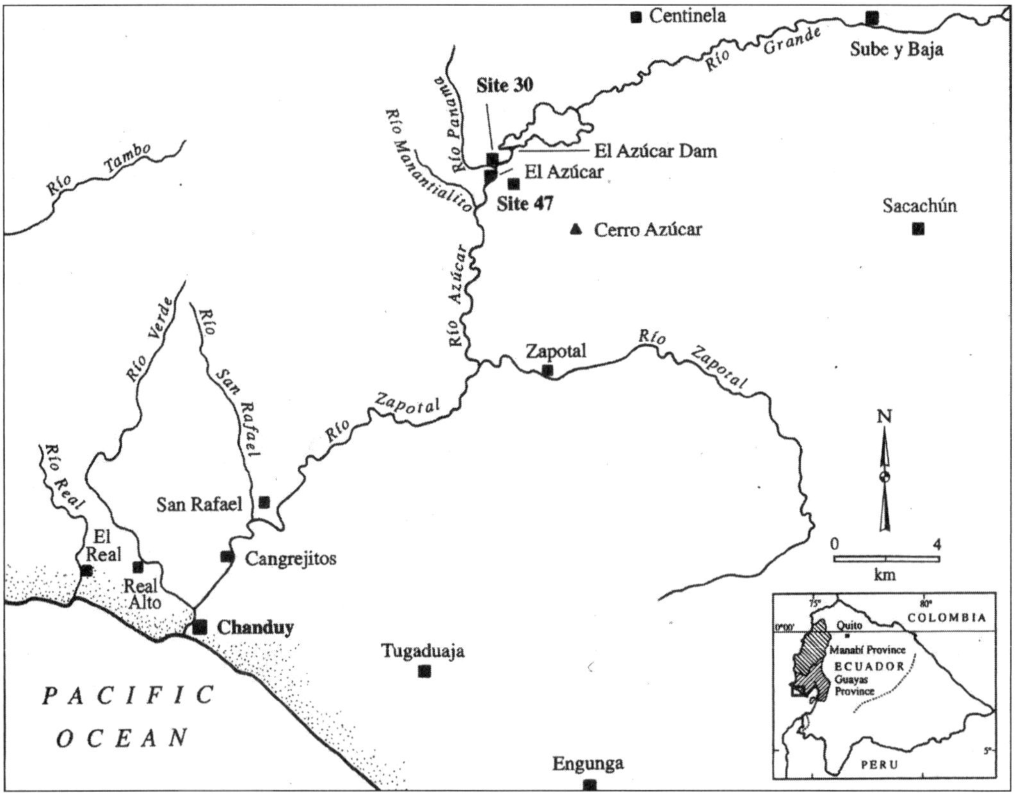

FIGURE 3.1. Chanduy Valley research sites (after Reitz and Masucci 2004:figure 2.1).

likely cultural and even political distinction between the central and southern coastal societies (Marcos 2013). Jorge Marcos (1995a, 1995b) even describes the existence of a Guancavilca State for the southern coast, distinct from, although likely related to, the central coastal Manteño society. Addressing basic questions of ceramic provenance and technology is a useful first step toward the evaluation of these alternative models and a more nuanced understanding of the Manteños and Guancavilcas of late prehispanic coastal Ecuador.

If the M-G style black finewares found at southern sites are nonlocal products and are not related to local traditions, then some form of movement of vessels would best explain the presence of the blackware ceramics. Furthermore, if technological traditions in other ceramic types show continuity at the southern sites and only the blackwares appear as a nonlocal product, then some form of trade or movement of these particular vessels is more likely than movement of peoples such as movement by nonlocal elites or even establishment of a foreign colony. For comparison, in the expansion of the Inka Empire it has been shown that a standardized technology for local production of imperial wares was exported rather than the vessels themselves (see Szilágyi and Szakmány 2009). Also, the obtaining of nonlocal ideologically or politically powerful items or materials rather than local copies could also be a strategy employed by rising local elites.

Background and Sample Description

This study of M-G ceramics is imbedded within Masucci's larger program of research of Ecuadorian coastal ceramics, which to date includes thin-section analysis of 348 ceramic samples, 62 clay samples, and 39 rock samples. The thin sections are archived at Drew University, with the ceramic samples stored in El Azúcar and type col-

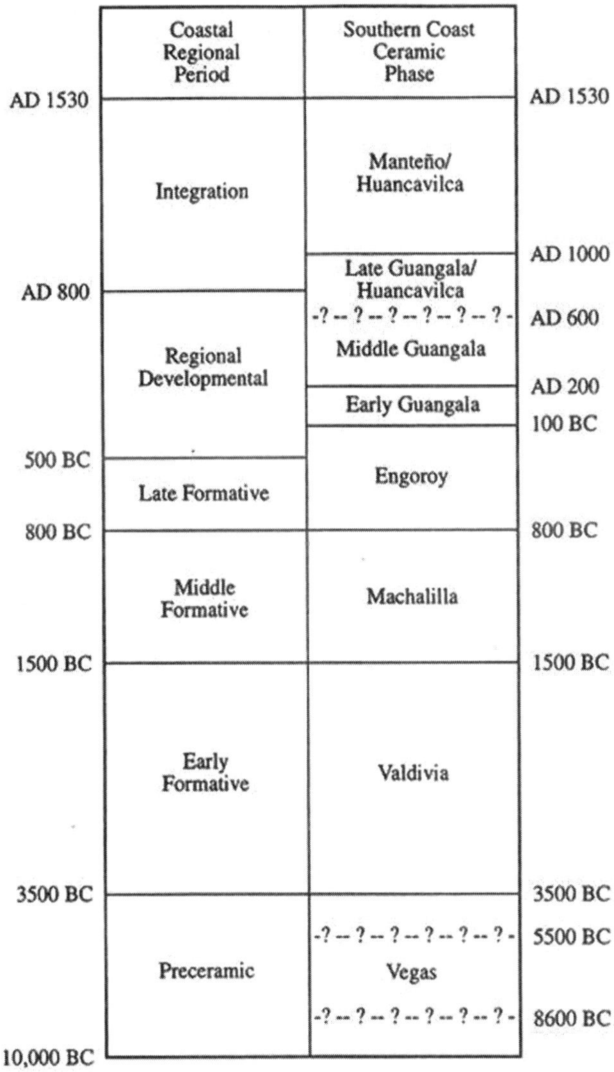

FIGURE 3.2. Prehispanic Ecuadorian coastal regional chronology and southern coast ceramic cultural phases.

lections at the Instituto Nacional del Patrimonio Cultural, Region 5, Guayaquil, Ecuador. A subset of 249 ceramic samples and 54 clay samples has been analyzed by NAA (Masucci and Macfarlane 1997; Masucci and Neff 1997, 2006). Both ceramic and raw material samples are drawn primarily from the Chanduy Valley system (Zapotal, El Azúcar, and Río Grande valleys, fig. 3.1), but additional samples have been obtained through resource surveys by Masucci and Macfarlane and contributions from other investigators and communities throughout Guayas, Santa Elena, and southern Manabí Provinces. Temporally the sample is weighted toward the Guangala Phase (200 BC–AD 600–800; fig. 3.2), but the sample includes Late Formative through colonial ceramics. The sample pertaining to Late Guangala and M-G consists of 126 ceramic thin sections, with 20 of these from blackware M-G style vessels discussed here (figs. 3.4 and 3.5 and table 3.1). The Late Guangala through M-G ceramic samples were selected to cover the range of traditional diagnostic types and the typological and chronological types recovered

FIGURE 3.3. Manteño-Guancavilca vessels from the collection of the Museo Antropológico y de Arte Contemporáneo (MAAC), Guayaquil, Ecuador. Photos courtesy of Karen Stothert (*a*) MAAC GA-1-95-76; (*b*) MAAC GA-3-833-78.

from excavated contexts dated to Late Guangala through M-G by Masucci throughout the Chanduy Valley system or obtained from colleagues in the southern or central coastal region (Masucci 1992; Masucci and Neff 1997, 2006). A subset of 110 of the total Late Guangala–M-G sample was included for NAA, including 16 blackware M-G style vessel sherds selected to cover the formal and regional range of the samples (table 3.1). The following discussion provides details on the petrographic analysis performed by Masucci and methods and statistical analyses of the NAA are provided elsewhere (Masucci and Macfarlane 1997; Masucci and Neff 1997, 2006).

Analytical Procedures

The combination of stylistic and chronological studies with petrography and NAA was selected to provide complementary data on provenance, style, technology, and organization of production for coastal Ecuadorian ceramics. Typological, chronological, and petrographic analyses were performed by Masucci (Masucci 1992, 1995, 2000; Masucci and Macfarlane 1997). NAA was performed independently at the University of Missouri Research Reactor by Hector Neff and Michael Glascock (Masucci and Neff 1997,

TABLE 3.1. Manteño-Guancavilca black fineware samples

Provenance	Petrographic fabric class	Chemical group
El Azúcar	Fabric Class 1	Anom MFP
Punta Carneiro	Fabric Class 1b	Core
San Marcos	Fabric Class 1	Core
Loma de los Cangrejitos	Fabric Class 1b	Fine Gray
Loma de los Cangrejitos	Fabric Class 1b	Fine Gray
Loma de los Cangrejitos	Fabric Class 1b	Fine Gray
Loma de los Cangrejitos	Fabric Class 1b	Fine Gray
Loma de los Cangrejitos	Fabric Class 1b	Fine Gray
El Azúcar	Fabric Class 1b	Fine Gray
Salango	Fabric Class 1	Slg 2
Salango	Fabric Class 1	Slg 2
El Azúcar	Fabric Class 1a	Slg 2
Tambo River	Fabric Class 1	Slg 2
San Rafael	Fabric Class 1	Slg 2
Sube y Baja	Fabric Class 1	Slg 2
Chanduy	Fabric Class 1a	
Agua Blanca	Fabric Class 1a	Seco
Agua Blanca	Fabric Class 1a	
Mar Bravo	Fabric Class 1a	
Mar Bravo	Fabric Class 1a	

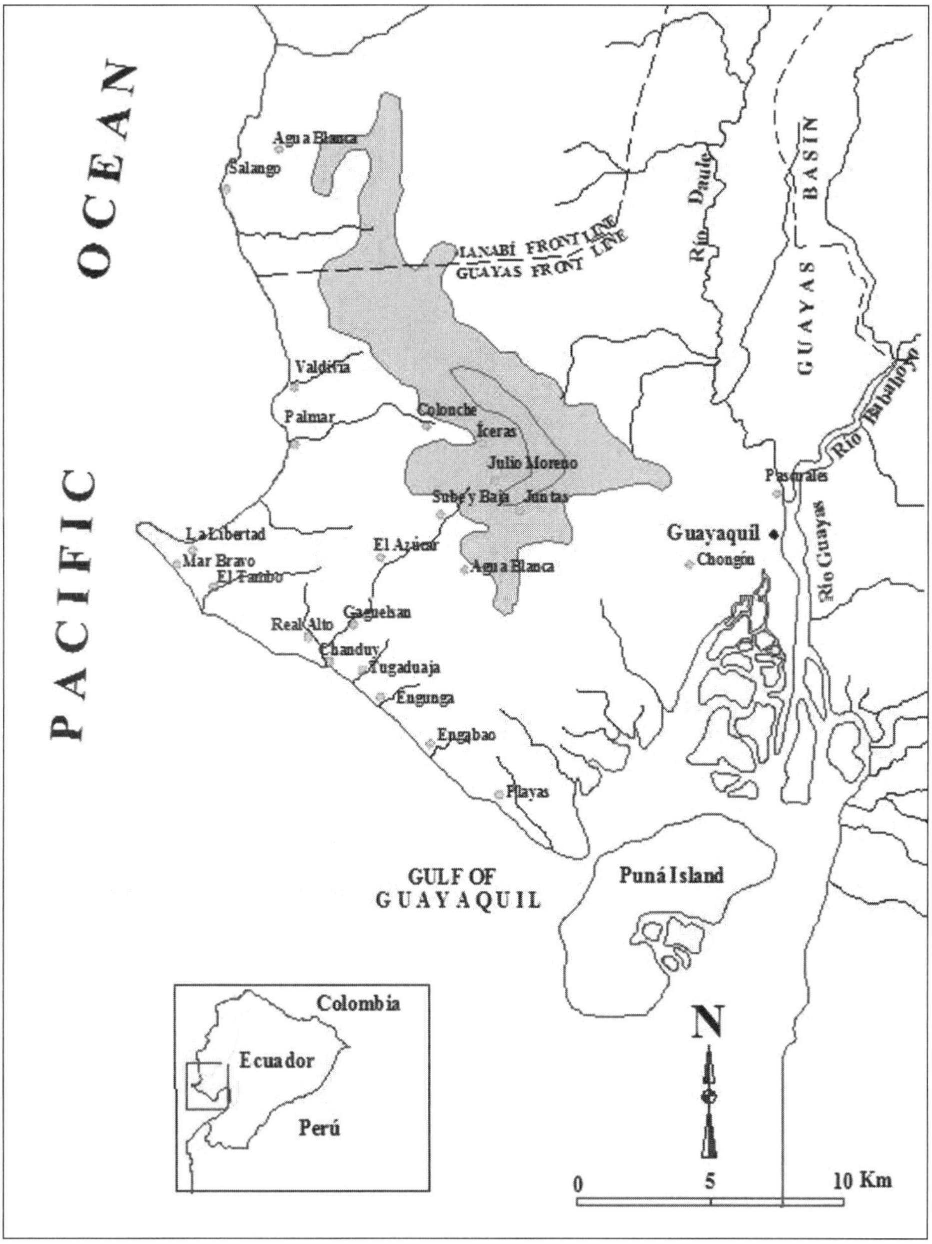

FIGURE 3.4. Ceramic and raw material sample locations. The Chongon-Colonche hill range is shaded.

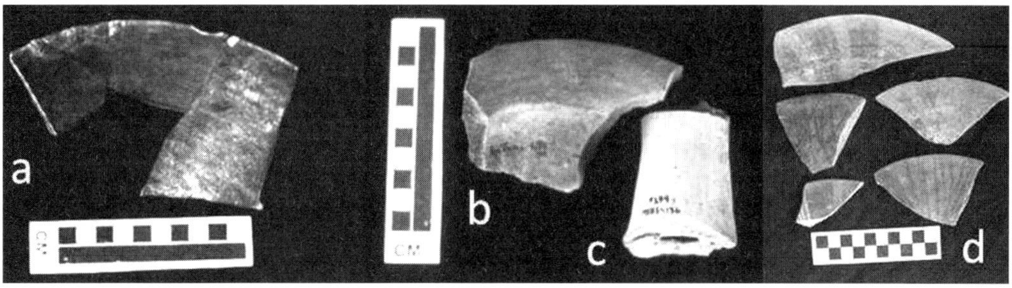

FIGURE 3.5. Manteño-Guancavilca black fineware rim and neck sherds: (*a*) Loma de los Cangrejitos; (*b*) Agua Blanca; (*c*) Sube y Baja; (*d*) Mar Bravo.

2006), including additional consultation with Jeff Speakman (see Glascock 1992 for the methodology). These methods were used to define typological, chronological, and compositional paste groups. The paste groups were then examined in light of analysis of raw material samples collected from the region, published accounts of local geology, results of geological survey (Masucci and Macfarlane 1997), the provenance of the artifact samples, and the ceramic typological and chronological information available through ongoing research in the Chanduy Valley (Masucci 1992, 2000). The earlier research focused on the Guangala cultural phase, while the research presented here is the first attempt to examine the organization of M-G ceramic production. The sample of M-G ceramics discussed here is small but has the advantage of being imbedded in the larger study. It is anticipated that these initial results will be helpful in guiding further sampling and research. A preliminary version of this research was presented at the annual meeting of the Society for American Archaeology in 2006 (Masucci and Neff 2006).

Thin-Section Analysis

Thin-section analysis focused on textural and micromass description and identification of mineral inclusions in the samples for comparison of composition (Folk 1974; Kerr 1977) and microstructure (Courty et al. 1990; Peacock 1977; Whitbread 1989). The attributes recorded for each thin section were identity of inclusions, range of grain sizes and shapes present, estimation of predominant grain size and shape, estimation of frequency, modal percentage, sorting, and association of mineralogical constituents of the pastes, as well as the weathered state, degree of alteration, degree of vitrification, and orientation of minerals and void spaces. Estimates of frequency, modal percentage and sorting, and description of shapes present were made by comparison with visual charts (Bullock et al. 1985; Folk 1974; Pettijohn 1975; Quinn 2013; Whitbread 1989). In addition, at least four samples of each fabric group identified were point-counted to verify estimates based on visual charts. The point-counting followed a line count method with a minimum of 200 counts reached in each section (Middleton et al. 1985), including two typical examples of each group and one at each extreme of the range of variation. Therefore in this study point-counting was used only as a check on the groupings and estimates of qualitative attributes rather than for statistical analysis as a basis for compositional groupings (Stoltman 1989, 1991).

The intent of the procedure outlined above was to link fabrics with similar attributes. Composition was awarded initial priority in forming groups that were then further subdivided on the basis of texture. Minor variations in microstructure were noted but were not primary attributes for forming groups unless it appeared that the distinctions were attributable to differences in raw materials or fabricating techniques. The groups were then compared to the provenance and pottery class to consider source areas (Bishop et al. 1982; Peacock 1977). Success in such provenance studies, however, depends largely on distinctive inclusions and regional geological variability (Jones 1984).

INAA

Multivariate statistical analyses were used to identify similarities and differences between specimens and determine whether there were any archaeologically meaningful groups. PCA (principal components analysis) was used for pattern recognition through bivariate and trivariate plots to examine the correlations between the elements and identify groups as well as detect any outliers. The PCA is based on Mahalanobis distances (see Glascock et al. 2004). PCA confidence ellipses surround the groups discussed.

Results

Petrographic Fabric Class 1: Fine-Paste Pumice Tempered Wares

The petrographic analysis grouped all the 20 black polished fineware samples into Fabric Class 1, but 6 formed Fabric Class 1a, due to the addition of distinct igneous minerals, and 7 formed Fabric Class 1b based on textural differences. Fabric Class 1 was established in the previous study of Guangala ceramics and is defined

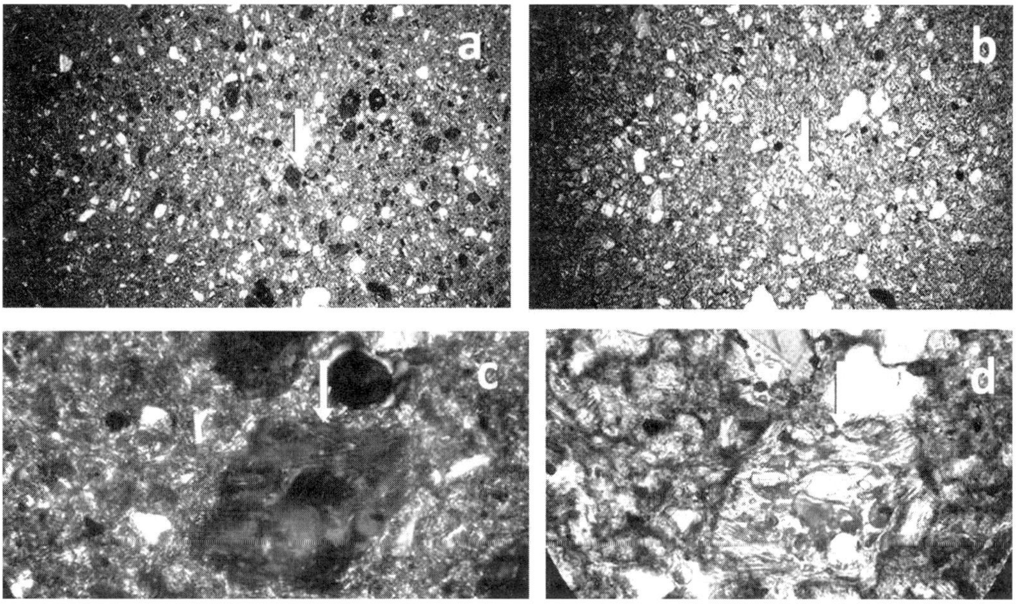

FIGURE 3.6. Photomicrographs of Fabric Class 1 with arrow indicating pumaceous ash temper fragment: (*a*) 40× xpl; (*b*) 40× ppl; (*c*) 400× xpl; (*d*) 400× ppl.

by a very homogeneous fine paste. It includes all fineware ceramics from Late Formative through M-G (Masucci 2000). The fabric is distinguished by the presence of pumice/pumaceous ash as the predominant inclusion in a fine-textured fabric (fig. 3.6). The pumice fragments are very consistent in size (very fine to fine silt-sized, 0.002–0.007 mm), shape (rounded), sorting (extremely well sorted), and textural and surface characteristics (consistent flow and bubble structure) with phenocrysts extremely rare to absent. The origin of this material is proposed to be trade, due to the consistency in characteristics across a long time span, the nature of the temper as unaltered pumice with clear flow structure, the lack of regular wind current systems that would deposit ash or pumice in the region, and the lack of identified tephra or related pyroclastic materials in the geology or stratigraphy of the region of the sites.

The 7 samples of Fabric Class 1 derive from a range of sites in the southern coast (El Azúcar *n* = 1, San Marcos *n* = 1, Tambo River Valley *n* = 1, San Rafael *n* = 1, Sube y Baja *n* = 1; table 3.1 and figs. 3.4 and 3.5) and a single site in southern Manabí Province (Salango *n* = 2). These sites range from Late Guangala–M-G storage sites to known Late M-G port and settlement sites.

Fabric Class 1a is consistent in fabric with Fabric Class 1, but the samples contain igneous inclusions. One sample includes basalt, which is typical of raw materials and sediments sampled to the north of the Chongon-Colonche hill range. The other members of Fabric Class 1a contain rare amounts of granodiorite, which is the predominant inclusion in Manteño Ordinario, the utilitarian wares characteristic of the M-G society after AD 1200. These 6 ceramics are from Agua Blanca (*n* = 2), a central Manabí Province M-G site with complex architecture that may have been a regional capital in the final portion of the period, El Azúcar (*n* = 1) and Mar Bravo (*n* = 2, fig. 3.4), and Chanduy (*n* = 1). These last two are late period port sites, on the southern coast. Granodiorite is not available in the southern coastal area, but a large outcrop is present immediately north of Guayaquil, with evidence of mining activity from the prehispanic period to the present.

Fabric Class 1b includes 3 samples from the pre–platform mound contexts at Loma de los Cangrejitos (fig. 3.5), which are dated with

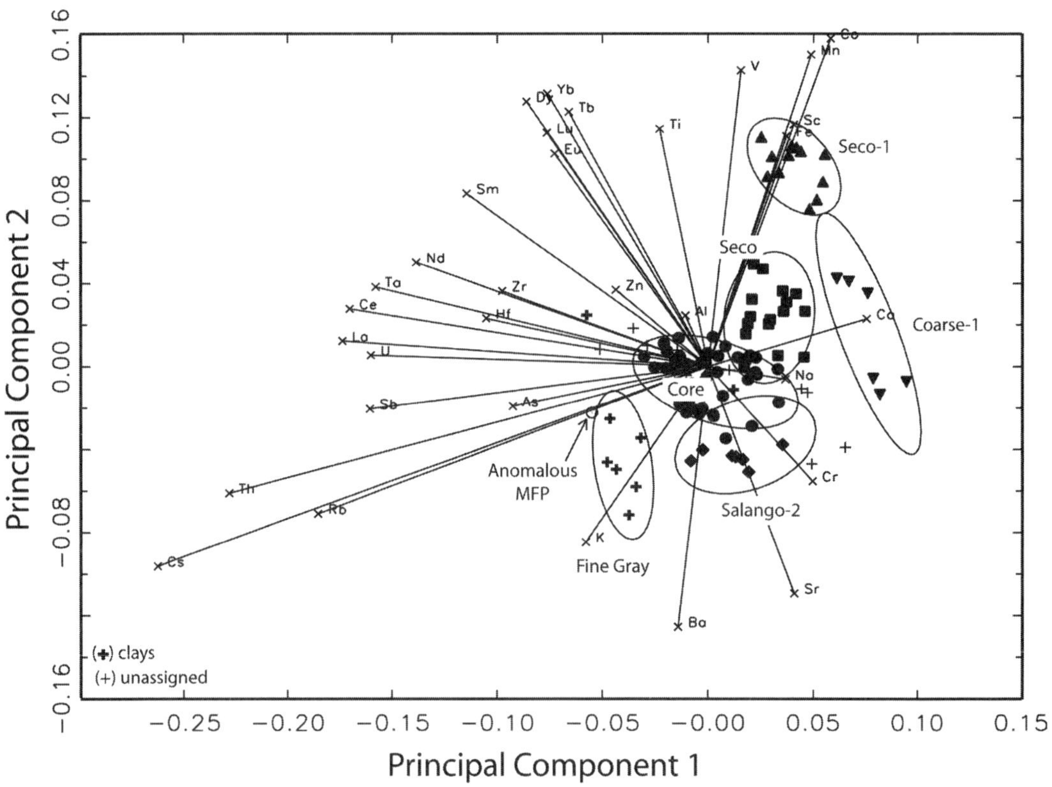

FIGURE 3.7. PCA scatter plot for NAA analysis.

radiocarbon assays to Late Guangala–Early M-G (AD 700–900), and 2 samples dating after AD 900. The remaining member of Fabric Class 1b from Punta Carneiro is estimated at Middle–Late M-G.

An earlier study of Guangala ceramics utilized electron microprobe analysis to source pumice fragments in the Guangala finewares (Masucci and Macfarlane 1997). Two distinct compositions were present that appear to have chronological significance, possibly related to a shift in source location. Analysis of samples of pumice from highland mines indicated the need for more sensitive instrumentation, as the preliminary microprobe analysis did not discern sufficient compositional diversity among the highland pumice deposits. Further analysis will include the M-G finewares, but these initial results were promising for examining patterns of interaction over time.

Elemental Compositional Groups

The elemental analysis included 16 samples of black polished fineware. These included 7 from Fabric Class 1, 1 of the Fabric Class 1a samples from Agua Blanca with granodiorite and another from El Azúcar, and all of Fabric Class 1b (table 3.1). Statistical analyses did not place these samples into a single group. Instead they have high probability of membership in two fineware groups defined in the earlier study of Guangala finewares and one new fine-textured elemental group, Fine Gray (fig. 3.7). These three groups are labeled as "Salango-2" (n = 6), "Anomalous MFP" (n = 1), and "Fine Gray" (n = 6). The one sherd from Agua Blanca of Fabric Class 1a grouped statistically with the M-G utilitarian wares from Agua Blanca and other Late M-G sites that contain ceramics with granodiorite. This group is labeled "Seco" in the elemental analysis. The blackware vessel sample from

Punta Carneiro and the vessel from San Marcos showed low probability of being members of the "Core" elemental group, which contains all Late Formative, Guangala, Late Guangala, and Early M-G utilitarian wares and clusters with the clay-rich sediments sampled for the southwest coast region. Petrographically these samples were members of Fabric Class 1.

The Fine Gray elemental group contains 6 samples of M-G fine-paste blackware. M-G fineware samples are the only members of this group. Of these, 5 are from Loma de los Cangrejitos, with 3 from the pre–platform mound contexts (ca. AD 700–900) and 2 associated with the platform mounds (ca. AD 900–1200). The remaining sample is from a Late Guangala–M-G site in the El Azúcar Valley (Site 531). Chronologically the contexts of these 6 samples would span the transition between Late Guangala to M-G and Early–Middle M-G, confirmed by radiocarbon dates between AD 640 and 1200 from Loma de los Cangrejitos. Therefore they are stylistically and technologically similar and were all recovered from the Chanduy Valley but potentially span the period from Late Guangala through Middle M-G. This elemental group also has internal variability, being enriched in a number of trace elements that are probably carried in the clay matrix (Masucci and Neff 2006). Their distinct trace-element patterns suggest that they represent distinct source zones. The source (or sources) for the Fine Gray group remains unknown, because many projections of the data do not show any of the raw clay samples falling within the range of variation of this group (fig. 3.7). To date, however, the raw material sampling is heavily weighted to the southern coastal Guayas and Santa Elena Provinces. Areas of Manteño-Guancavilca settlement outside the southern region obviously need to be sampled to investigate possible source areas for these ceramics (Masucci and Neff 2006). But the conclusion is drawn that this elemental group likely represents a range of production locations potentially outside the southern coastal region.

The single sample of M-G fineware that clusters in the Anomalous MFP group is from El Azúcar, while 3 other Guangala fineware polychrome sherds cluster with this sample: 2 from the southern coast Tambo Valley and 1 from the central coastal site of Salango.

The 6 samples that cluster in Salango-2 are from central coastal Salango (n = 2) and 1 sample each from Tambo, El Azúcar, San Rafael, and Sube y Baja, all in the southern coastal plain region. This group also includes Guangala fineware ceramics from this same range of sites.

Therefore the elemental analysis in general appears to parallel the divisions within Fabric Class 1 that were observable in the petrographic analysis but also detects additional variability not discernible petrographically in fineware pastes. This is one of the advantages of combining the methods, as petrographic analysis may not be able to detect the differences in fine paste wares. Both the petrographic analysis and elemental analysis indicate similar patterns in technology and production with the M-G finewares, following a pattern similar to that of earlier Guangala finewares. The petrographic analysis demonstrates that all M-G blackware vessels sampled continue a technological recipe dating from at least the Late Formative, utilizing pumice temper as the primary inclusion in a fine, highly refined and processed clay. In most cases the pumice is the only significant visible inclusion. The elemental analysis, however, demonstrates that vessels recovered from individual sites were likely from multiple production locales. This is the same pattern seen in earlier Guangala finewares (Masucci and Macfarlane 1997).

In style, technology, and cultural practices, as suggested by the patterns of ceramic production, there is continuity between the earlier Guangala into the Late Guangala to M-G transition and into Early M-G, possibly through AD 1200. Evidence is mounting from other investigations that the real changes and possible development of what should most accurately be termed the Manteño Empire or Manteño-Guancavilca is post AD 1200 and after the development and main occupation of Loma de los Cangrejitos. As has been proposed based on settlement evidence, the Late Guangala–M-G transition appears to represent a reorganization of population and economy to nucleation around coastal centers and ports, while in the past it had been seen as settlement abandonment and cultural devolution (Paulsen 1970). Continuity in ceramic technology and organization of production further supports this, as do settlement data showing similar settlement pattern changes in southern Manabí Province (Martín 2009). Loma de los Cangrejitos is unique in the region in that it becomes a regional center with earthen platforms and an adjacent cemetery with high-status burials.

Provenance and Production

The Manteño-Guancavilca Fine Gray vessels sampled in this study continue a deep tradition of fineware production utilizing fine pastes tempered with pumice. Therefore the 20 samples included are similar in technology and style, but elemental analysis suggests that they are from a range of production locales. There is no clear correlation between find site and production region. For example, all 5 samples from Loma de los Cangrejitos cluster in a single chemical group, Fine Gray, but with internal variability in trace elements. These Manteño-Guancavilca black polished finewares at Loma de los Cangrejitos, however, occur alongside coarsewares that are both petrographically and chemically associated with local sediments and materials suggesting local production. Hence the coarseware assemblages are locally produced and suggest not the appearance of enclaves or colonies but populations continuing previous traditions, although with the inclusion of new utilitarian styles and forms associated with the development of Manteño-Guancavilca societies. A preliminary hypothesis based on the results of this study would be that the Manteño-Guancavilca style of black polished fineware vessels from Loma de los Cangrejitos and throughout the southern and central coastal region were both local products and vessels obtained from other southern as well as central coastal production locations. The circulation of vessels as part of rituals and gift exchanges has been proposed for Guangala (Masucci 2008), and further sampling is necessary to investigate this model. Now it is possible to include the Late Guangala and M-G period in this same pattern. The black fineware vessels likely represent new ritual practices that would correspond well with the cultural changes that would have come with a shift from dispersed farmsteads to nucleation near coastal and inland centers. Also, the ability to obtain and manipulate the blackware vessels may be related to the development of Cangrejitos as a regional center, possibly by a social group that altered the hilltop site from domestic and craft production functions to the addition of public ritual spaces. Therefore a preliminary model of the blackware vessels at least at Cangrejitos as representing elite or rising local elite manipulation of key ritual and symbolic material items obtained through trade is proposed.

Not surprisingly this corresponds very well to changes elsewhere in the northern and central Andes. The period from AD 600 to 1200 is one of major changes. The nucleation of populations around coastal centers on the Ecuadorian coast focused on production of shell, particularly *Spondylus*, and textiles for export could easily have been spurred and be part of increased demand from southern and highland neighbors. The period after AD 1200 would also mark a time of significant cultural changes throughout the region with the origins of the Inka Empire, which is known to have interacted and even attempted to conquer the coastal Manteño-Guancavilca region (McEwan 1992; McEwan and Delgado Espinoza 2008).

Discussion and Conclusions

The research reported here examines a small sample of M-G style black finewares. This research is imbedded in a larger data set that combines mineralogical and chemical analysis of both ceramics and raw materials correlated with detailed ceramic stylistic analyses and systematic settlement survey and excavations. The results demonstrate the potential of this combination of methods and data. Petrographic analysis in particular indicates the existence of a set of long-standing traditions of pottery making and raw material choices in coastal Ecuadorian ceramic making from the Late Formative through

Manteño-Guancavilca, including trade in key desired temper (pumice) alongside changes in style and form, while NAA shows both correlation and lack of correlation between find site and raw materials, suggesting circulation of fineware vessels. The term "circulation" is used at this point because no consistent patterns of find site to production region have been discerned to suggest establishment of production centers and organization of trade. The vessels may be "circulating" through social or ritual rather than economic practices.

The elemental analysis presented in this study does indicate, however, some changes or new source locations (for example in the separation of the Fine Gray elemental group that includes only Manteño-Guancavilca fineware samples). These samples are not distinguishable petrographically from earlier finewares. The analysis shows the continued use of pumice temper, but elemental analysis suggests that these samples may be from distinct, although related, production sources from the earlier Guangala finewares. It is tantalizing to consider that the settlement reorganization, and in many areas lack of settlement continuity between Guangala and Late Guangala–Early Manteño (of which settlements like Loma de los Cangrejitos are examples), is related to new intersite or interregional relationships. The Fine Gray elemental group may be the ceramic production evidence of changes in other areas and segments of the societies. The variation in the new elemental group may also be evidence of the intraregional settlement changes during Late Guangala–Early Manteño. The petrographic and elemental analysis of raw materials from the southern and central coast indicated distinct southern and central coastal raw material provinces with internal variation, although the internal variation was significantly less than between the two regions (Masucci and Macfarlane 1997). Shifts in settlement locations within each area between the Guangala and Manteño phases could have placed potters in contact with acceptable raw materials that allowed continuation of long-standing technologies but had minor compositional differences not affecting ceramic products. These differences and therefore the settlement change may be reflected in the results of the elemental analysis.

Organization of ceramic production during the Guangala phase was complex with ceramic classes exhibiting distinct production histories, traditional diagnostic ceramic types as imported items, and interregional interactions, with trade in raw materials and fineware types circulated among settlements and across the coastal regions (Masucci and Macfarlane 1997). The research reported here demonstrates that these patterns continued into the final portion of the Guangala Phase and into Manteño. Late Guangala–Early Manteño is marked by changes in settlement organization, continuation in ceramic technological traditions, and in situ change in coarseware ceramic style and form. The appearance of M-G style black finewares was a local development but may have initially involved obtaining the vessels from sites outside the region, possibly related to new ritual practices associated with economic and sociopolitical changes and/or the use of the vessels in the rise of local elites. This would correspond well with the sequence of development of the platform center at Loma de los Cangrejitos. Of the black fineware vessels sampled here, three were from contexts underlying the platform mound structures, while two were associated with the platforms. Settlement nucleation near the coast at Cangrejitos is accompanied ca. AD 700 by increased craft production, particularly *Spondylus* shell working. The earliest contexts include the small quantities of blackware that based on the analysis presented here may be obtained from sites outside the region. By ca. AD 900–1200 the site had been transformed into a regional center with raised earthen platform constructions and much higher quantities of black fineware ceramics.

The later samples of M-G black fineware from the range of sites sampled here suggest both local and nonlocal production, possibly due to circulation of vessels obtained through participation in central and southern coast-wide cultural developments. The more dramatic changes signaled in ceramic technology and production come in the coastal Manteño-Guancavilca society after AD 900 and particularly after AD 1200, marked by the dramatic increase in use of Manteño black finewares and the production of the ubiquitous Manteño Ordinario vessel forms.

Many more details of these cultural changes remain to be investigated, of course, but this preliminary look at the technology of Manteño-Guancavilca finewares utilizing both petrographic and elemental analyses has provided initial models of the nature and origin of these changes that can be tested. Further investigations must include a larger sample of black M-G fineware, particularly from northern sites along with sediment samples from that region. The resulting ceramic data can then begin to be placed into full context with detailed site data, only recently emerging from investigations throughout the northern and southern coastal regions. Earlier investigations had been shadowed by acceptance of models of a monolithic Manteño "league of merchants" or the "Señorio de Salangome," which was seen as controlling the coast through a trading economy, particularly related to *Spondylus* (Stothert 2013). Questioning of this model has sparked renewed interest, and evidence is emerging of a more complex system of developments that resulted in a diversity of cultural trajectories, identities, customs, and settlements of the Manteño-Guancavilca and the other coastal cultures recorded by the European settlers.

Acknowledgments

This work would not have been possible without the generous sharing of samples and ideas of colleagues. Samples have been provided by the Comuna de Agua Blanca, Karen Stothert, Richard Lunnis, José Chancay, Amelia Sanchez, Kimball Smith, and Elisabet Bordt. Conversations with Olaf Holm and Karen Stothert provided much impetus for these investigations. We owe a debt to the exchanges with these scholars. Assistance and collaboration by Jorge Marcos, Maritza Freire, Patricia Netherly, Javier Jagual, and colleagues at the Instituto Nacional de Patrimonio Cultural (INPC)—Franklin Fuentes, Telmo Lopez, and José Chancay—as well as my geologist co-investigator, Allison Macfarlane, have been invaluable. The local residents, field staff, and friends, in particular, Rita de Molina and Santa Rodriguez, are always central to the success of the research. Financial support has been key to the project. The support of the National Science Foundation for both analytical and field research is gratefully acknowledged. The NSF supported the Missouri Research Reactor NAA study. Field grants from the Heinz Foundation, American Philosophical Society, Brennan Foundation, and Banco Central del Ecuador as well as grants through Drew University from the Mellon Foundation and NASA have all allowed for different phases of the research. Finally, the initial opening of a new window into ceramic analysis that is at the core of this research was made possible by the opportunity given to me by Heather Lechtman and Dorothy Hosler to begin technological analysis at the Center for Materials Research in Archaeology and Ethnology, MIT, particularly with the guidance of Ian Whitbread. For this opportunity and its resulting transformation in our understanding of ceramics Maria Masucci will always be grateful.

References Cited

Benzoni, M. Jerónimo

1967 *La historia del mundo nuevo*. C. Radicati di Primeglio, Traducción Universidad de San Marcos, Lima.

Bishop, Ronald L., Robert L. Rands, and George R. Holley

1982 Ceramic Compositional Analysis in Archaeological Perspective. In *Advances in Archaeological Method and Theory*, 5, edited by Michael B. Schiffer, pp. 275–330. Academic Press, New York.

Bullock, Peter, Nicolas Federoff, A. Jongerius, Georges Stoops, and Tatiana Tursina

1985 *Handbook for Soil Thin Section Description*. Waine Research Publications, Wolverhampton, UK.

Bushnell, Geoffrey H. S.

1951 *The Archaeology of the Santa Elena Peninsula in Southwest Ecuador*. Occasional Papers of the Cambridge University Museum of Archaeology and Ethnology, No. 1. Cambridge University Press, Cambridge.

Cieza de León, Pedro

1864 [1532–1550] *The Travels of Pedro de Cieza de Leon, AD 1532–50*. Translated by Clements R. Markham. Hakluyt Society, London.

Courty, Maire-Agnes, Paul Goldberg, and Richard Macphail
1990 *Soils and Micromorphology in Archaeology.* Cambridge University Press, Cambridge.
Folk, Robert L.
1974 *Petrology of Sedimentary Rocks.* Hemphill Publishing Company, Austin, Texas.
Glascock, Michael D.
1992 Neutron Activation Analysis. In *Chemical Characterization of Ceramic Pastes in Archaeology,* edited by Hector Neff, pp. 11–26. Monographs in World Archaeology No. 7. Prehistory Press, Madison.
Glascock, Michael D., Hector Neff, and Kevin J. Vaughn
2004 Instrumental Neutron Activation Analysis and Multivariate Statistics for Pottery Provenance. *Hyperfine Interactions* 154(1–4): 95–105.
Jijón y Caamaño, Jacinto
1941–1946 *El Ecuador interandino y occidental antes de la Conquista Castellana.* 4 vols. Editorial Ecuatoriana, Quito.
Jones, Richard E.
1984 Greek Potters' Clays: Questions of Selection, Availability and Adaptation. In *Ancient Greek and Related Pottery,* edited by H. A. G. Brijder, pp. 21–30. Allard Pierson, Amsterdam.
Kerr, Paul F.
1977 *Optical Mineralogy.* McGraw-Hill, New York.
Marcos, Jorge G.
1995a El mullo y el pututo: La articulación de la ideología y el tráfico a larga distancia en la formación del Estado Guancavilca. In *Primer Encuentro de Investigadores de la Costa Ecuatoriana en Europa,* edited by A. Álvarez, S. Álvarez, C. Fauría, and J. Marcos, pp. 97–142. Ediciones Abya-Yala, Quito.
1995b El sistema de parentesco, de los habitantes de la Costa de la Provincia de Guayas. In *Primer Encuentro de Investigadores de la Costa Ecuatoriana en Europa,* edited by A. Álvarez, S. Álvarez, C. Fauría, and J. Marcos, pp. 467–486. Ediciones Abya-Yala, Quito.
2013 *La historia prehispánica de los pueblos Manteño Guancavilca de Chanduy.* Universidad Internacional del Ecuador, Quito, Ecuador.
Martín, Alexander J.
2009 The Domestic Mode of Production and the Development of Sociopolitical Complexity: Evidence from the *Spondylus* Industry of Coastal Ecuador. Ph.D. dissertation, Department of Anthropology, University of Pittsburgh, Pittsburgh.
Masucci, Maria
1992 Ceramic Change in the Guangala Phase, Southwest Ecuador: A Typology and Chronology. Ph.D. dissertation, Department of Anthropology, Southern Methodist University, Dallas, Texas. University Microfilms, Ann Arbor.
1995 Explaining Ceramic Change in the Late Prehistoric Ceramics of Coastal Ecuador, South America. In *Proceedings of the 8th CIMTEC—World Ceramics Congress and Forum on New Materials, Florence, Italy, June 29–July 4, 1994,* edited by P. Vincenzini. pp. 123 138. Techna Publishers, Faenza.
2000 Defining Ceramic Change and Cultural Interaction: Results of Typological, Chronological and Technological Analyses of Guangala Phase Ceramics *Andean Past* 6:175–208.
2008 Early Regional Polities of Coastal Ecuador. In *Handbook of South American Archaeology,* edited by Helaine Silverman and William H. Isbell, pp. 489–503. Springer, New York.
Masucci, Maria, and Allison Macfarlane
1997 Tracing Ancient Ecuadorian Trade Routes: Application of Geological Survey and Ceramic Petrology to Provenance Studies. *Geoarchaeology* 12(7):765–793.
Masucci, Maria, and Hector Neff
1997 Ceramic Production and Trade: Correlating Mineralogical and Chemical Analyses of Guangala Ceramics, Ecuador. Paper presented at the 62nd Annual Meeting of the Society for American Archaeology, Nashville, Tennessee.
2006 Ceramic Production and Societal Change in the Manteño of Coastal Ecuador. Paper presented at the 71st Annual Meeting of the Society for American Archaeology, San Juan, Puerto Rico.
McEwan, Colin
1992 Sillas de poder: Evolución sociocultural en Manabí-Costa del Ecuador. In *5000 años de ocupación, Parque Nacional Machalilla,* edited by Mariella García and Presley Norton, pp. 53–70. Ediciones Abya-Yala, Quito.

McEwan, Colin, and Florencio Delgado Espinoza
2008 Late Pre-Hispanic Polities of Coastal Ecuador. In *Handbook of South American Archaeology*, edited by Helaine Silverman and William H. Isbell, pp. 505–525. Springer, New York.

Middleton, Andrew P., Ian C. Freestone, and Morven N. Leese
1985 Textural Analysis of Ceramic Thin Sections: Evaluation of Grain Sampling Procedures. *Archaeometry* 27(1):64–74.

Muse, Michael
1989 Corrología regional y relaciones de territorialidad en la región oeste del área septentrional andina, sigo XVI. In *Relaciones interculturales en el área ecuatorial del Pacífico durante la época precolombina*, edited by Jean-François Bouchard and Mercedes Guinea, pp. 187–197. BAR International Series, Oxford.

Paulsen, Allison C.
1970 A Chronology of Guangala and Libertad Ceramics of the Santa Elena Peninsula in South Coastal Ecuador. Ph.D. dissertation, Department of Anthropology, Columbia University, New York. University Microfilms, Ann Arbor.

Peacock, David P. S.
1977 Ceramics in Roman and Medieval Archaeology. In *Pottery and Early Commerce: Characterization and Trade in Roman and Later Ceramics*, edited by David P. S. Peacock, pp. 21–33. Academic Press, London.

Pettijohn, Francis J.
1975 *Sedimentary Rocks*. 3rd ed. Harper and Row, New York.

Quinn, Patrick S.
2013 *Ceramic Petrography: The Interpretation of Archaeological Pottery and Related Artefacts in Thin Section*. Archaeopress, Oxford.

Reitz, Elizabeth, and Maria Masucci
2004 *Guangala Fishers and Farmers: A Case Study of Animal Use at El Azúcar, Southwestern Ecuador*. University of Pittsburgh Memoirs in Latin American Archaeology No. 14. University of Pittsburgh, Pittsburgh.

Saville, Marshall H.
1910 *The Antiquities of Manabí, Ecuador: Final Report*. Heye Foundation Contributions to South American Archaeology, Vols. 1 and 2. Irving Press, New York.

Stoltman, James B.
1989 A Quantitative Approach to the Petrographic Analysis of Ceramic Thin Sections. *American Antiquity* 54(1):147–160.
1991 Ceramic Petrography as a Technique for Documenting Cultural Interaction: An Example from the Upper Mississippi Valley. *American Antiquity* 56(1):103–121.

Stothert, Karen E.
2001 Manteño. In *Encyclopedia of Prehistory, Volume 5: Middle America*, edited by Peter N. Peregrine and Melvin Ember, pp. 303–327. Published in conjunction with the Human Relations Area Files at Yale University. Kluwer Academic/Plenum Publishers, New York.
2006 La cerámica de etiqueta de las tolas de Japotó, costa del Ecuador. In *Avances de investigación en el Ecuador prehispánico: Número temático*, edited by M. Guinea and J.-F. Bouchard. *Bulletin del'Institut Français d'Études Andines* 35, no. 3:265–284.
2013 The Peoples of the Coast of Ecuador Accommodate the Inca State. *Ñawpa Pacha* 33 (1): 71–102.

Szilágyi, Veronika, and Gyorgy Szakmány
2009 Comparison of Volcaniclastic-Tempered Inca Imperial Ceramics from Paria, Bolivia, with Potential Sources. In *Interpreting Silent Artifacts*, edited by P. S. Quinn, pp. 211–225. Archaeopress, Oxford.

Velasco, Juan de
1842 *Historia del reino de Quito en la América Meridional, año de 1789, tomo III, que contiene la historia moderna*. Imprinter del Gobierno, Quito.

Whitbread, Ian
1989 A Proposal for the Systematic Description of Thin Sections towards the Study of Ancient Ceramic Technology. In *Archaeometry, Proceedings of the 25th International Symposium*, edited by Yannis Maniatis, pp. 127–138. Elsevier, Amsterdam.

Zeidler, James, and Deborah M. Pearsall
1994 *Regional Archaeology in Northern Ecuador, Volume 1: Environment, Cultural Chronology, and Prehistoric Subsistence in the Jama River Valley*. University of Pittsburgh Memoirs in Latin American Archaeology, No. 8. University of Pittsburgh, Pittsburgh.

CHAPTER 4

Petrography and pXRF at San Pedro de Atacama, Northern Chile

Exploring Ancient Ceramic Production

Emily M. Stovel, Beatriz Cremonte, and Ester Echenique

Here we review two recent attempts to broaden our understanding of ancient San Pedro life through archaeometric studies of household and burial ceramic remains. We juxtapose recent petrographic studies of local ceramics with an ongoing effort to characterize local ceramics using portable chemical instrumentation. The present sample focuses on highly polished Brown, Black, and Gray ceramics and thick-walled, coarser samples from both household and burial contexts of the Middle Period (AD 500–900; table 4.1).[1] Through this comparison of ceramic paste mineralogy, we confirm a high level of homogeneity found in previous metric studies (Stovel 2005), contribute new understandings of production decisions made in the past, and reflect on combining portable x-ray fluorescence spectrometry (pXRF) and petrography for a robust understanding of ceramic recipes through time.

San Pedro de Atacama: A Persistent Ceramic Style

San Pedro is a small grouping of kin-based communities located at the confluence of two rivers on the eastern edge of the Atacama Desert (fig. 4.1). These communities, 2,400 m above sea level, are called *ayllus*, a term used in Quechua and Aymara to describe a family-based community with land rights and a sense of common history (see Bastien 1978; Isbell 1997). Throughout prehistory in San Pedro, ayllus carved out a series of agriculturally productive oases between two rivers and a distinctive common material culture tradition (with some differences among ayllus; Nado et al. 2012; Torres-Rouff 2011; Torres-Rouff et al. 2013), while engaging in trade and exchange with many communities in neighboring areas. This interaction sphere consolidated into a more general cohabitation of the south-central Andes (southern Bolivia, northwestern Argentina, and northern Chile) by communities linked through marriage and long-term kin-like relationships (Martínez 1998; Stovel 2008). During the Middle Period in particular, the entire region was influenced by Tiwanaku, a large polity centered in western Bolivia whose material culture is found from Cochabamba in eastern Bolivia to southern coastal Peru. Local graves reflect the possible adoption of larger religious practices (including the consumption of hallucinogenic snuff) and iconographic depictions of rayed heads, frontal staff-bearing figures, and profile anthropomorphized bird and feline attendants (the Southern Andean Iconographic Series: Isbell and Knobloch 2006, 2009). This influence produced a change in burial furnishings in San Pedro, some increase in violence, and possibly changes in power structures within the oases (Agüero and Uribe 2011; Bravo and Llagostera 1986; Hubbe et al. 2012; Torres-Rouff 2008, 2011; Torres-Rouff

TABLE 4.1. San Pedro ceramic chronology

Dates	Period	Phase	Associated ceramic styles
1500–1200 BC	Transition to Formative	Tajane	Los Morros
1200–350 BC	Formative	Tilocalar	Los Morros, A & B Loa Café y Rojo Alisados Rojo Pulido Negro Pulido (early variants)
350 BC–AD 100		Toconao	
AD 100–500		Sequitor	
AD 500–700	Middle Period	Quitor	Negro Pulido (late variants) Incisos, Negro and Rojo Gris Grueso Pulido
AD 700–900		Coyo	
AD 900–1000	Late Intermediate Period	Yaye-Solor	Gris Café Pulido Dupont Negro Pulido Rojo Violáceo Aiquina Café Rojizo Turi Rojo Variants
AD 1000–1300		Turi-Quitor	
AD 1300–1450		Toconce-Zápar	
AD 1450–1550	Late Period	Catarpe	Lasana Café Rojo Revestido

and Costa 2006; Uribe and Agüero 2001). The local ceramic aesthetic, however, did not change during this transition.

A distinctive San Pedro ceramic aesthetic developed over the Late Formative period (1200 BC–AD 500; table 4.1).[2] Ceramic production began with a sporadic, thick ceramic type characterized by large gray inclusions (Los Morros A and B), later accompanied by large smoothed vessels (Loa Rojo Alisado and Toconao urns) and early polished forms after AD 350 (Agüero 2005; Uribe 2004, 2006; Uribe and Ayala 2004). Late Formative and Middle Period burial red and black wares (Rojo Pulido and Negro Pulido) are joined by gray and brown variants in houses (Gris Grueso and Café Pulido; fig. 4.2),[3] in both San Pedro and neighboring Upper Loa River Valley sites (Sinclaire et al. 1997).[4]

Middle Period graves could also include large smoothed utilitarian jars and during the latter half of the period (AD 700–900) incised Coyo bowls (again, black and red; fig. 4.3). The only decoration found in local ceramic production throughout prehistory consists of large applied and incised anthropomorphic faces on large Late Formative smoothed open-necked jars; small, often schematic and abstract anthropomorphic faces on closed-neck Middle Period bottles (Negro Pulido Decorado); and the incisions filled with contrasting white gypsum of Coyo bowls. The black and red dichotomy is retained in the Late Intermediate Period (AD 900–1450) in the production of black polished Dupont and red polished Aiquina bowls (Agüero et al. 1997; Uribe 2002). This stylistic continuity is echoed in metric consistency showing high levels of homogeneity, particularly in the Late Formative and early Middle Periods (Stovel 2005, 2013b). Thus we have a longstanding ceramic tradition of red and black polished wares used as grave furnishings (with a broader range of colors present in houses) that persists for over 1,000 years despite interaction with communities throughout Bolivia, northern Chile, and northwestern Argentina.

The adherence to a local standardized aesthetic despite wide-ranging interaction with neighboring groups and the regular consumption of nonlocal ceramics for local burial purposes, especially during a time of regional political changes tied to the rise of Tiwanaku and Inka states, has prompted an interesting discussion of identity maintenance in an articulated multicultural or multiethnic region (Agüero et al. 1997,

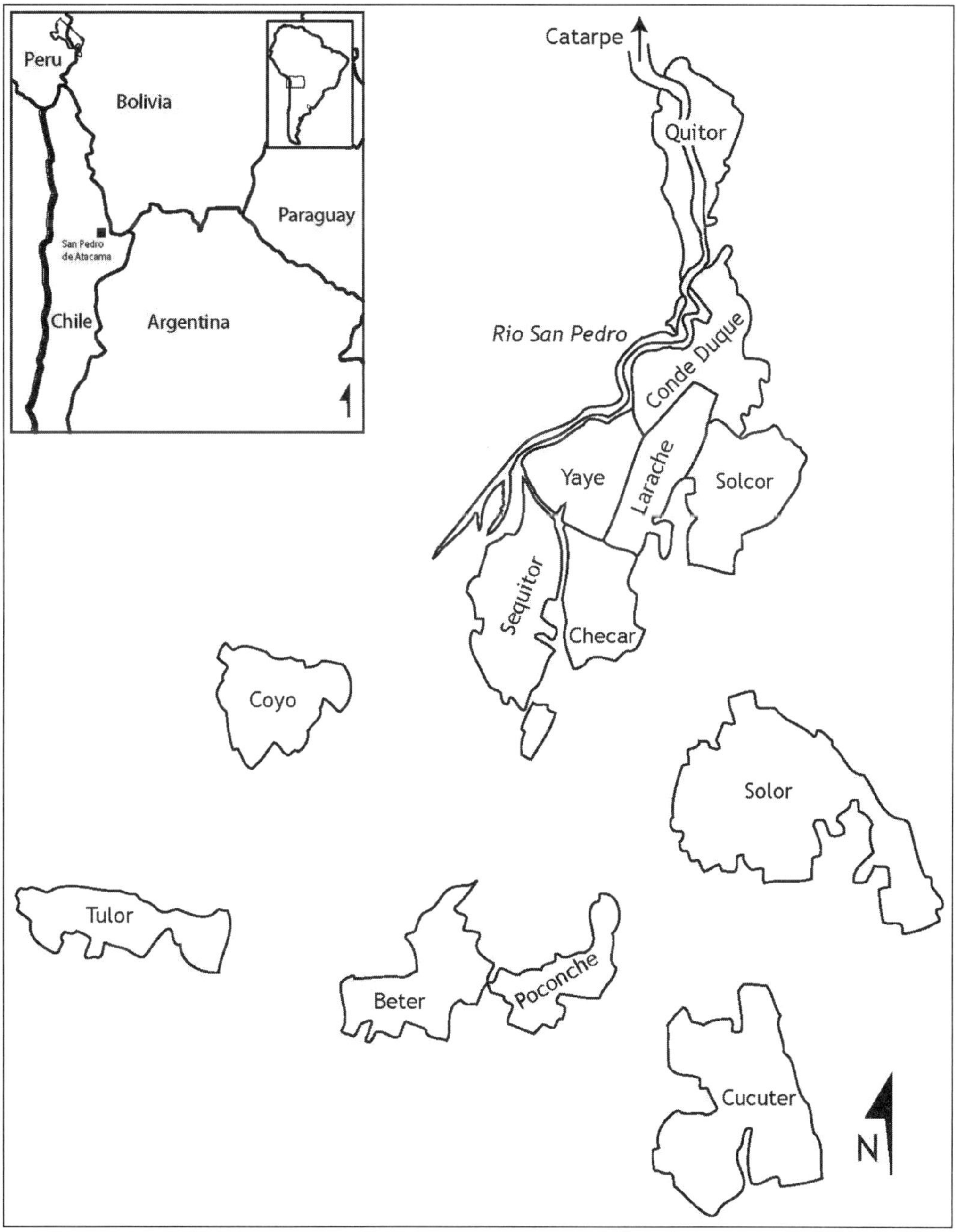

FIGURE 4.1. Map of San Pedro ayllus (prepared by W. Whitehead).

1999; Stovel 2008). Part of our motivation for carrying out chemical and petrographic analysis of San Pedro ceramics was to identify and trace production differences in houses and/or tombs that may have coincided with these political changes and to examine how these changes (if any) might have articulated with stylistic continuity. A first step in understanding temporal, stylistic, and contextual variation, however, involves characterizing local ceramic mineralogy. The present chapter juxtaposes the results of two distinct ceramic petrographic studies

FIGURE 4.2. Common household polished wares, San Pedro de Atacama Middle Period: (*a*) Negro Pulido; (*b*) Gris Grueso; and (*c*) Café Pulido. (photographs from Instituto de Investigaciones Arqueológicas y Museo R. P. Gustavo Le Paige SJ, de la Universidad Católica del Norte).

FIGURE 4.3. Common burial polished and smoothed wares, San Pedro de Atacama Middle Period: (*a*) Negro Pulido bottle; (*b*) Loa Rojo Alisado urn (photographs from Instituto de Investigaciones Arqueológicas y Museo R. P. Gustavo Le Paige SJ, de la Universidad Católica del Norte).

that explore the ceramic pastes of polished and smoothed vessels from the houses and graves of the Middle Period. This comparison is followed by a review of current portable X-ray fluorescence spectrometry research on similar material, in an effort to reflect on the potential for combining both methods in the consideration of prehistoric ceramic technology and social life.

Petrographic Analyses

The two petrographic studies of San Pedro ceramics summarized here involved:

1. a 60-sample study of three ware types from a Late Formative to Middle Period habitation site, Coyo Aldea (Echenique 2012; see also Núñez 2005);
2. a 40-sample comparative study of fine to utilitarian types from various domestic and burial contexts (Stovel et al. n.d.).

Through the review and juxtaposition of these two studies we present here petrographic data for three polished wares (Negro, Café, and Gris Pulido) and utilitarian remains from household contexts and black polished and utilitarian vessel fragments from burial contexts, an almost full complement of ceramic styles from the Middle Period. The less frequent and less well understood red polished ware (Rojo Pulido) was not included, as we have yet to confirm its suspected role as an exclusively burial ceramic (Uribe 2006). Nor were Late Middle Period Coyo incised fragments included here. The results of these studies provide a strong comparative characterization of Middle Period ceramic mineralogy, in particular of the diagnostic San Pedro Negro Pulido.

Petrographic analysis was conducted on thin sections polished to 30 microns by Spectrum Petrographics, Inc., and the Taller de Petrotomia of the Instituto de Geología y Minería at the Universidad Nacional de Jujuy, respectively. The fragments are now housed in the Instituto de Investigación Arqueológica y Museo Gustave Le Paige (IIAM) in San Pedro de Atacama, Chile. Because they remain part of active investigation programs, the thin sections analyzed by Echenique are housed at the University of Arizona and those analyzed by Cremonte are currently held at the Instituto de Geología y Minería of the Universidad Nacional de Jujuy. Copies of reports, theses, and published articles are provided to the IIAM library in San Pedro de Atacama, while raw data are in the hands of the co-authors of this paper and are available upon request. More raw data are also provided in the original analyses reported and compared here.

Petrographic characterization in both studies involved polarized microscopic examination of thin sections (1) to identify the mineralogical composition of the pastes, any paste groupings, and possible raw material sources, and (2) to investigate paste recipes and technological aspects of ceramic manufacturing. In the first study Ester Echenique (2012) employed a qualitative and semiquantitative approach for the identification of mineralogical and technological components. Each rock and mineral was first identified, followed by a description of modal shape, size, and approximate frequency. Shape was estimated using visual guides for determining degree of roundness (rounded, subrounded, subangular, and angular). Mineral sizes were recorded in approximate measures: large grain: > 0.30 mm; medium grain: 0.10 to 0.30 mm; small grain: < 0.10 mm. The approximate amount of each mineral type was recorded as abundant (> 10 percent), moderate (between 5 and 10 percent), or sparse (< 5 percent) (Matthew et al. 1991). Technological variables of ceramic manufacturing were also recorded, such as clay color and the use of one or more clays, clay birefringence, porosity, and surface treatments. In the second study (Stovel et al. n.d.) Beatriz Cremonte employed point-counting in addition to these methods to determine the relative frequency of the individual mineral types (Middleton and Freestone 1991; Quinn 2009; Whitbread 1989), the character of matrix structures (Curtois 1976), and the sizes of inclusions and voids or pores (Orton et al. 1997) in microns (100 measurements per thin section).

The multiple intercept point-counting method adopted for the current study measures an average of 250 points per thin section. This quantity of points provides good data for further

FIGURE 4.4. Echenique's paste groupings: (*a*) local, (*b*) nonlocal.

statistical comparisons and the classification of paste groupings. A mechanical stepping stage was attached to a Leica DM750P polarizing microscope that moves the thin section from left to right, establishing 1 mm intervals across the entire surface of the sample. The intersection of the crosshairs in the microscope's eyepiece at 60× magnification creates points where inclusions and pores above 15 μm and matrix (clay and inclusions below 15 μm) are recorded. This combination of qualitative and quantitative variables allowed further discussion of paste recipes and characterization of ceramic manufacturing processes as well as technological choices inherent to San Pedro polished ceramic production.

Echenique (2012) explored the petrography of three common domestic types: highly polished black, gray, and brown vessels (Negro, Gris and Café Pulido; see also Sinclaire et al. 1997; Uribe 2006). Cremonte (Stovel et al. n.d.) compared the mineralogy and elemental profiles of polished black, rough utilitarian, and incised ceramic types. Both authors identified a predominance of ignimbrite and volcanic rocks in the pastes they examined. Although neither study included the analysis of raw materials, both argued that ceramic paste mineralogy corresponded strongly with the geology of the immediate region surrounding San Pedro and reflected the exploitation of local resources, particularly the chain of volcanoes located immediately to the east of the oases.

Echenique (2012) found one general paste type characterizing all three polished domestic black, gray, and brown wares (table 4.2, fig. 4.4). The Coyo Aldea sherds revealed a paste characterized by abundant (20–50 percent of matrix), large, poorly sorted inclusions of ignimbrites and volcanic rock (andesite or dacite), and less frequently chert, sandstone, and undifferentiated hypabyssal rock (the latter only in graywares: Echenique 2012:42). Major and minor rock-

TABLE 4.2. San Pedro paste components (Echenique analysis)

Paste group label	Number of samples	Rock fragments	Major rock-forming minerals	Minor rock-forming minerals	Opaque minerals
Fabric 1 (local)	59	ignimbrite volcanics chert sandstone undiff. hypabyssal rock	quartz plagioclase	chalcedony nepheline biotite hornblende oxyhornblende orthopyroxene volcanic glass epidote	magnetite hematite
Fabric 2 (nonlocal)	1	None	quartz plagioclase	hornblende cummingtonite aegirine-augite epidote	magnetite

forming minerals identified in this paste group were plagioclase, feldspar, quartz, hornblende, biotite, magnetite and hematite, and less commonly chalcedony, nepheline, aegirine-augite, orthopyroxene, clinopyroxene, epidote, and volcanic glass.

The mineralogical data, in addition to offering explorations of surface treatments and porosity, point to a "distinctive mineralogical and technological homogeneity" (Echenique 2012: 59), a common recipe shared by all polished wares under study. Thus surface color differences were a production decision executed most likely at the firing stage of the manufacturing process rather than being the product of a discrete slip or pigment.[5] Interestingly, the sherds revealed a high proportion of large angular inclusions, indicating either the use of temper or the decision not to refine poorly sorted clays, a quality usually suggestive of cooking pots (Tite et al. 2001). Echenique (2012:66) argues that the presence of "fine walls, high concentrations of inclusions, and high porosity" in these polished domestic wares is consistent with their use as cooking pots (Tite et al. 2001; West 1992). Although the paste characteristics seem to indicate cooking functions, further evidence such as contextual data and functional analyses (use wear and experimental analyses) is needed.

Echenique (2012) also identified one sherd with a different mineralogical profile, which she interprets to be either nonlocal or manufactured with nonlocal materials (table 4.2, fig. 4.4). This sherd is composed of high amount of inclusions (50 percent or more) and poorly sorted fabric of quartz, plagioclase, and magnetite as major mineral inclusions and amphiboles (common amphiboles and commingtonite), pyroxene (aegirine-augite), and epidote as minor inclusions. What distinguishes this sherd from the rest is the absence of volcanic rock fragments and biotite and the presence of rare amphiboles (cummingtonite) not found in local pastes, along with the striking weathering of plagioclase and amphiboles.

Cremonte (Stovel et al. n.d.), in turn, identified five paste groups in her sample of 40 fine polished and utilitarian samples from San Pedro houses and graves (table 4.3, fig. 4.5). These five groups represent variants of the general paste type identified by Echenique with different quantities and combinations of the same mineralogical components. Although five paste groups were identified, they do not co-vary with surface treatment, smoothing or color, or thickness. The entire sample reflects a high degree of petrographic homogeneity. In other words the two classificatory systems of style (determined according to long-established shape and surface treatment classes) and mineralogy do not coincide, implying that that style decisions did not involve changes to paste recipes. Echoing Echenique's work, all pastes were characterized by the major rock-forming minerals quartz

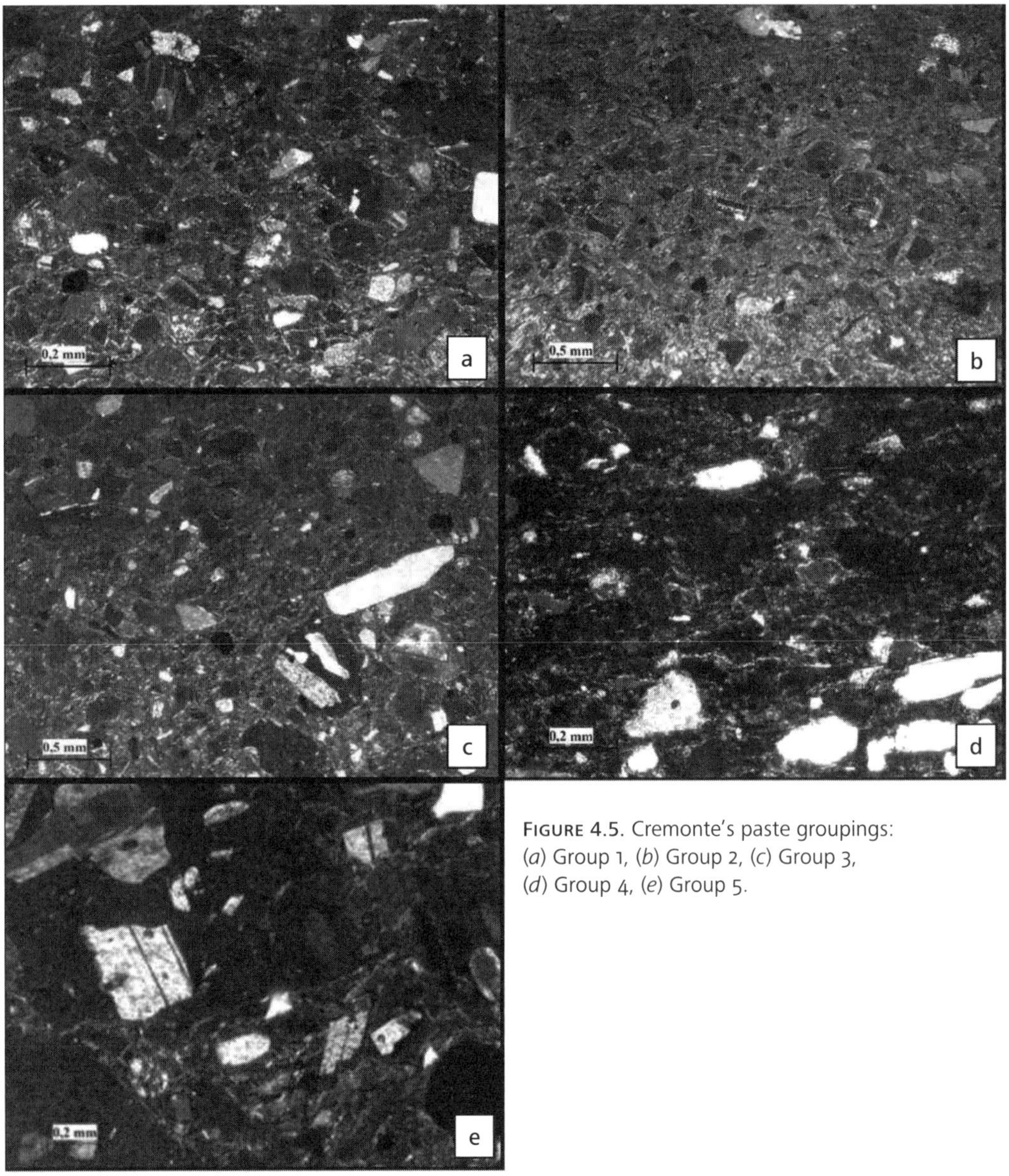

FIGURE 4.5. Cremonte's paste groupings: (*a*) Group 1, (*b*) Group 2, (*c*) Group 3, (*d*) Group 4, (*e*) Group 5.

and plagioclase in addition to the minor minerals hornblende, biotite, and orthopyroxene, followed in lower frequency by clinopyroxene, zircon, apatite, and titanite. Cremonte also noted the abundant ignimbritic inclusions.

Cremonte's Group 3 is the most common, combining household and burial samples and fine and utilitarian fragments. Group 4 includes four samples from the sites Coyo Aldea and Coyo 5 located near the ayllu of Coyo (table 4.4, fig. 4.1), but samples from these sites also appear in Groups 3 and 5, thus belying the idea of some distinctive paste types according to ayllu. Unlike Echenique, however, Cremonte (Stovel et al. n.d.) mentions the presence of pumice in the pastes that she analyzed and that it is more likely to be found in grave ceramics (which Echenique did not analyze). Even so, Cremonte also identified cases of high pumice presence in sherds from both burial and domestic contexts.

TABLE 4.3. San Pedro paste components (Cremonte analysis)

Paste group label	Number of samples	Petrographic components	Quantities (%)
Group 1	5	quartz	11–14
		volcanics	< 8
		ignimbrite	7–16
		pumice	3–16
		plagioclase	2.5–7
		biotite	1.5–2.3
		pyroxene	1–2
		quartzite	0.6–2.5
		amphibole	0–3
Group 2	3	ignimbrite	< 6
		quartz	5.47–11.57
		plagioclase	3.26–4.13
		amphibole	1.36–3.26
		biotite	1.36–2.17
		volcanics	0.83–6
Group 3	26	ignimbrite	14–26
		quartz	7–14
		plagioclase	3–8
		quartzite	< 3
		pumice	< 3
		volcanics	< 3
		biotite	1–4
Group 4	4	ignimbrite	17–25
		quartz	5–8
Group 5	2	ignimbrite	32–35
		quartz	10–14

Future work can shed light on this difference and explore its potential to distinguish tomb and household ceramic production practices (table 4.4).

Finally, both analysts tie these paste characteristics to the local geological landscape and thus to the exploitation of local resources for ceramic manufacturing. The principal components of these paste types are ignimbrites and volcanic rocks (such as andesite/dacite), while other mineral elements such as quartz and plagioclase may have originated from the same volcanic sources through weathering. This component may also have occurred naturally in sediments found in the area. The large size and predominance of angularity of the inclusions, however, suggest that they were included as an intentional temper to the clay matrix or that potters were using unrefined clays naturally mixed with fresh sediments (Echenique 2012:46–49). The geology of the region (fig. 4.6) is dominated by large volcanoes and volcanic deposits, an obvious source for the ignimbrites that so dominate these ceramic sherds. These results are echoed by early petrographic work by César Cortelezzi (appendix to Tarragó 1976), who examined four polished burial ceramics and found a predominance of plagioclase, orthoclase, quartz, hornblende, and

Table 4.4. Data set analyzed in Stovel et al. 2012

Artifact number	Site	Context	Surface treatment	Source
A9	Poconche 12	household	polished	Agüero 2005
A10	Poconche 12	household	polished	Agüero 2005
A83	Tchaputchayna	household	polished	Agüero 2005
A142	Tchaputchayna	household	polished	Agüero 2005
A31	Poconche 12	household	polished	Agüero 2005
A42	No provenance	burial	polished	
A44	Larache	burial	polished	Benavente et al. 1986
A45	No provenance	burial	polished	
A47	Sequitor Alambrado Oriental	burial	polished	Le Paige 1964, 1977; Tarragó 1989
A48	Quitor 8	burial	polished	Le Paige 1964, 1977; Tarragó 1989
A51	No provenance	burial	polished	
A52	No provenance	burial	polished	
A54	No provenance	burial	polished	
A56	No provenance	burial	polished	
A59	Quitor 7	burial	polished	Le Paige 1964, 1977; Tarragó 1989
A61	No provenance	burial	polished	
A556	Coyo 5	household	polished	Llagostera and Costa 1999
A566	Coyo 5	household	polished	Llagostera and Costa 1999
A568	Coyo 5	household	polished	Llagostera and Costa 1999
A569	Coyo 5	household	polished	Llagostera and Costa 1999
A613	Coyo Aldea	household	thick smoothed	Núñez 2005
A614	Coyo Aldea	household	thick smoothed	Núñez 2005
A615	Coyo Aldea	household	thick smoothed	Núñez 2005
A616	Coyo Aldea	household	thick smoothed	Núñez 2005
A617	Coyo Aldea	household	thick smoothed	Núñez 2005
A618	Coyo Aldea	household	thick smoothed	Núñez 2005
A619	Coyo Aldea	household	smoothed	Núñez 2005
A620	Coyo Aldea	household	smoothed	Núñez 2005
A621	Coyo Aldea	household	smoothed	Núñez 2005
A622	Coyo Aldea	household	smoothed	Núñez 2005
A623	No provenance	burial	smoothed	
A624	No provenance	burial	smoothed	
A625	No provenance	burial	smoothed	
A626	No provenance	burial	smoothed	
A627	No provenance	burial	burnished red	
A628	No provenance	burial	smoothed	
A629	No provenance	burial	fine smoothed	
A630	No provenance	burial	burnished red	
A631	No provenance	burial	burnished gray	
A632	No provenance	burial	burnished red	

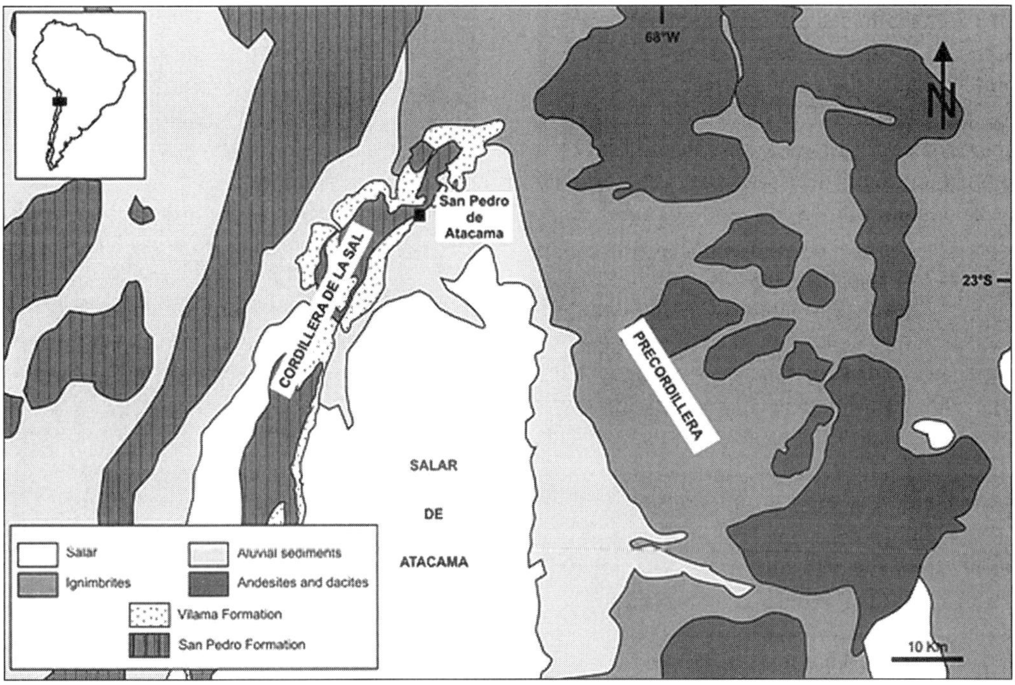

FIGURE 4.6. Geological map of San Pedro and environs (prepared by G. Maro).

biotite, and the abundance of large angular nonplastic inclusions.

Thus although pumice appears to play a larger role in Cremonte's analysis, a comparison of the two studies reveals the identification of the same mineralogical components, similar paste structures, the extraordinary petrographic homogeneity of the ceramic sample, and the conviction that local style ceramics are the product of local raw materials. Both analysts confirm the lack of co-variation of paste mineralogy with surface treatment, surface color, context of use, or wall thickness. There is clearly a correspondence in the pastes recovered in San Pedro houses and graves, regardless of style or functional type. Rather than using thin-walled polished pots as cooking vessels, it may be that fine and utilitarian jars were produced from the same generalized raw material source or recipe. We have evidence that fine polished and utilitarian vessels of all colors show similar paste structures and components, while there is very little evidence of discrete production practices for household and mortuary ceramics.

Ceramic production practices therefore remained unchanged among the various ayllus found in the oases of San Pedro throughout the Middle Period, while interaction with communities throughout Bolivia and northwestern Argentina continued and increased, particularly in the case of the regional polity of Tiwanaku. It is interesting that a distinctive ceramic aesthetic, bolstered by a specific ceramic recipe shared by ceramic types from graves to houses, from fine polished pots to large smoothed urns, should remain clearly unchanged while local inhabitants accentuated wealth differences and social tension (Torres-Rouff 2011; Torres-Rouff and Costa 2006) and adopted new iconography, religious practices, and cranial deformation (Bravo and Llagostera 1986; Lessa and Mendonça de Souza 2004, 2006; Torres-Rouff 2002, 2008).

These social differences are even reflected within and between different ayllus during this period (Nado et al. 2012; Torres-Rouff 2011). Interestingly, nonlocal individuals buried in the oasis during this period were buried with local goods and a local burial identity (Nado et al. 2012; Torres-Rouff and Knudson 2007). This interplay of social polarization with ritual material homogeneity is a strong indicator of the intentional development of a local identity in response to regional changes, perhaps even an indication of the production of local ceramics as emblematic of an ethnic identity (Stovel 2013a; Wiessner 1983, 1984) during the Middle Period. Certainly black polished San Pedro ceramics (in particular ritual burial forms) are the dominant San Pedro ceramic style found traded to or used in communities in southern Bolivia and northern Argentina (Llagostera et al. 1988; Nielsen 2004; Nielsen et al. 1999; Nielsen et al. 2006; Tarragó 1976). The inclusion of household ceramics in this mineralogical homogeneity provides perhaps even more compelling evidence.

pXRF Elemental Data

While this chapter seeks to present the combined results of petrographic studies of key ceramic types in San Pedro, the recent application of pXRF analyses on similar materials gives us an additional line of evidence to consider the ceramic production scenario previously presented. Ongoing elemental analysis of household and burial ceramic remains in San Pedro has occurred since 2007 using a hand-held X-Ray Fluorescence Innovx Alpha 4500 analyzer. Portable instrumentation allowed us to perform the first concerted chemical analyses on San Pedro ceramic materials in the storage area of the museum rather than relocate samples within or outside of Chile. Portable equipment offered us a more generous range of samples than would be permitted through the extraction of objects or sherds from the country. That said, concerns arise with portable devices that are powered by batteries that are less powerful and lack the detection potential of larger models. In addition, these instruments are not calibrated for ceramic analysis and provide algorithms that were designed for environmental contamination analysis or mining.

With these caveats in mind, we developed the following analytical protocol: readings were gathered in Soil Mode for 120 seconds, including a 60-second analysis under normal element identification and 60 seconds under a light element analysis protocol (LEAP). Soil mode was chosen because the algorithm provided by the Innovx has provided archaeologically relevant results in previous analyses (Stovel et al. n.d.). The internal INNOVX algorithm provides data in parts per million for 18 elements, reduced to 12 for statistical analyses to avoid those vulnerable to taphonomic variation (for example, Cl and Ca can fluctuate as salts suspended in water and deposited in sherds postdepositionally) and those at the upper limit of detection for our instrument (such as Co). Instrument calibration was achieved through National Institute of Standards and Technology standards and internal standards: two sherds from San Pedro that we repeatedly analyzed throughout the period of analysis. Readings from standards were checked to monitor and mitigate potential instrument drift. More of our analytical protocol can be found in Stovel et al. (n.d.). Elements useful in maximizing differences are evident in univariate and bivariate plots (Baxter and Jackson 2001), logged to normalize large natural differences among elements, and submitted to principal components (PCA) and cluster analyses. While PCAs were conducted on both elemental and petrographic data, the results of elemental PCAs are presented here, because they provide the most striking graphic representation of very slight differences among groups (figs. 4.7 and 4.8). Hence differences between archaeological ceramic categories are accentuated by both the data cleaning process and the statistical model presented. These pXRF studies used samples from different San Pedro sites, although fragments from Coyo Aldea appear in both (table 4.4).

The structural similarities between fine and utilitarian ceramics such that thin-walled polished vessels look like cooking pots from the perspective of their fabrics lead us to argue that fine and utilitarian vessels should look similar chemically because no structural differences exist at the level of porosity and the size, frequency, and type of inclusions. At each stage of analysis (Stovel and Deibel in press; Stovel et al. n.d.) we have confirmed the overall elemental homogeneity of San Pedro ceramics. The lack of elemental

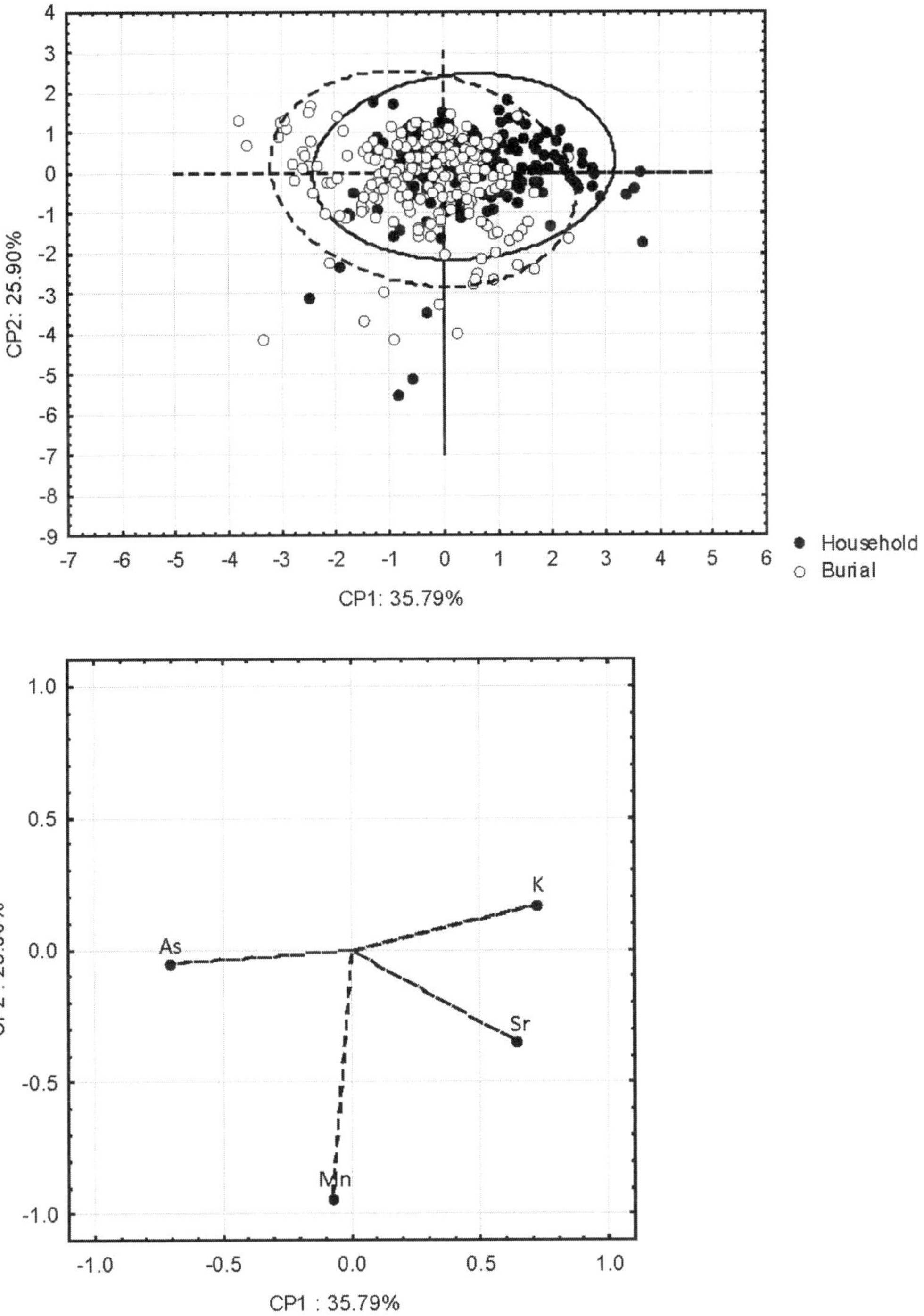

FIGURE 4.7. PCA of chemical data from house versus burial context ceramics.

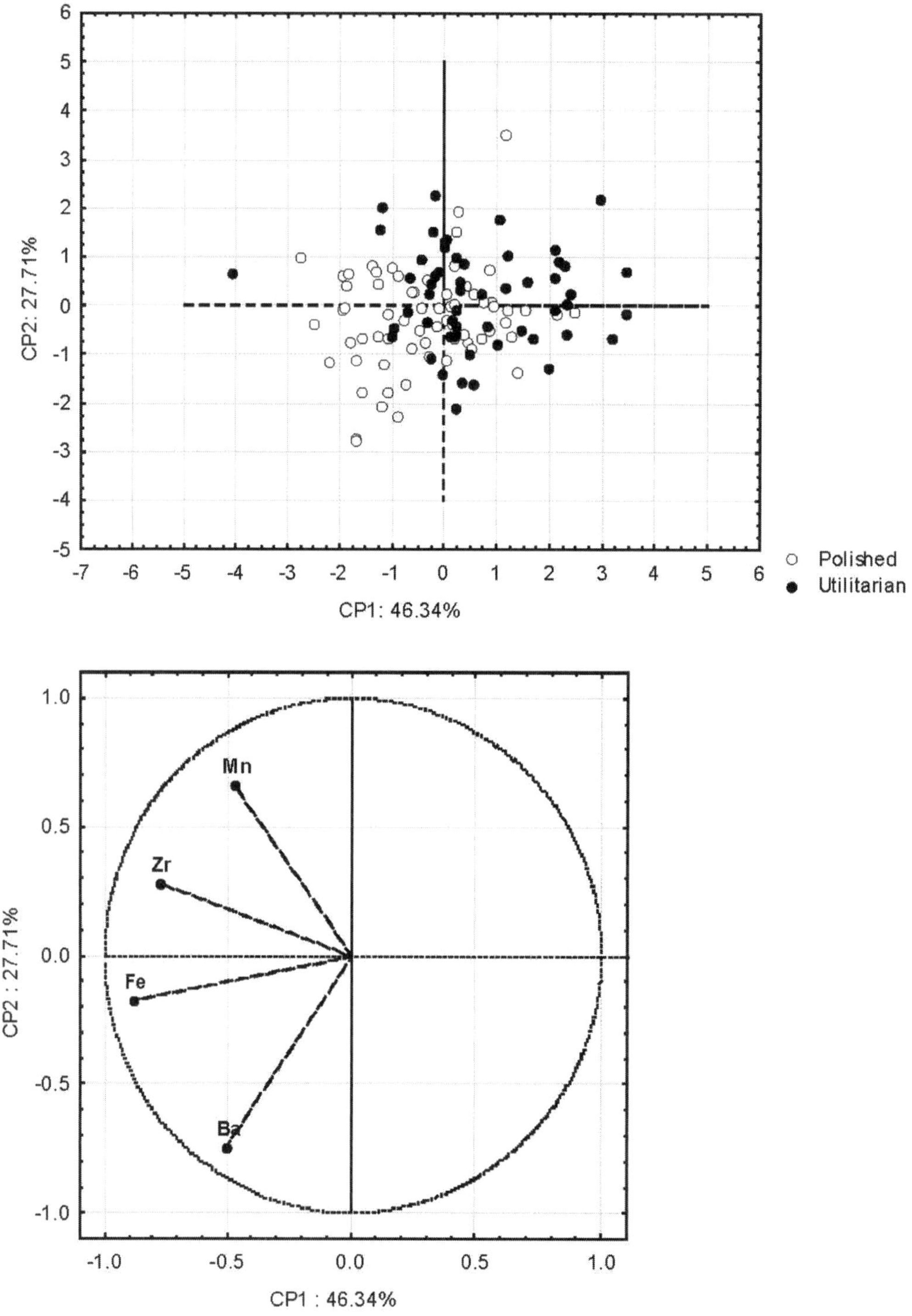

FIGURE 4.8. PCA of chemical data from fine versus utilitarian ceramics.

differences so far found among the various domestic polished ware types (black, brown, and gray; Stovel and Deibelin in press) confirms in general terms the petrographic arguments for homogeneous paste recipes used across functional and ware types. Some slight variations have appeared within this commonality, however, but it is unclear at present whether these are statistically significant as statistical tests are still underway.

A tendency toward differences in Mn and As are found between household and burial ceramics (fig. 4.7) and in Ba, Fe, and Zr between domestic fine and utilitarian vessels (fig. 4.8). The data sets overlap notably in multivariate space that seeks to exaggerate differences, while the eigenvalues are very low for each factor in the analysis, suggesting that any differences between archaeological types are slight. It is unclear what such chemical differences might mean about differences in the ceramic pastes, in particular because we have no knowledge of the source of elemental variation. They might be connected to the variability in mineralogical components that define the five paste variants already described, but these do not vary according to context or ware type. In pastes with such similar types and ranges of inclusions, elemental differences are more likely found in differently sourced clays. Of course, until raw materials are tested and compared with archaeological remains, we cannot determine the baseline variability of local clays, which may be elevated due to strong alluvial action that can mix clays and minerals.

Future work will explore the possible causes of chemical variation through laser ablation inductively coupled plasma mass spectrometry (LA-ICP-MS) or scanning electron microscopy with energy dispersive spectrometry (SEM-EDS), which allow us to pinpoint chemical variation within the ceramic matrix and identify where higher values for certain elements are introduced and how (whether this is a geological source or recipe issue). In other words, it behooves us to compare specific inclusions across these samples to see if the various important components of San Pedro pastes are chemically homogeneous and whether chemical variation is due to the clay matrix.

Social Change, Homogeneity, and Archaeometry

This chapter reviews recent archaeometric research in San Pedro de Atacama, northern Chile, on ceramic materials from the Middle Period (AD 500–900). As this period was characterized by coeval increases in quality of life, social tension, wealth distinctions, symbols of power, and new iconography and burial goods, a consideration of ceramic production practices can speak to the differences found in the burial and household contexts and the persistent production of a distinctive aesthetic style. This is particularly urgent given the lack of evidence of ceramic production in regional surveys. The results presented here speak to a conscious adherence to local ceramic styles and burial identities through the Middle Period, although some chemical differences may exist that warrant future work to tease out distinguishing characteristics of burial ceramic production and fine vessels from contemporary utilitarian urns. Petrographic evidence presented here suggests any differences would be very slight and unlikely to contradict the general homogeneity of ceramic pastes found in a wide range of morphological, decorative, and functional types.

It is clear that the combination of optical mineralogy and handheld chemical instrumentation in this case offers great promise for the exploration of ceramic production and the development of complex hypotheses about human life in the past. These hypotheses can be developed with a small investment in time and resources and can help focus future archaeometric research. The key to this process is to approach a large sample at the outset to minimize the impact of chance data contamination by large inclusions in the use of pXRF. In addition, optical mineralogy provides a vital source of explanation for variation identified in handheld analysis and the two methods have been combined usefully in the past (for example, Burley and Dickinson 2010; Goren et al. 2011). In our case we found interesting intersections between petrographic and chemical data without the aid of comparative samples of local raw materials.

We have corroborating support for a homogeneous volcanic paste recipe that was produced from locally available materials and was used to make a wide range of vessel types from polished black, brown and gray pots to large smoothed jars for use in both houses and graves. Nonplastic inclusions tend to be large and angular and were more likely added as temper because of these characteristics, although further work and comparison with samples of raw materials are needed to refine this hypothesis. Some slight chemical differences are evident that are proposed to be caused by differences in the clay matrix: some categories of vessels were produced with clays from different sources to which these nonplastic inclusions could have been added, thus mitigating chemical differences across the entire sample. It is possible that we are facing discrete production workshops that have access to different clays but use the same sources for nonplastic inclusions. If further work probes this issue more closely, we may be able to identify a community-wide adherence to the use of specific inclusions to "make our pots more San Pedro." This hypothesis, if confirmed, would provide important support for the idea that San Pedro ceramics were produced as an emblematic symbol of community identity in the face of social changes that defined the Middle Period. It is interesting that during the Late Intermediate, or Regional Developments, Period, members of a neighboring community chose to produce similar ceramic styles, using different raw materials (Stovel et al. n.d.). Thus the San Pedro recipe was shared by ancient producers in different ayllus, some characterized by internal social inequality.

Acknowledgments

Parts of this research were funded by a VRIDT Project (162/2011) through the Universidad Católica del Norte (Antofagasta, Chile), Proyecto Anillos ACT-96, the Tinker Foundation, a Reicker grant, the Emil W. Haury Education Fund, and the William and Nancy Sullivan Scholarship. We would like to thank the editors of this volume for their invitation to explore the results of two similar studies that constitute essentially a blind test of both our results and for their comments on earlier drafts of this paper. In addition we would like to thank Michael Deibel, Mark Fitzwater, David Killick, Guadalupe Maro, and William Whitehead. Without help from the staff and administration of the Instituto de Investigación Arqueológica y Museo Gustavo Le Paige in San Pedro de Atacama, this chapter would not have been possible.

Notes

1. Rice (1987:138) distinguishes between burnishing and polishing in that "polishing is done on a dry surface and gives a uniform luster without the pronounced parallel facets produced by burnishing leather-hard clay." Although such parallel facets are often visible on San Pedro pots, the long tradition of using the term "polished" (for example, Negro Pulido) in this area leads to its predominance over "burnished" in English-language publications.
2. For general discussions of San Pedro ceramic types, see Berenguer et al. (1986), Tarragó (1976, 1989), Thomas et al. (1984), Uribe (2002, 2006), and Uribe and Agüero (2001).
3. Considering the confusion surrounding the term "wares" in ceramic studies (Rice 1987:286–87), the term is used here to refer to the strikingly different colors of local polished ceramics.
4. Sinclaire et al. (1997:288–290, 301) group gray and brown household variants found in neighboring Alto Loa into a group Séquitor Gris–Café Pulido (Polished Gray–Brown Sequitor), equivalent to a wide range of Negro Pulido and Gris Grueso Pulido forms found in San Pedro de Atacama.
5. Echenique (2012:30) used transmitted and reflected light to explore the presence or absence of a slip on the surface of polished ceramics in her study. Reflected light, in particular, helps with the identification of minerals commonly used as a surface pigment in ceramics (such as hematite). Echenique, however, found no added layer of minerals, no color or textural discontinuities, and no differences in the orientation of surface particles indicative of a slip. This is contrary to the findings of César Cortelezzi (appendix to Tarragó 1976), who argued on the basis of petrographic analysis of two samples that a thin layer of hematite may have been applied to

a bisque fired vessel as a paste mixed with water and borax (Tarragó 1976:41).

References Cited

Agüero, Carolina
2005 Aproximación al asentamiento humano temprano en los oasis de San Pedro de Atacama. *Estudios Atacameños* 30:29–60.

Agüero, Carolina, and Mauricio Uribe
2011 Las sociedades Formativas de San Pedro de Atacama: Asentamiento, cronología y proceso. *Estudios Atacameños* 42:53–78.

Agüero, Carolina, Mauricio Uribe, Patricia Ayala, and Bárbara Cases
1997 Variabilidad textil durante el período intermedio tardío en el Valle de Quillagua: Una aproximación de la etnicidad. *Estudios Atacameños* 14:263–290.
1999 Una aproximación arqueológica a la etnicidad y el rol de los textiles en la construcción de la identidad cultural en los cementerios de Quillagua. *Gaceta Arqueológica Andina* 25:167–198.

Bastien, Joseph W.
1978 *Mountain of the Condor: Metaphor and Ritual in an Andean Ayllu.* West Publishing Company, St. Paul.

Baxter, Michael J., and Caroline M. Jackson
2001 Variable Selection in Artefact Compositional Studies. *Archaeometry* 43:253–268.

Benavente Aninat, Maria-Antonia, Claudio Massone Mezzano, Carlos Thomas Winter
1986 Larrache, evidencias atipicas. ¿Tiahuanaco En San Pedro de Atacama? *Chungara Revista de Antropología Chilena* 16/17:67–73.

Berenguer, José, Ángel Deza, Álvaro Roman, Agustín Llagostera
1986 La secuencia de Myriam Tarragó para San Pedro de Atacama: Un test por termoluminiscencia. *Revista Chilena de Anthropologia* 5:17–54.

Bravo, Leandro, and Agustín Llagostera
1986 Solcor 3: Un aporte al conocimiento de la cultura San Pedro. Período 500 al 900 d.C. *Chungara Revista de Antropología Chilena* 16/17:323–332.

Burley, David V., and William R. Dickinson
2010 Among Polynesia's First Pots. *Journal of Archaeological Science* 37:1020–1026.

Curtois, Liliane
1976 *Examen au microscope pétrographique des céramiques archéologiques.* Notes et Monographies Techniques 8. CRA, CNRS, Paris.

Echenique, Ester
2012 Ceramic Production in San Pedro de Atacama, Chile: A Petrographic Study of Pottery from Coyo Aldea. Master's thesis, Department of Anthropology, University of Arizona, Tucson.

Freestone, Ian C.
1991 Extending Ceramic Petrology. In *Recent Developments in Ceramic Petrology*, edited by Andrew Middleton and Ian C. Freestone, pp. 399–410. Occasional Paper No. 81. British Museum, London.

Goren, Yuval, Hans Mommsen, and Jörg Klinger
2011 Non-destructive Provenance Study of Cuneiform Tablets Using Portable X-Ray Fluorescence (pXRF). *Journal of Archaeological Science* 38:684–696.

Hubbe, Mark, Macarena Oviedo, and Christina Torres-Rouff
2011 Estado de conservación y contextualización cronológica de la Colección Osteológia "Gustavo Le Paige." *Estudios Atacameños* 41:29–44.

Hubbe, Mark, Christina Torres-Rouff, Walter Alves Neves, Laura M. King, Pedro Da Gloria, and María Antoinetta Costa
2012 Dental Health in Northern Chile's Atacama Oases: Evaluating the Middle Horizon (AD 500–1000) Impact on Local Diet. *American Journal of Physical Anthropology* 148:62–72.

Isbell, William H.
1997 *Mummies and Mortuary Monuments: A Postprocessual Prehistory of Central Andean Social Organization.* University of Texas Press, Austin.

Isbell, William H., and Patricia J. Knobloch
2006 Missing Links, Imaginary Links: Staff God Imagery in the South Andean Past. In *Andean Archaeology, Volume 3: North and South*, edited by William H. Isbell and Helaine I. Silverman, pp. 307–351. Kluwer Academic/Plenum Publishers, Springer, New York.
2009 SAIS: The Origin, Development, and Dating of Tiahuanaco-Huari Iconography. In *Tiwanaku: Papers from the 2005 Mayer Center Symposium at the Denver Art Museum*, edited by Margaret Young-Sánchez, pp. 163–210. Frederick and Jan Mayer Center for Pre-Columbian and Spanish Colonial Art at the Denver Art Museum, Denver.

Le Paige, Gustavo
1964 *El precerámico en la cordillera atacameña y los cementerios del período Agro-Alfarero de San Pedro de Atacama*. Universidad Católica, Santiago, Chile.
1977 Recientes descubrimientos arqueológicos en la zona de San Pedro de Atacama. *Estudios Atacameños*:109–124.

Lessa, Andrea, and Sheila M. Mendonça de Souza
2004 Violence in the Atacama Desert during the Tiwanaku Period: Social Tension? *International Journal of Osteoarchaeology* 14: 374–388.
2006 Broken Noses for the Gods: Ritual Battles in the Atacama Desert during the Tiwanaku Period. *Memórias do Instituto Oswaldo Cruz* 101:133–138.

Llagostera, Agustín, and María Antonietta Costa
1999 Patrones de asentamiento en la época agroalfarera de San Pedro de Atacama (norte de Chile). *Estudios Atacameños* 17:175–206.

Llagostera, Agustín, María Antoinetta Costa, and Francisco Téllez
1988 Interacciones transandinas en la formación y consolidación de la cultura San Pedro. Informe Proyecto FONDECYT 1016/86. Unpublished report.

Martínez, José Luis
1998 *Pueblos del chañar y el algarrobo: Los atacamas en el siglo XVII*. Dirección de Bibliotecas, Archivos y Museos, Santiago, Chile.

Matthew, Anthony J., Ann J. Woods, and C. Oliver
1991 Spots before the Eyes: New Comparison Charts for Visual Percentage Estimation in Archaeological Material. In *Recent Developments in Ceramic Petrology*, edited by Andrew P. Middleton and Ian C. Freestone, pp. 211–264. Occasional Paper No. 81. British Museum, London.

Middleton, Andrew P., and Ian C. Freestone, eds.
1991 *Recent Developments in Ceramic Petrology*. Occasional Paper No. 81. British Museum, London.

Nado, Kristin L., Sara J. Marsteller, Laura M. King, Blair M. Daverman, Christina Torres-Rouff, and Kelly J. Knudson
2012 Examinando las identidades sociales locales a través de los patrones de variación biológica y cultural en el ayllu de Solcor, San Pedro de Atacama, Chile. *Chungara Revista de Antropología Chilena* 44:341–357.

Nielsen, Axel E.
2004 Aproximación a la arqueología de la frontera tripartita Bolivia-Chile-Argentina. *Chungara Revista de Antropología Chilena* 36:861–878.

Nielsen, Axel E., José Berenguer, and Cecilia Sanhueza
2006 El Qhapaqñan entre Atacama y Lípez. *Intersecciones en Antropología* 7:217–234.

Nielsen, Axel E., María M. Vázquez, Julio C. Avalos, and Carlos I. Angiorama
1999 Prospecciones arqueológicas en la reserva "Eduardo Avaroa" (sud Lipez, dpto. Potosí, Bolivia). *Relaciones de la Sociedad Argentina de Antropología* 24:95–124.

Núñez, Lautaro
2005 La naturaleza de la expansión aldeana durante el formativo tardío en la cuenca de Atacama. *Chungara Revista de Antropología Chilena* 37:165–193.

Orton, Clive, Paul Tyers, and Alan Vince, eds.
1997 *La cerámica en arqueología*. Edición Crítica, Barcelona.

Quinn, Patrick Sean, ed.
2009 *Interpreting Silent Artefacts: Petrographic Approaches to Archaeological Ceramics*. Archaeopress, Oxford.

Rice, Prudence M.
1987 *Pottery Analysis: A Sourcebook*. University of Chicago Press, Chicago.

Sinclaire, Carole, Mauricio Uribe, Patricia Ayala, and Josefina González
1997 La alfarería del Período Formativo en la región del Loa Superior: Sistematización y tipología. *Actas del XIV Congreso Nacional de Arqueología Chilena, tomo II, Contribuciones Arqueológicas* 5:285–314.

Stovel, Emily M.
2005 The Archaeology of Identity Construction: Ceramic Evidence from Northern Chile. In *Global Archaeological Theory: Contextual Voices and Contemporary Thoughts*, edited by Pedro Paulo A. Funari, Andrés Zarankin, and Emily Stovel, pp. 145–166. Springer, New York.
2008 Interaction and Social Fields in San Pedro de Atacama, Northern Chile. In *The Handbook of South American Archaeology*, edited by Helaine I. Silverman and William H. Isbell, pp. 979–1002. Springer, New York.
2013a Concepts of Ethnicity and Culture in

Andean Archaeology. *Latin American Antiquity* 24:3–20.
2013b Prehistoric Atacameño Ceramic Styles and Chronology Reassessed. *Chungara Revista de Antropología Chilena* 45(3):371-385.

Stovel, Emily M., Beatriz Cremonte, Guadalupe Maro, and Michael A. Deibel
2012 Comparative pXRF and Petrography: Ancient Household and Burial Ceramics, San Pedro de Atacama, Northern Chile. Manuscript in possession of lead author, in preparation for submission to the *Journal of Archaeological Science*.

Stovel, Emily M., and Michael A. Deibel
In press San Pedro de Atacama, Northern Chile: The Domestic Ceramics of the Late Formative and Middle Periods. In *The Southern Andean Iconographic Series*, edited by William H. Isbell, Mauricio Uribe, and Anne Tiballi. Cotsen Institute of Archaeology, University of California, Los Angeles.

Tarragó, Myriam
1976 Alfarería típica de San Pedro de Atacama (norte de Chile). *Estudios Atacameños* 4:37–64.
1989 Contribución al conocimiento arqueológico de las poblaciones de los oasis de San Pedro de Atacama en relación con los otros pueblos puneños, en especial el sector septentrional del valle de Calchaquí. Ph.D. dissertation, Universidad Nacional del Rosario, Argentina.

Thomas, Carlos, Claudio Massone, and M. Antonia Benavente
1984 Sistematización de la alfarería del área de San Pedro de Atacama. *Revista Chilena de Antropología* 4:49–119.

Tite, Michael S., Vassilis Kilikoglou, and Giorgos Vekinis
2001 Strength, Toughness and Thermal Shock Resistance of Ancient Ceramics, and Their Influence on Technological Choice. *Archaeometry* 43:301–324.

Torres-Rouff, Christina
2002 Cranial Vault Modification and Ethnicity in Middle Horizon San Pedro de Atacama, Chile. *Current Anthropology* 43:163–171.
2008 The Influence of Tiwanaku on Life in the Chilean Atacama: Mortuary and Bodily Perspectives. *American Anthropologist* 110:325–337.
2011 Hiding Inequality beneath Prosperity: Patterns of Cranial Injury in Middle Period San Pedro de Atacama, Northern Chile. *American Journal of Physical Anthropology* 146:28–37.

Torres-Rouff, Christina, and María Antonietta Costa
2006 Interpersonal Violence in Prehistoric San Pedro de Atacama, Chile: Behavioral Implications of Environmental Stress. *American Journal of Physical Anthropology* 130:60–70.

Torres-Rouff, Christina, and Kelly J. Knudson
2007 Examining the Life History of an Individual from Solcor 3, San Pedro de Atacama: Combining Bioarchaeology and Archaeological Chemistry. *Chungara Revista de Antropología Chilena* 39:235–257.

Torres-Rouff, Christina, Kelly J. Knudson, and Mark Hubbe
2013 Issues of Affinity: Exploring Population Structure in the Middle and Regional Developments Periods of San Pedro de Atacama, Chile. *American Journal of Physical Anthropology* 152:370–382.

Uribe, Mauricio
2002 Sobre alfarería, cementerios, fases y procesos durante la prehistoria tardía del desierto de Atacama (800–1600 DC). *Estudios Atacameños* 22:7–32.
2004 Alfarería, arqueología y metodología: Aportes y proyecciones de los estudios cerámicos del Norte Grande de Chile. Master's thesis, Department of Anthropology, Universidad de Chile.
2006 Sobre cerámica, su origen y complejidad social en los Andes del desierto de Atacama, norte de Chile. In *Esferas de interacción prehistóricas y fronteras nacionales modernas: los Andes sur centrales*, edited by Heather Lechtman, pp. 449–502. Instituto de Estudios Peruanos de Lima and Institute of Andean Research, New York.

Uribe, Mauricio, and Carolina Agüero
2001 Alfarería, textiles y la integración del Norte Grande de Chile a Tiwanaku. *Boletín de Arqueología Pontificia Universidad Católica del Perú* 5:397–427.

Uribe, Mauricio, and Patricia Ayala
2004 La alfarería de Quillagua en el contexto formativo del Norte Grande de Chile (1.000 aC–500 dC). *Chungara Revista de Antropología Chilena* 36:585–597.

West, Steven Michael
1992 Temper, Thermal Shock and Cooking Pots: A Study of Tempering Materials and Their Physical Significance in Prehistoric and Traditional Cooking Pottery. Master's thesis, Department of Materials Science and Engineering, University of Arizona, Tucson.

Whitbread, Ian K.
1989 A Proposal for the Systematic Description of Thin Sections towards the Study of Ancient Ceramic Technology. In *Archaeometry: Proceedings of the 25th International Symposium*, edited by Yannis Maniatis, pp. 127–138. Elsevier, Amsterdam.

Wiessner, Polly
1983 Style and Social Information in Kalahari San Projectile Points. *American Antiquity* 48:253–276.
1984 Reconsidering the Behavioral Basis for Style: A Case Study among the Kalahari San. *Journal of Anthropological Archaeology* 3:190–234.

CHAPTER 5

Technical Comparisons of Halaf and Ubaid Sherds from Tell Ziyadeh

A Pilot Study

Yukiko Tonoike

This chapter presents a pilot project focused on the Halaf and Ubaid period sherds excavated at Tell Ziyadeh, a site located in the Khabur River basin of northern Syria, in order to achieve a better understanding of the prehistoric pottery from Tell Ziyadeh for an upcoming larger project encompassing the entire Khabur region. The period relevant to this study is the time of early agricultural villages, characterized by Halaf and Ubaid cultures, dating to the fifth millennia BC in northern Mesopotamia. The Halaf culture flourished primarily in northern Mesopotamia approximately 5900–5300 BC and was characterized by distinctive painted pottery (Akkermans and Schwartz 2003:115). The sites attributed to the Halaf culture also tend to share other similarities such as round structures called tholoi and the use of rain-fed agriculture as well as stylistically similar figurines and seals (Campbell 1992).

The late Ubaid culture at Tell Ziyadeh dates to approximately 4900–4600 cal BC and the post-Ubaid to 4600–4300 cal BC. The Ubaid culture originated as a southern Mesopotamian phenomenon, but the early Ubaid period does not appear in northern Mesopotamia. Instead the late Ubaid seems to have replaced the Halaf culture by 5300 cal BC (Hole 2001). The Ubaid period is characterized by larger villages consisting of rectangular buildings, early public architecture, and buff pottery with a greenish tinge and simple geometric designs.

While Tell Ziyadeh is primarily an Ubaid and post-Ubaid site, more than 50 painted Halaf sherds were found out of context in the basal layers of the site (fig. 5.1). Presumably there once was a small Halaf settlement, most of which has disappeared, possibly as a result of flooding. Although in general radiocarbon dating suggests some overlap between the Halaf and Ubaid societies, in the Khabur region it is commonly accepted that the late Ubaid was intrusive and was not a development out of the earlier cultures.

The study of ceramic manufacturing techniques through macroscopic and microscopic analyses has the potential to help us understand the relationship between Halaf and Ubaid period occupations by focusing on the practice and process of making ceramic vessels. Through analytical methods such as petrography and X-ray diffraction, we should be able to identify similarities or differences in resource use and production techniques. If potters of both periods utilized local clays, any observable differences in aspects such as paste, temper, firing, and surface treatment would relate to cultural preferences rather than to availability of source materials. Along with other lines of evidence, this will help us address the question of whether the Ubaid period at Tell Ziyadeh is a gradual development

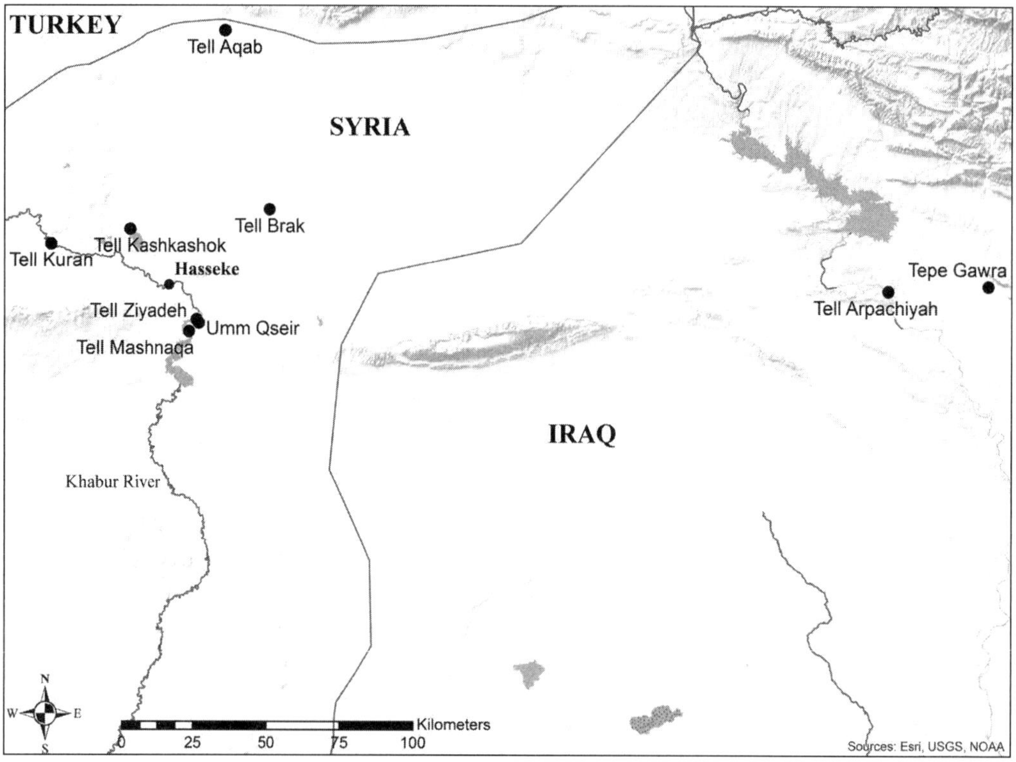

FIGURE 5.1. Map showing location of Tell Ziyadeh, Syria.

from the previous Halaf period or whether it is a later intrusion.

Broader Study

Due to the extensive survey and excavation carried out by Yale University in the Khabur region of northeastern Syria, artifacts from some 8,000 years of history are available for study. The main goals of the Yale projects "were to document the early prehistory" and "to better understand the relationship between environmental changes and human responses" (Hole 2000:609). Based on results from the current pilot study of the Halaf and Ubaid period sherds, we should be able to document differences in ceramic technology, uses of different raw material sources, and evidence of exchange over this long span of time for the greater Khabur region. These results will inform about aspects of interactions within and outside the Khabur Basin, which, in turn, will provide insight into the changing social and political milieu. While the early cultures operated in limited spheres, at least from the third millennium BC sites in the Khabur Basin were linked through political and commercial activities to a much broader region (Hole, personal communication, 2013). Data from this analysis are expected to delineate when boundaries of interaction expanded and contracted and to suggest tangible links to social and political events known from historic sources.

Tell Ziyadeh

The site of Tell Ziyadeh falls into the area influenced by the Halaf and Ubaid period cultures. Tell Ziyadeh is located approximately 12.5 km southeast of the northeastern town of Hasseke, Syria, on the middle Khabur River. It is currently a mound 8 m high above the sterile river terrace, measuring approximately 120 m by 90 m. This is substantially smaller than its original size, however, as the river has eroded much of the southern and southeastern part of the mound in addition to an undetermined amount of erosion from the top of the mound (Hole 2000).

The first excavations were carried out by

FIGURE 5.2. Examples of Halaf painted pottery from Tell Ziyadeh (*left*: Z02; *right*: Z07).

the International Institute of Mesopotamian Area Studies for three years from 1988 to 1990 (Buccellati et al. 1991). Further excavations were carried out by Yale University for two seasons between 1995 and 1997 (Hole and Arzt 1998). The goals of the Yale excavations were to open a large Ubaid architectural exposure and to understand the archaeological sequence of the site, which covered two periods of occupation: one dating to the Ubaid through post-Ubaid periods and a later occupation dating to the early third millennium BC (Hole 2000, personal communication, 2013).

In addition, a few Halaf painted sherds dating to the late sixth millennium BC were found out of context near the basal Ubaid levels of the site, suggesting that the earliest settlement at the site was a Halaf occupation layer (fig. 5.2). No occupation layer dating to this period has been found, so it has been suggested that this layer may have been washed away by erosion from the river prior to the following settlement by the late Ubaid people (Hole 2000). A few hundred meters upstream from Tell Ziyadeh a middle to late Halafian site called Tell Umm Qseir has been located. Tell Umm Qseir was first excavated by the Yale team led by Hole and then by a Japanese team from Tsukuba University. Tell Umm Qseir has ceramics of the middle Halaf style and has been dated between 5600 and 5900 cal BC (Hole 2000). It was a very small settlement with tholos-like and rectangular structures and plant and animal remains suggesting a permanent agricultural settlement that utilized both local wild and domestic resources. The Halaf sherds from Tell Ziyadeh look stylistically different from the ones from Tell Umm Qseir, however, so it is suggested that the Halaf occupation at Tell Ziyadeh was slightly earlier (Hole, personal communication, 2013).

The main occupation layer covering most of Tell Ziyadeh dates to the late Ubaid to post-Ubaid periods. A succession of several buildings and features accumulated over a span of some 600 years, approximately dating between 4900 and 4300 BC. Ubaid ceramics in general are characterized by a buff-colored background with characteristic designs painted with dark paint (fig. 5.3). They have been found over a large area ranging from the Persian Gulf across the Mesopotamian Plains, the western Zagros Mountains, southern Anatolia, and even along the Mediterranean (see Carter and Philip 2010 for more discussion on the Ubaid in general). Despite regional variations, Ubaid ceramics from Tell Ziyadeh are comparable with those from other sites located on the Khabur River, such as Tell Mashnaqa, Tell Kashkashok, and Tell Kuran (Hole 2000). The faunal and botanical evidence suggests that the Ubaid people at Tell

FIGURE 5.3. Examples of Ubaid painted pottery from Tell Ziyadeh (*left*: Z16; *right*: Z23).

Ziyadeh carried out wheat agriculture as well as sheep and goat husbandry and relied heavily on local game from the steppe and riverine resources. There are remarkably few exotic materials or artifacts in the Ubaid occupation layer (Hole 2000).

Ceramic Analysis

Although substantial work has been carried out on the stylistic aspects of painted prehistoric pottery in the Near East, especially for the Halaf painted pottery (see Frankel 1979), very little else is known. Compared to stylistic studies, a very limited number of archaeometric studies have focused on these ceramics (for the Halaf ceramics: Clop Garcia et al. 2004; Davidson 1981; Davidson and McKerrell 1976, 1980), especially on the Ubaid ceramics (Berman 1989; Davidson and McKerrell 1980; Eiland and Williams 2000). The main period of settlement at Tell Ziyadeh was the Ubaid period, so the first goal was to characterize the Ubaid pottery. The Halaf painted pottery was also characterized as a means of comparison. No previous archaeometric studies have been carried out on the Tell Ziyadeh pottery or on the prehistoric pottery of the Khabur region in general.

The macroscopic and microscopic observations help answer technical questions such as whether the manufacturing processes for the Halaf and Ubaid vessels were the same. Various aspects such as paste, temper, firing, surface treatment, and hardness, based on the Moh's hardness scale, were considered for this purpose. In addition, observations were made on the decorated vessel surfaces, revealing differences in the approach to painting, such as the quality of line, the type of brush used, and the general care in the manufacture of the vessels. The sherds from the two periods have a distinctively different feel and look, so we hypothesized that differences in these variables would aid in objectively characterizing the pottery from these two periods and help to get closer to answering the question of whether the potters living in Tell Ziyadeh in the Halaf and Ubaid periods used the same methods.

The main method utilized was petrographic analysis accompanied by careful macroscopic observations of the paste and the vessel surface using a hand lens and a Dino-Lite Pro AM413ZTAS digital microscope. Petrographic analysis was chosen due to its unique ability to focus on the microstructure of the ceramics in addition to its accessibility and low cost. Furthermore, I had used this method successfully in studying other Near Eastern ceramics in a previous study (see Tonoike 2010 for details and more discussion on the advantages of petrographic analysis). A small-scale X-ray diffraction study focusing on the clay particles was also carried out. The traditional assumption has been that vessels from both periods were made from the local clay, based on availability. If this is true, any variation seen in the sherds of the two periods would be a result of cultural preferences in the manufacturing techniques or clay source selection or a change in the clays themselves (for example, the deposition of a new clay source as a result of changes in the river) or a combination of these. X-ray diffraction, in combination with petrographic analysis, enabled me to consider such aspects, by allowing us to compare the clay fraction of the vessels of the two periods.

From Hole's collection of sherds from Tell Ziyadeh, currently housed in the Department of Anthropology at Yale University, 10 Ubaid painted sherds were chosen for thin sectioning in addition to 7 Halaf painted sherds for comparison. At the end of analyses, all sherds will be donated to the Peabody Museum of Natural History at Yale University, together with copies of any publications. Actual data can be obtained from the author. These sherds were all carefully selected from the assemblage in order to cover the variation in vessel type, macroscopically visible paste variation, and vessel wall thickness. I carried out thin sectioning at the Department of Anthropology at Yale University. The initial cut of all of the sherds was made longitudinally in order to observe as much of the vessel profile as possible. The petrographic analysis was done in the Yale University Archaeology Laboratories (YUAL) at the Department of Anthropology, using an Olympus BX60 microscope set up with transmitted light, with objective lenses with magnifications of ×10, ×20, and ×40 with a ×10 eyepiece.

Much petrographic research has focused on sourcing ceramics (for example, Quinn et al. 2010; Stoltman 1991) based on the nature of the site and the style of ceramics. But this pilot project was carried out under the traditional assumption that the Halaf and Ubaid painted pottery from Tell Ziyadeh was manufactured locally, until proven otherwise. Therefore more emphasis was placed on microstructural analysis by carrying out a thorough observation of all aspects of the sherds in order to understand their variability for future fabric classification as well as manufacturing technology. To this end, textural analysis, such as the qualitative and semi-quantitative description of various aspects of the ceramic microstructure, "the shapes and arrangement of voids, micromass (material < 0.1 mm in size) and inclusions," became crucial (Whitbread 1989:129).

When describing ceramic thin sections, especially texture, it is important to use standardized terms so that the results can be compared to other related studies (Shepard 1985; Whitbread 1989). Therefore, following Whitbread (1989, 1995), modifiers and other descriptive terms borrowed from sedimentary petrography and soil micromorphology, such as those proposed by Bullock et al. (1985), FitzPatrick (1984), and Kemp (1985), were used to describe the various features characterizing the fabric of the sherds. The inclusions and voids were counted using a combination of the area counting and point-counting methods, as appropriate (see Middleton et al. 1985). Some samples were reanalyzed in order to review the data for consistency.

Following this methodology, the fine/coarse inclusion size boundary was considered to be 62 microns (0.062 mm) and the groundmass refers to anything smaller than 10 microns in size. Size modifiers were used to divide the size classes of inclusions and voids. "Mega" refers to voids and inclusions larger than 2 mm and "macro" refers to voids and inclusions between 0.5 and 2 mm. For voids, "meso" refers to sizes between 0.05 and 0.5 mm and "micro" refers to the visible range below 0.05 mm (50 microns). For inclusions, "meso" refers to sizes between 0.0625 (62.5 microns) and 0.5 mm. Fine inclusions are synonymous with the micro size range.

Macroscopic and Petrographic Results

Ubaid Sherds

The 10 Ubaid painted sherds selected for this pilot project were chosen with a careful attempt to include the range of variation in qualitative criteria (macroscopically visible paste variation and vessel wall thickness) of the fine painted wares such as fine cups (n = 5), goat bowls (n = 3), and eye bowls (n = 2).

In general, Ubaid painted sherds from Tell Ziyadeh have simple dark painted geometric designs on a buff background (table 5.1). They have a chalky feel and leave a film of white dust on the fingers when handled. No clear slip layer was visible macroscopically or with a hand lens. This was also confirmed by petrographic analysis. The wall thickness of the Ubaid painted sherds varies greatly, ranging between 2.7 mm and 10.3 mm. Z17 is an extreme example at 10.3 mm, with the next thickest Ubaid painted sherd having a wall thickness of 6.8 mm. The average wall thickness is 5.3 mm; but the average wall thickness measures 4.7 mm if Z17 is omitted. Even with such a large variation in vessel wall thickness, however, the surface hardness is extremely uniform and fairly soft at 3 on the Moh's hardness scale for all 10 sherds.

Petrographically, the Ubaid painted vessels from Tell Ziyadeh are also fairly consistent (see table 5.1). They generally are evenly fired with reddish brown or gray groundmass with varying degrees of optical activity (Z16 and Z21 are optically inactive), almost no inclusions measuring more than 0.5 mm, and only a small number of voids measuring more than 0.5 mm. Z21 and Z24 are extreme, with no inclusions and almost no voids measuring larger than 0.5 mm. Z16 is a little different, having a few macro voids following the general trend but also some inclusions, measuring approximately 0.5 mm, which is the lowest end of the macro size range.

The percentage of all voids in the area of field of view (coarse inclusion:fine inclusion:groundmass:void: c:f:g:v, which adds up to 100 percent) is fairly consistent, ranging from 1 to 2 percent, with only Z19 having 3 percent. The types and sizes of the voids vary from sherd to sherd, however, although there are some patterns. For example, mega-sized voids (larger than 2 mm) are almost nonexistent (only Z21 has one channel measuring 2.5 mm) and the majority of the voids are meso- to micro-sized (under 0.5 mm). Macro-sized voids (between 0.5 and 2 mm) range from 3 to 7 percent of all voids in all Ubaid painted pottery analyzed (fig. 5.4). Channel voids, indicating the use of chaff temper, are the least abundant in all analyzed Ubaid painted pottery. Several Ubaid samples show secondary calcite developing along the void walls.

The inclusions in the Ubaid samples show a clear bimodal distribution. Coarse inclusions, larger than 0.62 mm (62 microns), are rare. The percentage is fairly consistent, ranging from 0.2 to 1 percent of the field of view (c:f:g:v). Fine inclusions, between 10 to 62 microns, are also very rare, with less than 1 percent for all (except Z24, which still only had 2 percent). The mineral components of the coarse inclusions are mainly quartz and feldspar with varying but fairly small amounts of hornblende, biotite, muscovite, augite, calcite, grog, and mudstone in all samples. Most of the coarse inclusions are angular or semiangular, suggesting that they were not part of the clay deposit but were more likely intentionally added as temper. But the grog (which often is accompanied by a separation void) and mudstone tend to be rounded. The fine inclusions are of similar minerals. They are rounded, however, suggesting that they were a natural component of the clay. In addition, although ranging slightly in amounts from 0.2 to 2 percent of field of view (not counted when c:f:g:v is discussed), all Ubaid painted pottery has amorphous concentration features such as small amounts of black opaques in the micro to meso size range (smaller than 0.5 mm), which are probably hematite or magnetite. The slightly larger red opaques, identified as hematite-stained clay pellets in the meso size range (0.05–0.5 mm) are apparent in all but two of the Ubaid samples (Z23 and Z25). Argillaceous clay pellets are also visible in many of the Ubaid samples (Smith 2008). An interesting feature in three Ubaid samples (Z19, Z24, Z25) was the possible evidence of clay mixing.

Halaf Sherds

In addition to the Ubaid painted sherds, seven Halaf painted sherds were examined for comparative purposes, selected out of a total of over 50 Halaf painted sherds found in a mixed context

TABLE 5.1. Summary of macroscopic observations and petrographic results for sampled sherds

<table>
<tr><th colspan="2"></th><th>Halaf Painted</th><th>Ubaid Painted</th></tr>
<tr><td rowspan="5">Macroscopic observations</td><td>Sample size</td><td>7</td><td>10</td></tr>
<tr><td>Stylistic description</td><td>Dark complex decorative motifs on buff to salmon background</td><td>Dark simple geometric designs on buff background</td></tr>
<tr><td>Thin-section direction</td><td colspan="2">Longitudinal</td></tr>
<tr><td>Wall thickness (mm)</td><td>3.1–8.6 (average 5.4)</td><td>2.7–10.3 (average 4.7, if 1 outlier omitted)</td></tr>
<tr><td>Surface hardness (Moh's)</td><td>3–5 (majority 3)</td><td>3</td></tr>
<tr><td rowspan="14">Petrographic results</td><td>Groundmass</td><td>Noncalcareous red clay with fairly even firing</td><td>Evenly fired with reddish brown or gray groundmass</td></tr>
<tr><td rowspan="4">Inclusions</td><td colspan="2">Almost no inclusions over 0.5 mm</td></tr>
<tr><td>Coarse inclusions are 0.25–1% of field of view, except 2 samples over 7%; semiangular to angular</td><td>Coarse inclusions are 0.2–1% of field of view; semiangular to angular</td></tr>
<tr><td>Fine inclusions are less than 0.5%, except Z24 with 2%; rounded</td><td>Fine inclusions are less than 2%; rounded</td></tr>
<tr><td>Amorphous concentration features</td><td>Amorphous concentration features; argillaceous clay pellets</td></tr>
<tr><td>Mineral components</td><td>Mainly quartz and feldspar, with smaller amounts of mudstone, hornblende, muscovite, calcite, biotite, grog, and pyroxenes</td><td>Mainly quartz and feldspar, with small amounts of hornblende, biotite, muscovite, augite, calcite, grog, and mudstone</td></tr>
<tr><td rowspan="6">Voids</td><td>Some voids over 0.5 mm, especially Z04 with 36% of all voids</td><td>Very few voids over 0.5 mm</td></tr>
<tr><td>Voids are 1–2% of field of view</td><td>Voids are 1–3% of field of view</td></tr>
<tr><td colspan="2">Voids over 2 mm are almost nonexistent</td></tr>
<tr><td>Types of voids vary greatly</td><td>Channel voids are least abundant</td></tr>
<tr><td>Channel voids are not frequently present</td><td>Channel voids are present in many samples</td></tr>
<tr><td colspan="2">Secondary calcite along some voids</td></tr>
<tr><td colspan="2" rowspan="2">Special notes</td><td>Dark opaque paint</td><td>Chalky feel; no evidence of slip layer</td></tr>
<tr><td>Calcareous slip layer visible on some</td><td>Possible clay mixing in 3 samples</td></tr>
</table>

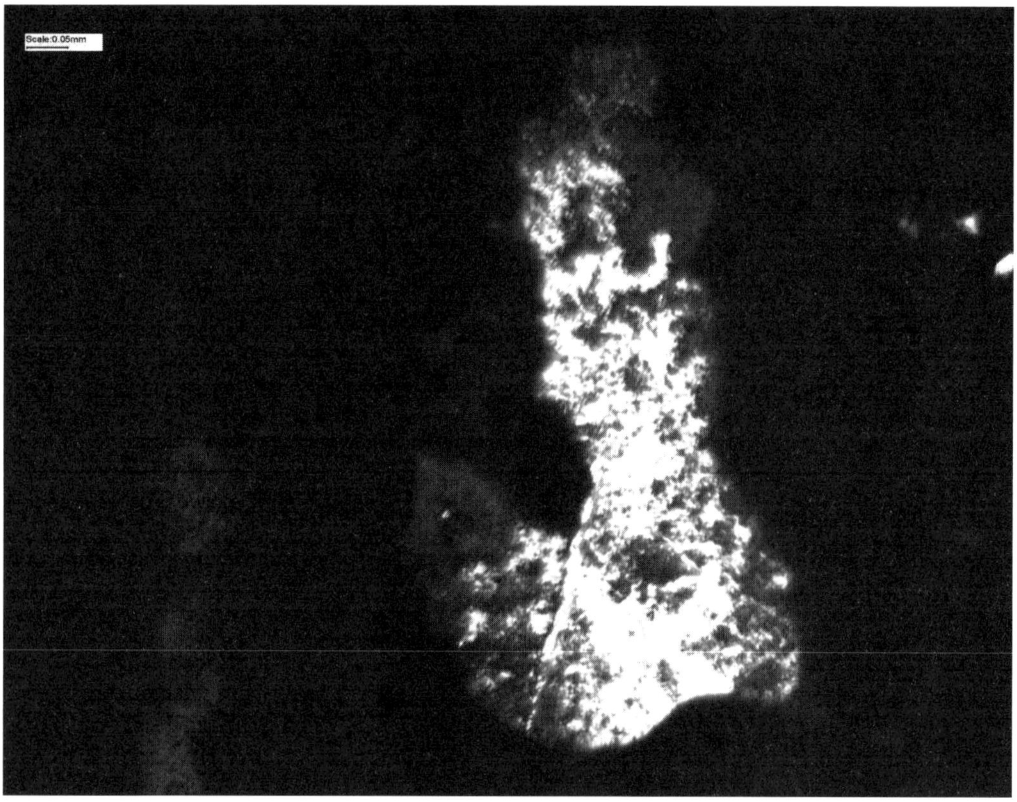

FIGURE 5.4. Example of a macro-sized void in an Ubaid painted sherd (Z20), XPL (the thin section was left thicker than usual to ensure that large void size is not a result of overpolishing).

in the basal levels of Tell Ziyadeh. The selected sherds include the variety in paste and wall thickness observed macroscopically and with a hand lens for the Halaf sherds from the site. They were all body sherds, although the decorative motifs and vessel shape allowed for the general orientation of the sherds. The Halaf painted pottery from Tell Ziyadeh was identified by its distinctive complex decorative motifs following the Halafian tradition, painted on a buff to salmon background with brown to black paint.

The wall thickness ranged from 3.1 mm to 8.6 mm, with an average wall thickness of 5.4 mm. Z07 was much thicker than the others at 8.6 mm, with the next thickest sherd (Z03) measuring 6.5 mm (see table 5.1). It is interesting to note that Z07 also stood out with a much grittier feel than the others. The hardness of the sherd surfaces ranged from 3 to 5 on the Moh's scale, with Z03 being the hardest at 5 and the majority measuring 3.

The seven Halaf painted sherds share many petrographic characteristics, such as a red noncalcareous clay, preferred orientation of elongated voids and inclusions parallel to the vessel walls, fairly even firing, bimodal distribution of inclusions, and very small amounts of inclusions and voids (see table 5.1). A very thin layer of calcareous slip is visible on two of the samples (Z03 and Z05) and dark opaque paint is visible on some others (Z02 and Z04). The red noncalcareous clay included amorphous concentration features such as the micro-sized opaque black rounded minerals, most likely magnetite or hematite (1–2 percent of the field of view), and the macro-sized rounded red opaque hematite stained clay pellets (0–2 percent of the field of view).

Coarse inclusions in general are rare (0.25–1.5 percent of the field of view, c:f:g:v = 100 percent), except for Z06 and Z07, which had 7 percent and 12 percent, respectively. Coarse inclusions larger than 0.5 mm are almost nonexistent and tend to be angular to semiangular in shape, as is the case with the Ubaid painted pottery. The most common coarse inclusions are quartz and feldspars, with lesser amounts of mudstone, hornblende, muscovite, calcite, biotite, grog, and pyroxenes. The fine inclusions also have the same mineral components.

Voids are rare as well: only 1–2 percent in all observed samples. In addition, voids large than 0.5 mm are also not very abundant (0–11 percent of all voids, except for Z04, which had 36 percent). But the types of voids tend to vary greatly between sherds. For example, some have many channels and few vesicles (for example, Z04 has 70 percent of all voids as channels and only 0.3 percent as vesicles), others have a fairly even mix (for example, Z02 has 21 percent channels, 44 percent vughs, and 35 percent vesicles), and some have mostly vughs and no channels (for example, Z07 has 0 percent channels, 68 percent vughs, and 32 percent vesicles). Vughs tend to be fairly common within all Halaf sherds, ranging from 30 to 68 percent. Some of the Halaf sherds, like the Ubaid sherds, show a development of secondary calcite along the void walls.

X-Ray Diffraction Results

In addition to the petrographic analysis, I carried out a small-scaled preliminary powder X-ray diffraction project on 4 sherds and 3 soil samples to achieve a better understanding of the clay fraction. I compared 2 Halaf painted sherds (Z02 and Z07) and 2 Ubaid painted sherds (Z17 and Z19) to 3 different soil samples (ZS01, ZS02, ZS03) collected from excavated areas during the Yale excavations at Tell Ziyadeh. All samples were finely powdered by hand using a pestle and mortar. Of the soil samples, 2 (ZS02b and ZS03b) were further fired at 900°C for approximately two hours in order to observe the effects of firing on the diffraction pattern. I did the analysis at the Yale University Chemical and Biophysical Instrumentation Center (CBIC) using a Bruker AXS D8 Focus. Although the results of the analysis are still preliminary, the diffraction patterns do not contradict the assumption that the 4 sherds were produced using locally available clays from around the site.

Interpretation and Conclusions

The petrographic and XRD analyses of a total of 17 sherds and 3 soil samples support the hypothesis that the Halaf and Ubaid painted pottery at Tell Ziyadeh was manufactured using locally available clays and tempering materials. Many aspects of the Halaf and Ubaid pottery seem to share similarities, with few differences except for surface treatment and the use of chaff as temper, which was indicated in some of the Halaf sherds by the channel voids. Although the time scale is not controlled well in this assemblage, the variability of the pottery within each ceramic tradition suggests that each household potter manufactured pottery by hand, adding different inclusions to the various clays, and manipulated the clay in different ways, perhaps as a result of the nature of the clay that was chosen.

Very few inclusions were observed. Due to their angularity compared to the finer inclusions, the coarser inclusions were most likely added to the clay as temper, perhaps from natural sandier deposits, as they were all fairly fine. Small amounts of grog were occasionally also added as temper. Due to the same mineral composition observed for the fine and coarse inclusions, no evidence supports a nonlocal source for the clays or the tempers. As stated earlier, the preliminary results of the XRD analysis also support this interpretation.

Although Ubaid pottery in general is well known to have been chaff tempered, voids were not frequently observed in Ubaid pottery at Tell Ziyadeh, while they were more frequently observed in Halaf pottery. The chaff inclusions were very fine, measuring approximately 0.15 mm in diameter and 0.3 to 2 mm in length (fig. 5.5). This suggests that it was not chaff from grasses obtained by normal husking activities, which would tend to be much thicker in diameter (Hole, personal communication, 2013). The chaff produced by harvester ants, a sample of which was collected at the site of Tell Ziyadeh, is of the general size range of the voids in the ceramics (fig. 5.6). Therefore I suggest that the channel voids are a result of the pottery being tempered by chaff similar to the chaff discarded by harvester ants.

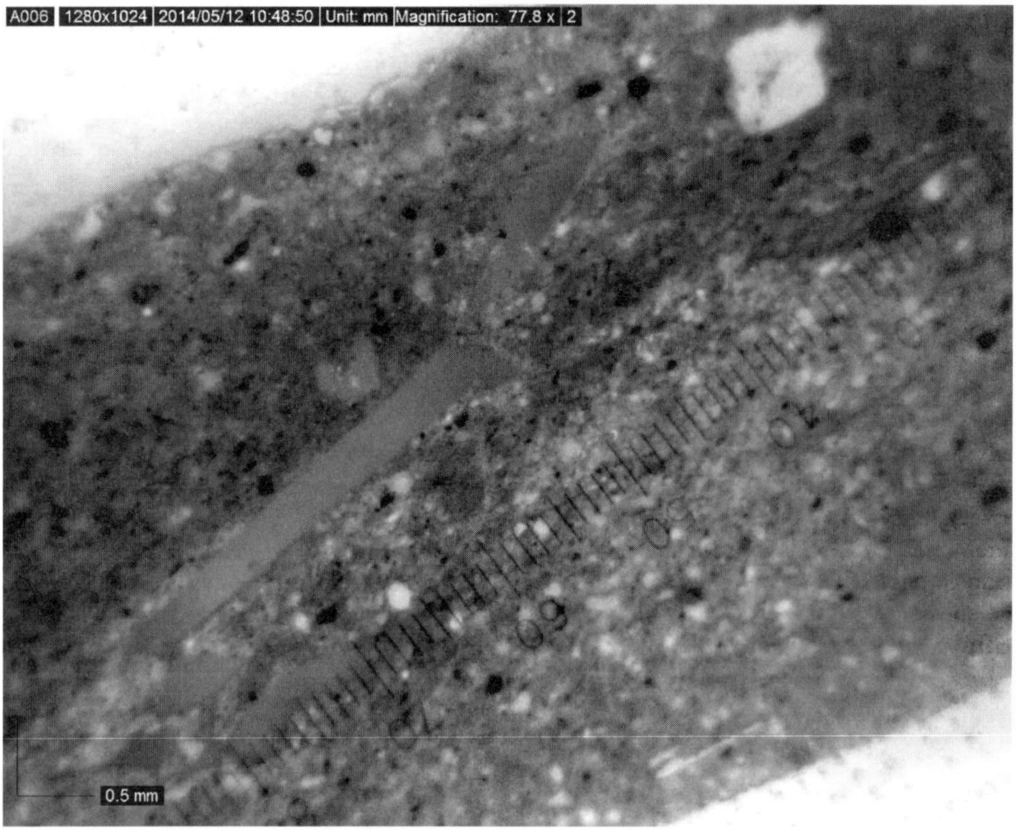

FIGURE 5.5. Example of a channel void indicating chaff inclusion in a Halaf painted sherd (Z04), taken with Dino-Lite Pro AM413ZTAS.

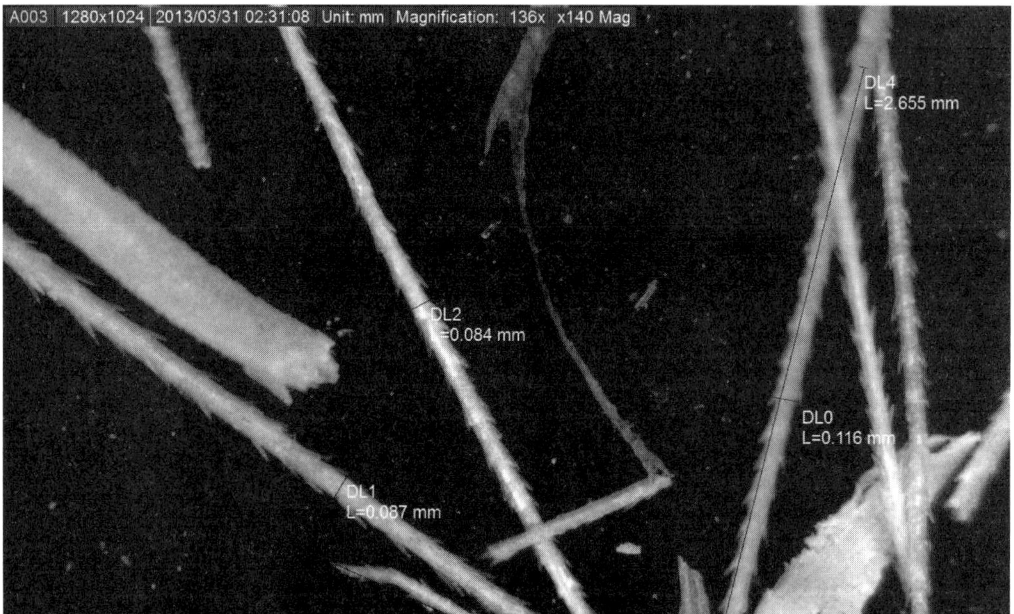

FIGURE 5.6. Examples of chaff collected by harvester ants, taken with Dino-Lite Pro AM413ZTAS.

This pilot project has provided a baseline understanding for the Halaf and Ubaid cultures at Tell Ziyadeh. There were no drastic changes in the materials that were used in the pottery manufactured between the two periods. The manufacturing techniques also did not show a major difference, other than the unexpected frequent use of chaff temper in the Halaf period and the unexpected lack of chaff temper in the Ubaid periods. Other Ubaid regions' pottery is commonly chaff tempered and fired in fairly high temperatures, as evidenced by the greenish tinge on the buff background, hard vessel walls, and frequently vitrified dark paint (Campbell and Fletcher 2010; Huot 1989; Oates 1960). The Ubaid pottery at Tell Ziyadeh is fairly soft, leaving a white residue on the fingers, with no evidence of vitrification of the paint, suggesting that it was fired at lower temperatures. This may indicate regional differences in Ubaid pottery to be explored in further study.

Earlier limited archaeometric studies of Halaf and Ubaid sherds, such as Davidson and McKerrell (1976, 1980) on those from the Khabur headwaters, Tell Arpachiyah, and Tepe Gawra and Davidson (1981) on sherds from Tell Aqab, have shown that Halaf sherds indicate a combination of local production and nonlocal imports from larger production centers. But there seem to be regional differences in ceramics from these periods. For example, the Ubaid sherds at Tell Aqab and Tepe Gawra, as well as at Tell Brak (Eiland and Williams 2000), seem to have been made locally using the same clays as the locally made Halaf sherds, whereas the clays used to make Ubaid pottery at Tell Arpachiyah were different from those used to make the Halaf sherds. Furthermore, the presence and amount of temper used in Halaf sherds vary according to site. The results of this study of the Halaf and Ubaid sherds from Tell Ziyadeh fit into this general picture of local production and resulting regional variation. Due to the limited number of sherds analyzed for this study, however, it was not possible to identify whether there were nonlocal imports in the Halaf period assemblage at Tell Ziyadeh.

The results of this pilot study suggest that by using these techniques on a large sample we should be able to acquire a better grasp of these two ceramic traditions and more readily discern differences in the ceramics from the remaining 8,000 years of the ceramic sequence in the region.

Acknowledgments

I would like to express my great appreciation to Dr. Frank Hole for the opportunity to study these materials and for providing the necessary background information on the site. Any inaccuracies are my own responsibility. I would also like to thank the Department of Anthropology and the Chemical and Biophysical Instrumentation Center at Yale University for the access to the necessary instrumentation.

References Cited

Akkermans, Peter M. M. G., and Glenn M. Schwartz
2003 *The Archaeology of Syria: From Complex Hunter-Gatherers to Early Urban Societies (c. 16,000–300 BC)*. Cambridge University Press, Cambridge.

Berman, Judith C.
1989 Ceramic Production and Its Implications for the Sociopolitical Organization of the Susiana Plain during the Late 'Ubaid. In *Upon This Foundation—The 'Ubaid Reconsidered*, edited by Elizabeth F. Henrickson and Ingolf Thuesen, pp. 257–280. Museum Tusculanum Press, Copenhagen.

Buccellati, Giorgio, Daniela Buia, and Stephen Reimer
1991 Tell Ziyadeh: The First Three Seasons of Excavation (1988–1990). *Bulletin of the Canadian Society for Mesopotamian Studies* 21:31–62.

Bullock, Peter, Nicholas Fedoroff, A. Jongerius, Georges Stoops, and Tatiana Tursina
1985 *Handbook for Soil Thin Section Description*. Waine Research Publications, Wolverhampton, UK.

Campbell, Stuart
1992 The Halaf Period in Iraq: Old Sites and New. *Biblical Archaeologist* 55(4):182–187.

Campbell, Stuart, and Alexandra Fletcher
2010 Questioning the Halaf-Ubaid Transition. In *Beyond the Ubaid: Transformation and*

Integration in the Late Prehistoric Societies of the Middle East, edited by Robert A. Carter and Graham Philip, pp. 69–83. University of Chicago Press, Chicago.

Carter, Robert A., and Graham Philip, eds.
2010 *Beyond the Ubaid: Transformation and Integration in the Late Prehistoric Societies of the Middle East*. University of Chicago Press, Chicago.

Clop Garcia, Xavier, Aureli Alvarez Perez, and Frédéric Hatert
2004 Characterization Study of Halaf Ceramic Production at Tell Amarna (Euphrates Valley, Syria). In *Tell Amarna (Syrie)—La période de Halaf*, edited by Önhan Tunca and Miquel Molist, pp. 201–212. Peeters, Leuven.

Davidson, Thomas E.
1981 Pottery Manufacture and Trade at the Prehistoric Site of Tell Aqab, Syria. *Journal of Field Archaeology* 8:65–77.

Davidson, Thomas E., and Hugh McKerrell
1976 Pottery Analysis and Halaf Period Trade in the Khabur Headwaters Region. *Iraq* 38(1):45–56.
1980 The Neutron Activation Analysis of Halaf and 'Ubaid Pottery from Tell Arpachiyah and Tepe Gawra. *Iraq* 42(2):155–167.

Eiland, Murray L., and Quentin Williams
2000 Infra-Red Spectroscopy of Ceramics from Tell Brak, Syria. *Journal of Archaeological Science* 27:993–1006.

FitzPatrick, Ewart A.
1984 *Micromorphology of Soils*. Chapman and Hall, London.

Frankel, David
1979 *Archaeologists at Work: Studies on Halaf Pottery*. Trustees of the British Museum, London.

Heidke, James M., and Elizabeth J. Miksa
2000 Correspondence and Discriminant Analyses of Sand and Sand Temper Compositions, Tonto Basin, Arizona. *Archaeometry* 42:273–299.

Hole, Frank
2000 Tell Ziyadeh on the Middle Khabur, Syria. In *Proceedings of the First International Congress on the Archaeology of the Ancient Near East, Rome, May 18th–23rd 1998*, edited by Paolo Matthiae, Alessandra Enea, Luca Peyronel, and Frances Pinnock, pp. 609–619. Università degli Studi di Roma "La Sapienza," Rome.
2001 A Radiocarbon Chronology for the Middle Khabur, Syria. *Iraq* 63:67–98.
2013 Verbal communication, February 20, 2013.

Hole, Frank, and Jennifer Arzt
1998 Excavations at Tell Ziyadeh, 1996–97. *Orient Express* 98(3):64–66.

Huot, Jean-Louis
1989 'Ubaidian Villages of Lower Mesopotamia: Permanence and Evolution from 'Ubaid 0 to 'Ubaid 4 as seen from Tell el'Oueili. In *Upon This Foundation—The 'Ubaid Reconsidered*, edited by Elizabeth F. Henrickson and Ingolf Thuesen, pp. 19–40. Museum Tusculanum Press, Copenhagen.

Kemp, Rob A.
1985 *Soil Micromorphology and the Quaternary*. Quaternary Research Association Technical Guide 2. Quaternary Research Association, Cambridge.

Middleton, Andrew P., Ian C. Freestone, and Morven N. Leese
1985 Textural Analysis of Ceramic Thin Sections: Evaluation of Grain Sampling Procedures. *Archaeometry* 27(1):64–74.

Oates, Joan
1960 Ur and Eridu, the Prehistory. *Iraq* 22:32–50.

Quinn, Patrick, Peter Day, Vassilis Kilikoglou, Edward Faber, Stella Katsarou-Tzeveleki, and Adamantios Sampson
2010 Keeping an Eye on Your Pots: The Provenance of Neolithic Ceramics from the Cave of the Cyclops, Youra, Greece. *Journal of Archaeological Science* 37(5):1042–1052.

Shepard, Anna O.
1985 *Ceramics for the Archaeologist*. Volume 609, Carnegie Institution, Washington, DC.

Smith, Michael S.
2008 Petrography (Chapter 6). In *Woodland Pottery Sourcing in the Carolina Sandhills*, edited by Joseph M. Herbert and Theresa E. McReynolds, pp. 142–170. Research Report. Research Laboratories of Archaeology, University of North Carolina at Chapel Hill, Chapel Hill.

Stoltman, James B.
1989 A Quantitative Approach to the Petrographic Analysis of Ceramic Thin Sections. *American Antiquity* 54(1):147–160.
1991 Ceramic Petrography as a Technique for Documenting Cultural Interaction: An Example from the Upper Mississippi Valley. *American Antiquity* 56(1):103–121.

Tonoike, Yukiko
2010 Beyond Style: Petrographic Analysis of Dalma Ceramics from Two Regions in Iran. Ph.D. dissertation, Yale University.

Van Der Plas, Leendert, and Alexander C. Tobi
1965 A Chart for Judging the Reliability of Point Counting Results. *American Journal of Science* 263:87–90.

Whitbread, Ian K.
1989 A Proposal for the Systematic Description of Thin Sections towards the Study of Ancient Ceramic Technology. In *Archaeometry: Proceedings of the 25th International Symposium*, edited by Yannis Maniatis, pp. 127–138. Elsevier, Amsterdam.
1995 *Greek Transport Amphorae: A Petrological and Archaeological Study.* Fitch Laboratory Occasional Paper 4. British School at Athens, Oxford.

CHAPTER 6

Petrography in the Age of Instrumental Characterization

An Example from Honey Bee Village, Pima County, Arizona

JAMES M. HEIDKE

Anna Shepard's insights regarding the necessity of characterizing the temper in larger lots of sherds than it is practical to analyze petrographically often seem to have been lost on the discipline, especially since the advent of instrumental characterization studies. Shepard repeatedly stressed the idea that temper must be accurately characterized in the majority of the sherds undergoing analysis and that petrographic and reflected light temper characterization results must be in agreement (Shepard 1936, 1942, 1965). For example, her famous study of Rio Grande glaze paint ware involved the reflected light temper characterization of 16,738 sherds recovered from 170 sites as well as the examination of 679 thin sections (Shepard 1942:140). She then relied on temper data from 5,494 rim sherds in her final statistical analysis (Shepard 1942:figures 10, 12, 14). Individual, site-based data sets were combined to form 14 spatially distinct districts. Time was partitioned into early, middle, and late divisions. Considered together, those spatial and temporal variables yielded 32 data batches with an average of 172 sherds per batch.

The current study is based on Shepard's approach. Here petrographic and binocular microscopic studies of the pottery from Honey Bee Village in southeastern Arizona are shown to complement one another, allowing questions regarding the interrelationship of time, space, technology, and pottery function to be addressed. Correspondence, linear regression, and multiple linear regression analyses are shown to be especially useful statistical tools for recognizing and quantifying those relationships. The findings indicate that potters acting at specific times and in specific places practiced distinct paste recipes in order to produce different wares, each with their own intended functions.

Sample, Variables, and Methods

The ceramic sample was recovered from features excavated at Honey Bee Village, a Hohokam settlement located in the middle Cañada del Oro Valley on the *bajada* (alluvial fan along the mountain front) of the Tortolita Mountains just north of Tucson, Arizona. The site overlooks the broad Sausalito Wash–Big Wash floodplain, whose braided channels still flow intermittently through the Sonoran Desert landscape during the winter and summer rainy seasons. This area would have been an optimal location to practice floodwater farming prehistorically (Huckleberry 2012).

The central core of the site is ringed by trash mounds and contains both a Preclassic (AD 500–1150) ballcourt and an early Classic (AD 1150–1300) compound. Portions of the site located

TABLE 6.1. Phase and ware systematics (including abbreviations) used in tables and figures

Phase		Ware (Abbreviation)
Name (Abbreviation)	Date (AD)	Plain (PW), Red (RW), Red-on-brown (Dec or Rdbn)
Tortolita (Tort)	650–700	PW, RW
Snaketown (Sntwn)	700–750	PW
Cañada del Oro (CdO)	750–850	PW, Dec
Rillito (Rill)	850–950	PW, Dec/Rdbn
Rillito to Early Rincon (Rill–ER)	850–1000	PW, Dec
Early Rincon (ER)	950–1000	PW, Dec
Early Rincon to Middle Rincon 1 (ER–MR1)	950–1040	PW, Dec/Rdbn
Middle Rincon (MR)	1000–1100	PW, RW, Dec/Rdbn, Rincon Red (RinRed or RW), Rincon Polychrome (RinPoly)
Transitional Middle Rincon 3/Late Rincon (MR3/LR)	1090–1110	PW, Dec/Rdbn
Late Rincon (LR)	1100–1150	PW, RW, Dec/Rdbn
Transitional Late Rincon/Tanque Verde (LR/TV)	1140–1190	PW, RW, Dec/Rdbn
Tanque Verde (TV)	1150–1200	PW, Dec

outside the core area were excavated prior to residential development (Wallace 2012b). Feature types encountered include houses, trash mounds, roasting pits, and mortuary deposits. Occupation began during the Tortolita phase (AD 500–700), with an estimated 47 persons living at the site in 7–11 households, grew to a maximum of approximately 90 people living in 11–23 households during the Late Rincon phase (AD 1100–1150), and ended sometime early in the Tanque Verde phase (AD 1150–1300) with 22 people living in 2–6 households (Wallace 2012a).

Excavation of the nonmortuary features yielded a total of 49,526 sherds, representing portions of at least 9,299 vessels—a figure that includes 134 reconstructible vessels that range from 25 to 100 percent complete. Most of those vessel fragments were recovered from contexts that accumulated over long spans, such as trash mounds, or showed signs of temporal mixing. This chapter focuses on 3,966 vessels recovered from typologically well-dated contexts. Overall 96.1 percent of the temporally diagnostic pottery types in those deposits are attributable to the assigned phase, 2.5 percent to an earlier phase, and 1.4 percent to a later phase.

Relationships among five variables are examined. The first variable is time. The long occupation of Honey Bee Village is divisible into twelve sequential temporal units spanning the period AD 650–1200 (table 6.1). The second variable is ceramic ware. Four wares produced in the greater Tucson area are included in the study: plain brown, red-slipped, red-on-brown bichrome, and black-and-white-on-red polychrome. One to four of those wares are present in each of the temporal units (see table 6.1).

The third variable is temper source. Sand temper provenance was characterized with reference to the current Tucson Basin Petrofacies model (see the discussion in the introduction). That actualistic model is based on petrographic point-count data recorded from 296 wash sand samples. Statistical analysis of these data by petrographer Elizabeth Miksa (2011; Miksa et al. 2012) defined 38 distinct sand temper compositions available to potters in the area (fig. 6.1).

Pottery exhibiting a temper composition available within 3 km of Honey Bee Village is viewed as having been produced locally. The 3 km value represents the maximum distance traveled by foot to collect sand temper documented in 24 ethnographic and ethnoarchaeologic studies of traditional potter behavior

FIGURE 6.1. Current Tucson Basin Petrofacies map showing locations, letter designations, and names (prepared by Catherine Gilman).

(Heidke 2011a:table 4.10). Here sand-tempered pottery displaying compositions not available within 3 km of Honey Bee Village are considered nonlocal items. We documented 11 nonlocal compositions at distances ranging from 5.6 km to 31.5 km from the site.

The fourth variable is temper type, the type(s) of nonplastic material added to form a workable paste. Sand, disaggregated metamorphic rocks grading from schist to gneiss, muscovite mica, grog (crushed sherd), and mixtures of those materials exhibiting varying degrees of processing were documented. Sherds assigned to the "schist/gneiss ± muscovite mica" temper source and temper type groups were treated as products of local manufacture, because numerous micaceous rocks were recovered from deposits at the site and many thin-sectioned sherds contain metamorphic rock and mineral tempers similar to those recovered from the archaeological deposits (Heidke 2012:225–227; Miksa et al. 2012:176).

The fifth variable is vessel function. Function was inferred by using an approach developed by David Braun (1980; Heidke 2006). It is based on a synthesis of ethnographic data from the American Southwest regarding pottery usage. Rim sherds and reconstructible vessels having measurable openings at their mouths and determinate forms were assigned to functional categories defined by attributes of containment security, frequency of access, and overall morphology (Shepard 1995).

Temper source and type were characterized by the author. Clean sherd edges were examined using a Unitron ZSM binocular microscope set at 15× magnification and fitted with a Stocker and Yale Lite Mite Series 9 circular illuminator. Binocular microscopic identification of the sand temper in a pottery fabric was facilitated through the use of grain boxes: small boxes containing representative rock and mineral grains, keyed to detailed written descriptions of the composition of each petrofacies (Heidke et al. 2002:159; Kamilli 1994:638). In addition, loose sand samples from each petrofacies were examined. They provided a good sense of the abundance of the sand-sized rocks and minerals present in each composition.

Temper assignments were verified and refined with the aid of 100 thin sections selected by the author. Sampling was not random; the judgmental sample was chosen to represent temper sources and types, ceramic wares, and temporal units in numbers proportionate to their abundance in the well-dated contexts (Heidke et al. 2002:160). Thin sections were prepared by Quality Thin Sections of Tucson, Arizona. Each sherd was saturated with epoxy, sectioned parallel to the vessel wall, stained for potassium and plagioclase feldspars, and cover-slipped (Miksa and Heidke 2001:209).

The thin-sectioned sherds were point-counted by Carlos Lavayen (Miksa et al. 2012:131) using the Gazzi-Dickinson method (Dickinson 1970; Gazzi 1966) with improvements to the nomenclature and procedure made by others (Graham et al. 1976; Ingersoll 1983; Ingersoll and Suczek 1979; Lombard 1987; Miksa and Heidke 2001) (see the introduction for a description of the method). Petrographers and I (acting as project ceramicist) were in agreement regarding temper source and type in 96 of the 100 cases (Heidke 2012:208–218; Miksa et al. 2012:158–172).

The physical project collection, including thin sections, is archived at the Arizona State Museum (ASM), located on the University of Arizona campus. Digital data analysis files are archived at Desert Archaeology, Inc. (DAI). Both the ASM and DAI are located in Tucson, Arizona.

The time depth and presence of multiple wares throughout most of the site's occupation meant that all patterning in the Honey Bee Village ceramic attribute data would be multidimensional. Therefore a method of data reduction and exploration was required to facilitate pattern identification. Correspondence analysis was chosen for that purpose and performed using Multi-Variate Statistical Package version 3.1 by Kovach Computing Services (2000). Although it is not as well known as some statistical routines, correspondence analysis has been used to examine ceramic compositional (Gihwala et al. 1985; Heidke and Miksa 2000; Jacobson et al. 1991; Jacobson et al. 1995; Jacobson et al. 2002), functional (Farry 2005; Pitts 2005; Robertson 1999), and decorative style and

TABLE 6.2. Temper source reported by phase and ware

Phase + Ware	Temper provenance[a]													Row total
	E3/Z	E3	LMT	E3 SS-599	E1	E2	J1	J2s	J2n	J3	K	M	MW	
TortRW	0	42	4	1	0	14	0	0	0	0	0	0	0	61
TortPW	2	77	2	39	2	5	0	0	0	0	0	0	0	127
CdODec	0	2	1	0	0	0	1	3	3	0	0	0	0	10
RillDec	0	2	54	0	0	0	20	12	3	6	0	0	0	97
RillPW	10	28	50	0	18	6	3	1	1	2	0	0	0	119
Rill–ERDec	0	1	2	0	0	0	13	1	1	3	0	0	0	21
Rill–ERPW	1	2	9	0	0	0	0	0	0	0	0	0	0	12
ERDec	1	1	6	0	0	0	65	39	6	2	0	0	0	120
ERPW	2	5	10	0	1	0	1	0	0	1	0	0	0	20
ER–MR1Dec	0	7	6	0	0	0	145	79	5	3	0	2	1	248
ER–MR1PW	36	33	103	1	14	0	6	2	1	0	0	0	0	196
MRDec	6	34	7	2	3	0	215	100	23	15	5	0	3	413
RinRed	0	4	1	1	0	1	59	2	0	1	1	2	0	72
RinPoly	0	0	0	0	0	0	17	0	0	0	0	0	0	17
MRPW	73	97	125	3	38	6	9	5	3	0	0	0	0	359
MR3/LRDec	1	2	1	0	0	0	11	11	2	0	2	0	0	30
MR3/LRPW	4	7	4	0	10	3	0	0	1	1	1	0	0	31
LRDec	58	102	15	8	8	17	45	88	141	32	6	7	17	544
LRRW	2	3	4	0	0	0	0	0	0	1	0	0	0	10
LRPW	77	110	69	4	34	15	6	7	17	5	0	0	4	348
LR/TVDec	23	58	5	2	4	3	10	35	57	13	8	3	9	230
LR/TVPW	29	73	39	3	14	10	1	4	8	1	3	1	0	186
TVDec	0	10	0	0	0	0	1	5	12	2	1	1	7	39
TVPW	3	1	11	1	0	1	0	1	0	0	0	0	0	18
Column total	328	701	528	65	146	81	628	395	284	88	27	16	41	3,328

[a] E3/Z is Eastern Tortolita/Big Wash, E3 is Eastern Tortolita, LMT is schist/gneiss ± muscovite mica, E3 SS-599 is Eastern Tortolita Sand Sample 599, E1 is Western Tortolita, E2 is Central Tortolita, J1 is Beehive, J2s is Twin Hills south, J2n is Twin Hills north, J3 is Wasson, K is Black Mountain, M is Rillito, and MW is Rillito West.

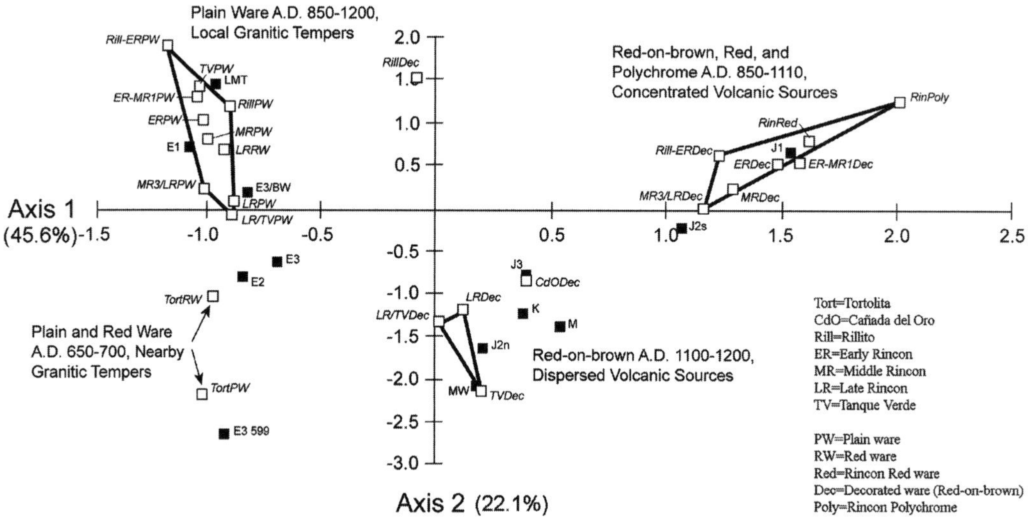

FIGURE 6.2. Correspondence analysis joint plot showing relationships between temper sources and wares.

design data (Müller 2009; Nováček 2009) and in ceramic seriation studies (Duff 1996; Pavúk 2010; Peeples and Schachner 2012; Smith and Neiman 2007).

Correspondence analysis is a principal components analysis method for the display of rows and columns of a two-way contingency table as points in a low-dimensional vector space (Carr 1990). The geometry of the rows, which in the Honey Bee Village data sets are represented by time plus ware cases, is related to the geometry of the columns, which represent analytical attributes (temper source, temper type, and vessel function), resulting in a correspondence between rows and columns (Greenacre 1984).

One of the primary reasons for choosing correspondence analysis as a method of data reduction and exploration is that the technique's requirements are easily met: (1) all values in a data matrix must be positive (zeros are acceptable); and (2) all row and column totals must be greater than zero (Hill 1979:10; Weller and Romney 1990:72). These are important assumptions whenever a database contains extremely heterogeneous attribute information, such as many of the Honey Bee Village ceramic data. It should be noted that the algebraic technique employed by correspondence analysis is purely deterministic (Ringrose 1992) and therefore provides little indication of the strength of any apparent relationships. Relationships identified during an analysis, however, can be used to develop testable hypotheses (Blasius 1994:23). Here linear and multiple linear regression analyses were used to model the dependence of a variable on one (or more) of the explanatory variables suggested by the correspondence analysis results.

Results

Temper Source

Figure 6.2 displays a scatterplot of the temper source, ware, and time data on the first two correspondence analysis axes. Rows and columns with fewer than 10 cases, potentially extrabasinal phyllite temper, and indeterminate provenance cases were deleted prior to the analysis, thereby decreasing the sample size from 3,966 to 3,328 cases in a 24 row by 13 column matrix (table 6.2). The first factor accounts for 45.6 percent of the variation, the second factor accounts for 22.1 percent, and the third factor accounts for 16.5 percent. The first three factors represent a total of 84.1 percent of the variation, indicating that the correspondence analysis was highly successful in reducing the dimensionality of the data. Each of the remaining nine factors accounts for less than 6 percent of the variation.

The first factor (Axis 1) is interpreted as a contrast between Tortolita Mountain granitic and metamorphic temper resources (points located to the left along Axis 1), some of which are located relatively close to the site, and the more distant nonlocal Tucson and Black Mountain volcanic and mixed volcanic compositions (points located to the right along Axis 1). The first factor is also interpretable in terms of wares and to a lesser extent time. All of the plainware data sets (regardless of time), the Tortolita Red, the Late Rincon redware, and the Rillito phase decorated ware data sets occur on the left side of Axis 1, while all of the remaining red-on-brown data sets, the Rincon Red, and the Rincon Polychrome data sets occur on the right side of Axis 1.

The second factor (Axis 2) is interpreted as a contrast between the compositions of an unusual Eastern Tortolita Petrofacies sand sample collected about 5.6 km north of Honey Bee Village (E3 599) and the Rillito West (MW) petrofacies (lowest Axis 2 negative values) and schist/gneiss ± muscovite mica tempers (LMT) (highest Axis 2 positive value). Note also that Axis 2 opposes the Eastern Tortolita Petrofacies Sand Sample-599 sand and schist/gneiss ± muscovite mica tempers, which had very similar negative scores on the first axis.

Distinct clusters are apparent when the variable and case data points are considered together on the first and second axes. The two data points representing the earliest pottery types recovered from the site, Tortolita phase plainwares and redwares plot close together (lower left). Most of the AD 650–700 Tortolita phase pottery is tempered with Eastern Tortolita (E3), Eastern Tortolita Sand Sample-599 (E3 599), or Central Tortolita (E2) petrofacies sands.

A large second cluster is formed by plainware produced from the Rillito phase through the Tanque Verde phase (AD 850–1200) as well as redware of the Late Rincon phase (AD 1100–1150) (upper left). Those cases always contain sherds tempered with Eastern Tortolita (E3) and Eastern Tortolita/Big Wash (E3/Z) petrofacies sands and schist/gneiss ± muscovite mica temper (LMT); pots tempered with Western Tortolita (E1) and Central Tortolita (E2) petrofacies sands often occur too.

A third cluster is composed of the painted and/or slipped pottery recovered from contexts spanning Rillito–Early Rincon through transitional Middle Rincon 3/Late Rincon times (AD 850–1110, upper right). This decorated pottery is dominated by Beehive (J1) Petrofacies sand temper, although Twin Hills south (J2s), Twin Hills north (J2n), Wasson (J3), and Eastern Tortolita (E3) sand tempers and schist/gneiss ± muscovite mica (LMT) temper occur at all points in time and most of the other source areas occur at least sporadically. Rincon Red displays a similar range of sources, although no sherds tempered with the Twin Hills north sand composition were documented. All of the Rincon Polychrome pottery recovered from Honey Bee Village is tempered with Beehive (J1) Petrofacies sand.

The final cluster consists of decorated pottery recovered from contexts spanning Late Rincon through Tanque Verde times (AD 1100–1200, lower right). The decorated collections from Late Rincon onward have a greater abundance of pottery tempered with nonlocal Twin Hills north (J2n) sand and local Eastern Tortolita (E3) and Eastern Tortolita/Big Wash (E3/Z) sands than any of the decorated collections from earlier times.

The Rillito phase (AD 850–950) decorated data point holds a unique position, as slightly more than half of the sherds in that data set are tempered with schist/gneiss ± muscovite mica (LMT), with most of the remaining sherds tempered with sands from volcanic Tucson Mountain sources (Twin Hills south [J2s], Twin Hills north [J2n], Wasson [J3], and especially the Beehive [J1] petrofacies).

The Cañada del Oro phase (AD 750–850) decorated pottery data point plots in an unusual, though easily explained, location. We would think that sand temper from the Wasson (J3) Petrofacies would be a common occurrence in the Cañada del Oro phase decorated pottery, given that their data points overlap. But none of it is tempered with sand from that source. Instead the compositional array consists of equal parts Twin Hills south (J2s) and Twin Hills north (J2n) sand tempers with subordinate amounts of Eastern Tortolita (E3), Beehive (J1), and schist/gneiss ± muscovite mica (LMT) tempers. That suite of sources forced the Cañada del Oro decorated pottery data point to plot between Twin Hills south and Twin Hills north with a deflection off the line toward Eastern Tortolita, a location also occupied by the Wasson Petrofacies data point.

These findings are easier to appreciate when the patterns identified are mapped onto the landscape. As shown in figure 6.3, plainware production was concentrated in the Tortolita Mountains (Petrofacies E1, E2, E3, and Z), with much of it occurring at Honey Bee Village. Preclassic red-on-brown, red, and polychrome ware production was concentrated in the southern Tucson Basin in the Beehive (J1) and Twin Hills south (J2s) petrofacies, whereas Classic Period red-on-brown production occurred over a much greater area (Petrofacies MW, M, J3, J2n, J2s, J1, and K). The dramatic change in painted pottery production location occurred over an interval of about seventy years (ca. AD 1040–1110). It likely reflects a reduction in both regional and basin-wide economic integration that occurred after the Preclassic ballcourt system was abandoned, "markets" or "fairs" (Haury 1976:78; Wilcox 1979:111) that accompanied the ritual ballgame ended (Heidke 2009; Heidke et al. 2002), and, as another result, the importation of Middle Gila buffware from outside the basin declined markedly (Heidke 2012:table 8.6).

Temper Type

Figure 6.4 displays a scatterplot of the temper type, ware, and time data on the first two correspondence analysis axes. Rows and columns with fewer than 10 cases, potentially extrabasinal phyllite temper, and indeterminate temper type

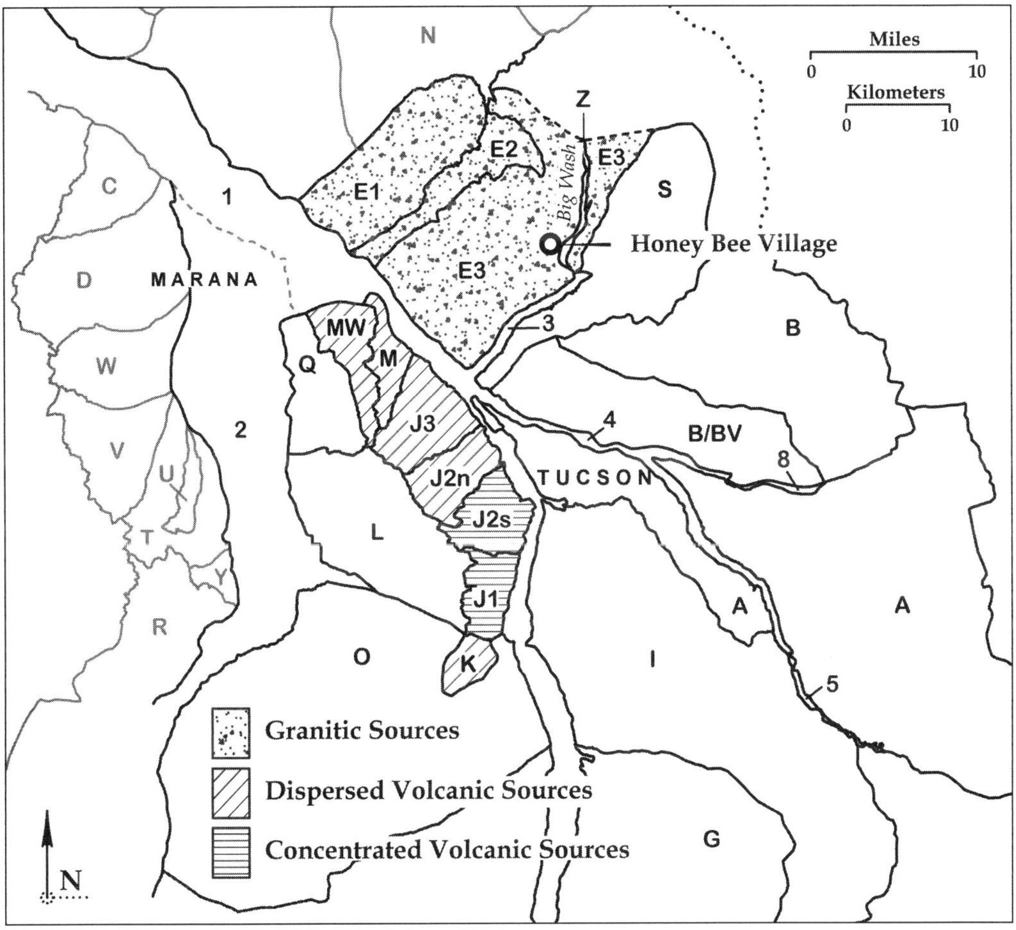

FIGURE 6.3. Visual summary of ceramic production and consumption patterns at Honey Bee Village (prepared by Catherine Gilman).

cases were deleted prior to the analysis, thereby decreasing the sample size from 3,966 to 3,785 cases in a 26 row by 9 column matrix (table 6.3). The first factor accounts for 45.7 percent of the variation, the second factor for 34.2 percent, and the third factor for 13.2 percent. The first three factors represent a total of 93.1 percent of the variation. Each of the remaining five factors accounts for less than 4 percent percent of the variation.

The first factor (Axis 1) is interpreted as a contrast between tempers that have abundant free mica (TT7, TT5, TT6, TT8), abundant pieces of schist/gneiss (TT1), or crushed sherd (TT9) (points located to the left along Axis 1) and tempers having lesser amounts of schist/gneiss (TT3, TT2) or just sand (TT4) (points located to the right along Axis 1). The first factor is also interpretable in terms of wares and to a lesser extent time. All of the plainware data sets from the Rillito phase through the Tanque Verde phase and the Late Rincon and transitional Late Rincon/Tanque Verde decorated and redware data sets are located to the left on Axis 1, while the Tortolita and Snaketown phase plainware, the Tortolita phase redware, the Cañada del Oro through transitional Middle Rincon 3/Late Rincon decorated ware, Rincon Red, and Middle Rincon polychrome (Rincon Polychrome and Rio Rico Polychrome) data sets are grouped to the right on Axis 1.

The second factor (Axis 2) is interpreted as a contrast between sand tempers lacking schist/gneiss (TT4) (lowest Axis 2 negative value) and

Table 6.3. Temper type reported by phase and ware

Phase + Ware	Temper type[a]									Row total
	TT1	TT2	TT3	TT4	TT5	TT6	TT7	TT8	TT9	
TortRW	2	0	0	65	0	0	2	0	0	69
TortPW	2	0	0	137	1	1	0	1	0	142
SntwnPW	0	0	0	10	0	0	0	0	0	10
CdODec	1	2	1	9	0	1	0	1	0	15
RillDec	60	12	14	10	0	1	6	1	0	104
RillPW	17	1	2	29	18	20	37	2	1	127
Rill–ERDec	4	7	5	6	0	1	0	1	0	24
Rill–ERPW	0	0	0	0	2	1	9	0	0	12
ERDec	11	29	50	31	0	1	1	2	0	125
ERPW	4	0	1	4	3	1	7	2	0	22
ER–MR1Dec	12	47	128	75	1	6	0	2	0	271
ER–MR1PW	28	4	8	15	61	16	74	6	0	212
MRDec	4	52	143	230	5	15	5	5	4	463
RinRed	1	3	29	47	1	0	0	2	1	84
RinPoly	0	0	7	10	0	0	0	0	0	17
MRPW	32	2	7	80	99	50	107	12	3	392
MR3/LRDec	0	0	2	30	3	1	1	0	0	37
MR3/LRPW	1	0	1	13	6	7	4	1	0	33
LRDec	1	4	13	477	38	98	11	8	1	651
LRRW	1	0	1	4	1	1	3	0	0	11
LRPW	19	3	1	157	84	72	47	15	3	401
LR/TVDec	0	1	9	195	19	46	3	3	3	279
LR/TVRW	1	0	0	8	1	0	0	0	0	10
LR/TVPW	2	1	1	98	41	19	27	2	5	196
TVDec	0	0	0	55	1	3	0	0	0	59
TVPW	1	0	0	4	8	0	5	1	0	19
Column total	204	168	423	1,799	393	361	349	67	21	3,785

[a] TT1 is high LMT (> 25% schist/gneiss), TT2 is high LMT/low sand (7–25% schist/gneiss), TT3 is low LMT/high sand (1–7% schist/gneiss), TT4 is high sand (< 1% schist/gneiss), TT5 is high muscovite mica (> 25% MUSC), TT6 is mixed sand and muscovite mica (1–25% MUSC), TT7 is schist/gneiss and muscovite mica (> 25% LMT + MUSC), TT8 is mixed sand, schist/gneiss, and muscovite mica (1–25% LMT + MUSC), and TT9 is sand and crushed sherd.

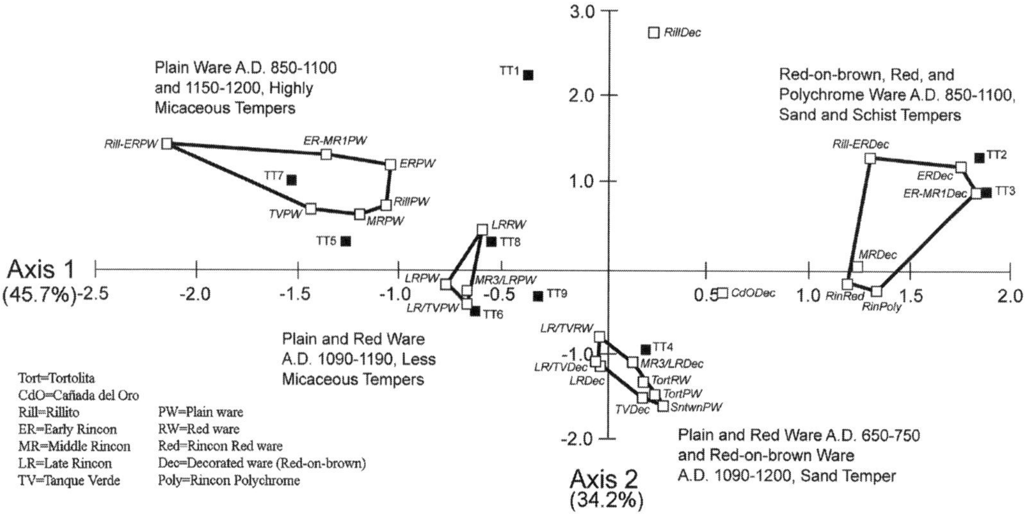

Figure 6.4. Correspondence analysis joint plot showing relationships between temper types and wares.

tempers with abundant schist/gneiss (TT1) (highest Axis 2 positive value).

Distinct clusters are apparent when the variable and case data points are considered together on the first and second axes. Data points representing plainwares recovered from Rillito through Middle Rincon phase contexts (AD 850–1100) as well as Tanque Verde phase (AD 1150–1200) plainware plot close together and near the highly micaceous temper types (TT5 and TT7).

Data points representing plainware recovered from transitional Middle Rincon 3/Late Rincon through transitional Late Rincon/Tanque Verde phase contexts and Late Rincon redware plot close together and near the mixed sand and mica temper types (TT6 and TT8) as well as sand and grog temper (TT9). Accordingly, this cluster represents plainware and redware production from AD 1090 to 1190.

A third cluster represents Tortolita and Snaketown phase plainware, Tortolita phase redware, decorated pottery from transitional Middle Rincon 3/Late Rincon through Tanque Verde contexts, and redware from transitional Late Rincon/Tanque Verde phase contexts tempered primarily with sand (TT4). This cluster represents early plain and redware production (AD 650–750), late decorated ware production (AD 1090–1200), and the latest redware production (AD 1140–1190) documented at Honey Bee Village.

The final cluster consists of data points representing decorated and/or red-slipped pottery recovered from deposits that began to accumulate in the Rillito phase and continued through Middle Rincon times (AD 850–1100). Decorated pottery from this span often contains sand temper (TT4) or mixed sand and schist/gneiss tempers (TT2 and TT3).

The Rillito phase (AD 850–950) decorated ware data point holds a unique position. Nearly 60 percent of the sherds are tempered with schist/gneiss (TT1) with another 25 percent exhibiting mixed sand and schist/gneiss tempers (TT2 and TT3). In the Honey Bee Village collection, the Rillito phase (AD 850–950) decorated pottery contains the highest proportion of sherds tempered with crushed schist/gneiss.

The Cañada del Oro phase (AD 750–850) decorated pottery data point also plots away from any of the other data points. As is the case with the pottery in the Middle Rincon cluster, most Cañada del Oro decorated pottery is sand-tempered (TT4), with most of the remaining sherds tempered with various amounts of schist/gneiss (TT1, TT2, and TT3). In the Cañada del Oro data set, however, the amount of sand-tempered pottery is greater.

Discussion

The correspondence analyses of the temper type and provenance data sets provided similar results; each analysis greatly reduced the dimensionality of its data set. Given that nearly half of the variation in each data set is summarized by the ordering of the cases along the first axis, examining the location of those cases on a scatterplot consisting of both correspondence analyses first axes seemed a logical, if novel, extension of the technique. After all, potters acting in specific places at specific times and practicing different approaches to paste formulation gave rise to the variation that each of the correspondence analyses helped to characterize.

Figure 6.5 displays the resulting scatterplot of the data points. The linear appearance of the points suggested that a regression model would provide a measure of the relationship between when and where a pot was made and the technology used to prepare its paste (Shennan 1990). Accordingly, temper provenance was viewed as the independent variable (x axis) and temper type the dependent variable (y axis). The value of r^2 in the regression analysis is 0.78, meaning that 78 percent of the total variation in the temper type data is accounted for by time of manufacture, temper provenance, and kind of ware.

To investigate function, a correspondence analysis was performed with function, ware, and time data (fig. 6.6). The analysis was limited to (1) reconstructible vessels and rim sherds with measurable openings at their mouths and determinate forms and (2) rows and columns with five or more cases. As most vessels are represented by body sherds, imposing those criteria lowered the sample size from 3,966 to 960 cases in a 16 row

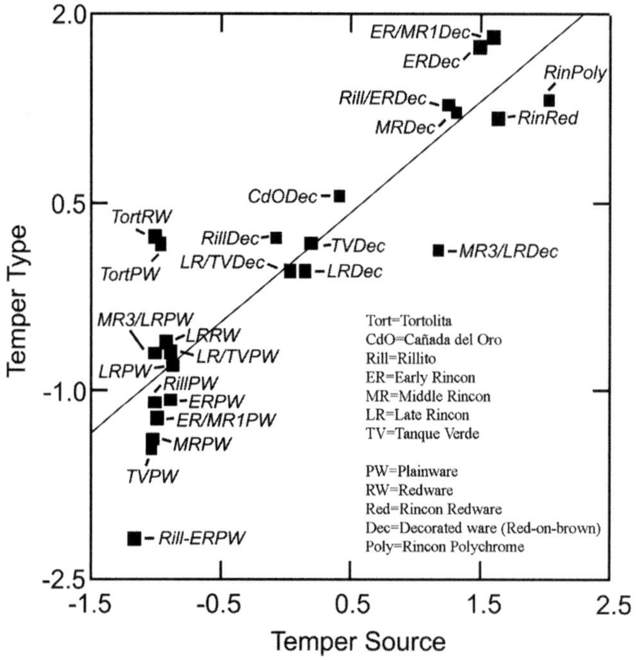

FIGURE 6.5. Linear regression model scatterplot showing the relationship of temper type to time, ware, and temper source ($r^2 = 0.78$).

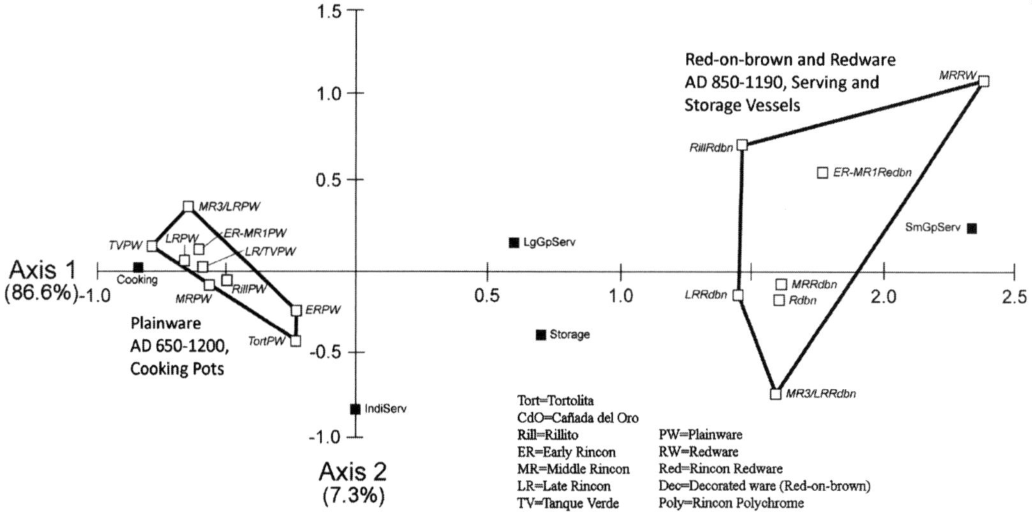

FIGURE 6.6. Correspondence analysis joint plot showing relationships between vessel functions and wares.

by 5 column matrix (table 6.4). The first factor accounts for 86.6 percent of the variation. The second factor accounts for only 7.3 percent, and the remaining two factors each account for less than 4 percent of the variation. The first factor is interpreted as a contrast between cooking vessels (point located to the left along Axis 1) and individual, small, and large group serving as well as storage vessels (points located to the right along Axis 1).

TABLE 6.4. Vessel function reported by phase and ware

		Serving[a]				
Phase + Ware	Cooking	IndiServ	SmGpServ	LgGpServ	Storage	Row total
TortPW	16	2	0	6	7	31
RillRdbn	0	0	3	9	1	13
RillPW	31	2	0	10	5	48
ERPW	6	0	0	2	3	11
ER–MR1Rdbn	0	0	9	13	4	26
ER–MR1PW	62	0	0	17	8	87
MRRdbn	0	0	16	20	23	59
MRRW	0	0	3	2	0	5
MRPW	143	3	0	30	31	207
MR3/LRRdbn	0	0	2	1	5	8
MR3/LRPW	8	0	0	3	0	11
LRRdbn	0	1	20	37	36	94
LRPW	165	0	0	33	25	223
LR/TVRdbn	0	1	13	15	18	47
LR/TVPW	53	0	0	12	10	75
TVPW	12	0	0	2	1	15
Column total	496	9	66	212	177	960

[a] IndiServ is an individual-sized serving vessel, SmGpServ is a small group-sized serving vessel, and LgGpServ is a large group-sized serving vessel.

The first factor is more readily interpretable in terms of wares. All of the plainware data sets from the Tortolita phase through the Tanque Verde phase (AD 650–1200) are grouped on the left, while the Rillito through transitional Late Rincon/Tanque Verde phase decorated ware and Rincon Redware data sets (AD 850–1190) are grouped on the right. Considered together, the results of this correspondence analysis show that throughout Honey Bee Village's occupation most storage, small group serving, large group serving, and, to a lesser extent, individual serving vessels were red-slipped or painted, whereas cooking vessels were not.

The importance of the time-ware-function analysis lies in the distinct clustering of plainware versus the red-slipped and red-on-brown ware cases, as we have already seen the exact same kind of clustering in the temper source and type data. A multiple linear regression model was run in order to provide a measure of the relationship between vessel function and when, where, and how pastes were prepared. The location of the cases on all three correspondence analyses first axes constituted the data set. Temper provenance and temper type were treated as independent variables and vessel function as the dependent variable. The result is a regression surface or plane (Berry and Feldman 1985) (fig. 6.7a).

But it is easier to label points in a two-dimensional scatterplot than on a three-dimensional surface. Figure 6.7b is a scatterplot of estimated (predicted) values by residuals (the difference between the observed and predicted values of the dependent variable *function*) and shows three distinct clusters of data points. Plainware data sets from the Tortolita phase through the Tanque Verde phase (AD 650–1200) are located on the left side of the plot, red-on-brown and redware data sets from AD 950–1110 contexts are on the right side, while the latest red-on-brown data sets (AD 1100–1190) are located at the top of the plot. The multiple linear regression analysis produced an r^2 value of 0.85, meaning that the model accounts for 85 percent of the total variation in the functional data. The model shows that potters acting at specific

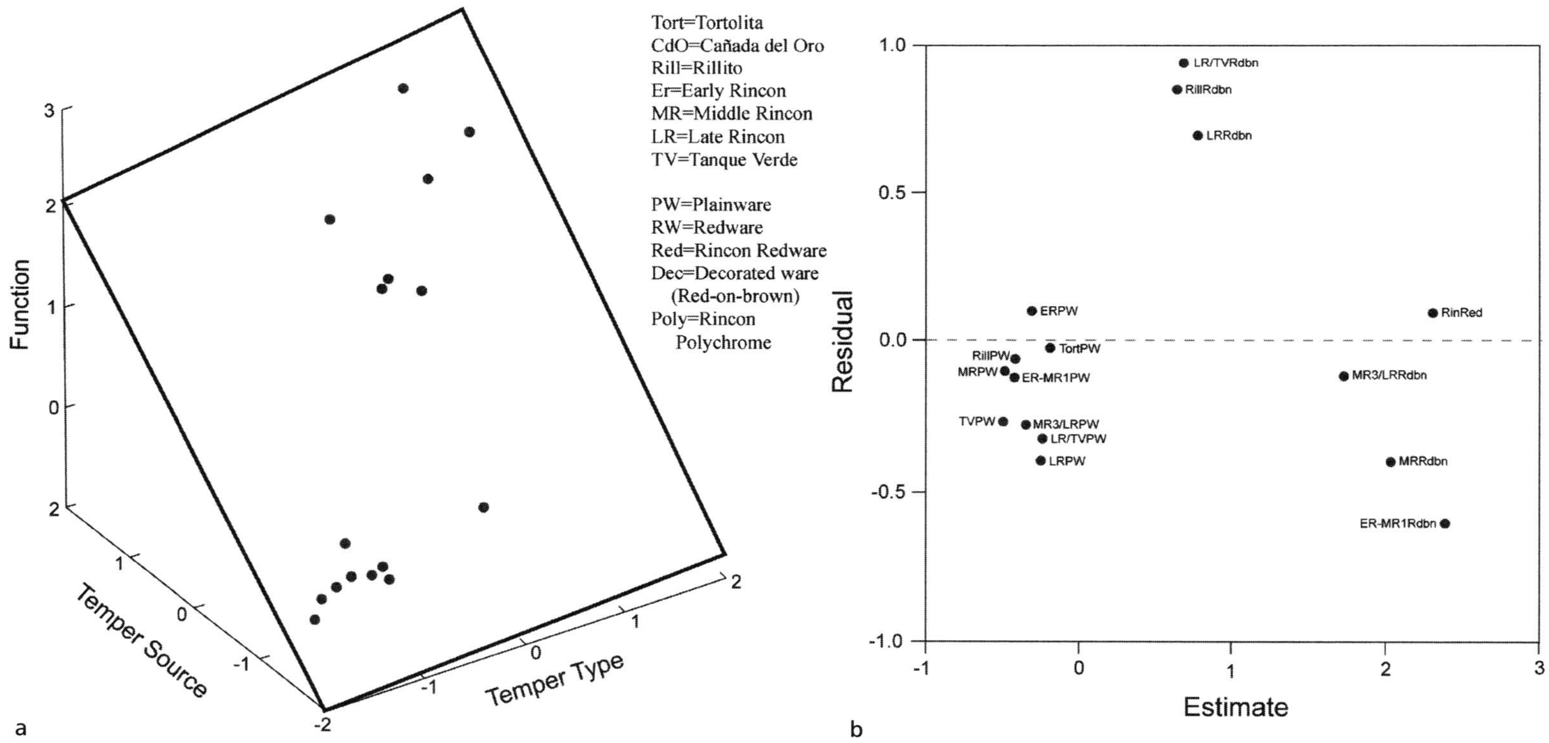

Figure 6.7. Multiple linear regression model showing the relationship of vessel function to temper source, temper type, time, and ware ($r^2 = 0.85$): (*a*) three-dimensional view; (*b*) two-dimensional view.

times and in specific places utilized distinct paste recipes (temper types) to produce different wares, each with their own intended functions.

While ethnographic studies from the Americas (Fontana et al. 1962:33–49, 67; Hagstrum 1989:figure 1.5; Mindling 2010:112; Russell 1980: 124–129; Shepard 1963:4–9, 14, 17–18), Africa (Arthur 2006:31–35, 44; David and Hennig 1972: 27; Herbich and Dietler 2008:231), and Asia (Longacre et al. 2000:277; Miller 1985:64, 142–148; Stark 1991:199) suggest that this should be the case, it is a pattern not often demonstrated with archaeological data. Having a large, petrographically calibrated ceramic sample provided the means to model these relationships.

Archaeological Conclusions

Temper provenance data recorded from pottery recovered from Honey Bee Village's well-dated contexts largely conform to known temporal trends related to ceramic production and distribution in the Tucson Basin. Tortolita and Snaketown phase pottery was usually made at or near the site where it was recovered. Community specialization in the manufacture of decorated pottery for basin-wide distribution began during the Cañada del Oro phase at a village located in the Twin Hills south Petrofacies. Subsequently manufacture shifted to three or four villages located within 3 km of the Beehive Petrofacies during the Early and Middle Rincon phases. Production at those settlements declined during the Late Rincon phase, with manufacture largely gone by the Tanque Verde phase (Heidke 2009, 2011b, 2012; Heidke et al. 2002). Indirect provenance evidence supports the idea that Classic Period ceramic production took place at settlements located throughout the greater Tucson Basin, with distinct distribution systems centered on each of the area's trunk streams.

It is important to have some understanding of the site's occupation in order to examine ceramic production there. Twenty-five house clusters were identified at Honey Bee Village, which span the site's Preclassic occupation (Lindeman 2012:table 17.1). A typical cluster consists of two to four contemporaneous structures arranged around a small courtyard. Eight of the house clusters were occupied and abandoned within a single phase and consist of just one courtyard. Twelve clusters were occupied and abandoned within a single phase but contain more than one courtyard group. The remaining five house clusters span multiple phases, contain multiple courtyard groups, and were occupied for many generations.

Overall, twelve of the house clusters yielded direct evidence of ceramic manufacture, generally ground stone pottery polishers exhibiting moderate wear (Adams and Stroud 2012:332–334). Only 40 percent of the single phase residential clusters yielded direct evidence of ceramic manufacture, whereas 80 percent of the multiphase clusters produced similar evidence. The ground stone tool and provenance evidence of ceramic production collected from Honey Bee Village therefore suggests that, while some households were largely self-reliant, the organization of other households, especially those that persisted for hundreds of years, is consistent with a model of individual specialization: "autonomous individuals or households producing for unrestricted local consumption" (Costin 1991:8).

Finally, functional characteristics of the ceramic collection recovered from Honey Bee Village exhibit little change over time, and those changes that are detectable reflect basin-wide changes. As is the case with all analyzed Tortolita phase pottery collections, the one from Honey Bee Village reflects a greater familiarity with the production of independent, restricted vessel forms when compared with the preceding Agua Caliente phase (AD 50–500) potters' emphasis on simple and dependent restricted forms. During the Tortolita phase potters throughout the basin began to make ceramic containers suitable for a wide variety of storage, food preparation, cooking, and serving tasks. The range of pottery functions established in the Tortolita phase continued throughout the rest of prehistory. This suggests that a single food use pattern (*sensu* Hally 1986:291) may have been established by the Tortolita phase, which continued to be practiced until at least AD 1450 (Heidke 2012:307–309).

The intent of this study has been to provide an example of the kinds of results it is possible to achieve when Shepard's approach is followed and petrography is used to verify temper characteristics recorded with the aid of a reflected light microscope. Furthermore, it highlights the important role that correspondence and regression analyses together can play in identifying patterns in large ceramic data sets and testing relationships of temporal, provenance, technological, and functional attributes of potter behavior.

Acknowledgments

Funding was provided by Pima County, Arizona, Vistoso Partners LLC, Canada Vista LLC (and Steve Solomon), Archaeology Southwest, and Desert Archaeology, Inc. Petrographers Elizabeth Miksa, Mary Ownby, and Carlos Lavayen provided critical insight into the tempering materials observed in the collection.

References Cited

Adams, Jenny L., and Amanda Stroud
2012 Ground Stone Artifacts and Ecofacts Recovered from Honey Bee Village, AZ BB:9:88 (ASM). In *Life in the Valley of Gold: Archaeological Investigations at Honey Bee Village, a Prehistoric Hohokam Ballcourt Village, Part 1*, edited by Henry D. Wallace, pp. 323–432. Anthropological Papers No. 48. Archaeology Southwest, Tucson.

Arthur, John W.
2006 *Living with Pottery: Ethnoarchaeology among the Gamo of Southwest Ethiopia*. University of Utah Press, Salt Lake City.

Berry, William D., and Stanley Feldman
1985 *Multiple Regression in Practice*. Sage Publications, Beverly Hills.

Blasius, Jörg
1994 Correspondence Analysis in Social Science Research. In *Correspondence Analysis in Social Science Research: Recent Developments and Applications*, edited by Michael Greenacre and Jörg Blasius, pp. 23–52. Academic Press, San Diego.

Braun, David P.
1980 Experimental Interpretation of Ceramic Vessel Use on the Basis of Rim and Neck Formal Attributes. In *The Navajo Project: Archaeological Investigations, Page to Phoenix 500 KV Southwestern Transmission Line*, edited by Donald C. Fiero, Robert W. Munson, Martha R. McClain, Suzanne M. Wilson, and Anne H. Zier, pp. 171–231. Research Papers No. 11. Museum of Northern Arizona, Flagstaff.

Carr, James R.
1990 CORSPOND: A Portable FORTRAN-77 Program for Correspondence Analysis. *Computers and Geosciences* 16:289–307.

Costin, Cathy L.
1991 Craft Specialization: Issues in Defining, Documenting, and Explaining the Organization of Production. In *Archaeological Method and Theory*, vol. 3, edited by Michael B. Schiffer, pp. 1–56. University of Arizona Press, Tucson.

David, Nicholas, and Hilke Hennig
1972 The Ethnography of Pottery: A Fulani Case Seen in Archaeological Perspective. In *McCaleb Module in Anthropology 21*, pp. 1–29. Addison-Wesley Publishing, Reading, Massachusetts.

Dickinson, William R.
1970 Interpreting Detrital Modes of Graywacke and Arkose. *Journal of Sedimentary Petrology* 40:695–707.

Duff, Andrew I.
1996 Ceramic Micro-Seriation: Types or Attributes? *American Antiquity* 61:89–101.

Farry, Andrew
2005 Regulars and "Irregulars": British and Provincial Variability among Eighteenth-Century Military Frontiers. *Historical Archaeology* 39:16–32.

Fontana, Bernard L., William J. Robinson, Charles W. Cormack, and Ernest E. Leavitt Jr.
1962 *Papago Indian Pottery*. University of Washington Press, Seattle.

Gazzi, Paolo
1966 Le arenarie del flysch sopracretaceo dell'Appennino Modenese: Correlazioni con il flysch di Monghidoro. *Mineralogica e Petrografica Acta* 12:69–97.

Gihwala, Dherendra, Leon Jacobson, Max Peisach, Carlos A. Pineda, and Hennie N. Vos
1985 Analysis of Chinese Porcelains and Ceramics. *South African Archaeological Bulletin* 40:96–99.

Graham, Stephan A., Raymond V. Ingersoll, and William R. Dickinson
1976 Common Provenance for Lithic Grains in Carboniferous Sandstones from Ouachita

Mountains and Black Warrior Basin. *Journal of Sedimentary Petrology* 46:620–632.

Greenacre, Michael J.
1984 *Theory and Applications of Correspondence Analysis*. Academic Press, New York.

Hagstrum, Melissa B.
1989 Technological Continuity and Change: Ceramic Ethnoarchaeology in the Peruvian Andes. Ph.D. dissertation, Department of Anthropology, University of California, Los Angeles.

Hally, David J.
1986 The Identification of Vessel Function: A Case Study from Northwest Georgia. *American Antiquity* 51:267–295.

Haury, Emil W.
1976 *The Hohokam: Desert Farmers and Craftsmen. Excavations at Snaketown, 1964–1965*. University of Arizona Press, Tucson.

Heidke, James M.
2006 Native American Pottery. In *Rio Nuevo Archaeology, 2000–2003: Investigations at the San Agustín Mission and Mission Gardens, Tucson Presidio, Tucson Pressed Brick Company, and Clearwater Site*, edited by J. Homer Thiel and Jonathan B. Mabry, pp. 7.1–7.94. Technical Report No. 2004-11. Desert Archaeology, Tucson.
2009 Multi-Village Specialized Craft Production and the Distribution of Hohokam Sedentary Period Pottery, Tucson, Arizona. In *Interpreting Silent Artefacts: Petrographic Approaches to Archaeological Ceramics*, edited by Patrick Sean Quinn, pp. 227–244. Archaeopress, Oxford, England.
2011a Prehistoric Pottery Containers from the Julian Wash Site, AZ BB:13:17 (ASM). In *Craft Specialization in the Southern Tucson Basin: Archaeological Excavations at the Julian Wash Site, AZ BB:13:17 (ASM): Part 1. Introduction, Excavation Results, and Artifact Investigations*, edited by Henry D. Wallace, pp. 263–294. Anthropological Papers No. 40. Center for Desert Archaeology, Tucson.
2011b Sedentary Period Ceramic Production and Distribution: New Evidence from the Julian Wash Site, AZ BB:13:17 (ASM). In *Craft Specialization in the Southern Tucson Basin: Archaeological Excavations at the Julian Wash Site, AZ BB:13:17 (ASM): Part 2. Synthetic Studies*, edited by Henry D. Wallace, pp. 529–552. Anthropological Papers No. 40. Center for Desert Archaeology, Tucson.
2012 Prehistoric Pottery from Honey Bee Village: Dating, Provenance, Typology, and Function. In *Life in the Valley of Gold: Archaeological Investigations at Honey Bee Village, a Prehistoric Hohokam Ballcourt Village, Part 1*, edited by Henry D. Wallace, pp. 183–321. Anthropological Papers No. 48. Archaeology Southwest, Tucson.

Heidke, James M., and Elizabeth J. Miksa
2000 Correspondence and Discriminant Analyses of Sand and Sand Temper Compositions, Tonto Basin, Arizona. *Archaeometry* 42: 273–299.

Heidke, James M., Elizabeth J. Miksa, and Henry D. Wallace
2002 A Petrographic Approach to Sand-Tempered Pottery Provenance Studies: Examples from Two Hohokam Local Systems. In *Ceramic Production and Circulation in the Greater Southwest: Source Determination by INAA and Complementary Mineralogical Investigations*, edited by Donna M. Glowacki and Hector Neff, pp. 152–178. Monograph No. 44. Cotsen Institute of Archaeology, University of California, Los Angeles.

Herbich, Ingrid, and Michael Dietler
2008 The Long Arm of the Mother-in-Law: Learning, Postmarital Resocialization of Women, and Material Culture Style. In *Cultural Transmission and Material Culture: Breaking Down Boundaries*, edited by Miriam T. Stark, Brenda J. Bowser, and Lee Horne, pp. 223–244. University of Arizona Press, Tucson.

Hill, Mark O.
1979 *DECORANA: A FORTRAN Program for Detrended Correspondence Analysis and Reciprocal Averaging*. Ecology and Systematics, Cornell University, Ithaca, New York.

Huckleberry, Gary
2012 Geomorphology of Honey Bee Village and the Middle Cañada del Oro Valley. In *Life in the Valley of Gold: Archaeological Investigations at Honey Bee Village, a Prehistoric Hohokam Ballcourt Village, Part 1*, edited by Henry D. Wallace, pp. 29–44. Anthropological Papers No. 48. Archaeology Southwest, Tucson.

Ingersoll, Raymond V.
1983 Petrofacies and Provenance of Late Mesozoic Fore-Arc Basin, Northern and Central

California. *American Association of Petroleum Geologists Bulletin* 67:1125–1142.

Ingersoll, Raymond V., and Christopher A. Suczek
1979 Petrology and Provenance of Neogene Sand from Nicobar and Bengal Fans, DSDP Sites 211 and 218. *Journal of Sedimentary Petrology* 49:1217–1228.

Jacobson, Leon, Warren S. Fish, and Willem A. van der Westhuizen
2002 XRF Analysis of Pottery from Mutokolwe, a Khami Settlement from the Soutpansberg Mountains, South Africa. In *Geochemical Evidence for Long Distance Exchange*, edited by Michael D. Glascock, pp. 215–228. Bergin and Garvey, Westport, Connecticut.

Jacobson, Leon, Thomas N. Huffman, Willem A. van der Westhuizen, and Hendrik de Bruiyn
1995 X-Ray Fluorescence and Microprobe Analyses on Pottery from the Northern and Eastern Transvaal, South Africa and Southern Zimbabwe. In *The Aim of Laboratory Analyses of Ceramics in Archaeology*, edited by Anders Lindahl and Ole Stilborg, pp. 49–61. KVHAA Konferenser 34, Stockholm.

Jacobson, Leon, Jannie H. N. Loubser, Max Peisach, Carlos A. Pineda, and Willem van der Westhuizen
1991 PIXE Analysis of Pre-European Pottery from the Northern Transvaal and Its Relevance to the Distribution of Ceramic Styles, Social Interaction, and Change. *South African Archaeological Bulletin* 46:19–24.

Kamilli, Diana C.
1994 A Petrographic Study of Pottery from the Schuk Toak District. In *Archaeological Studies of the Avra Valley, Arizona: Excavations in the Schuk Toak District: Vol. 2. Scientific Studies and Interpretations*, by Allen Dart, pp. 627–652. Anthropological Papers No. 16. Center for Desert Archaeology, Tucson.

Kovach Computing Services
2000 *MVSP Version 3.1*. Kovach Computing Services, Angelsey, Wales.

Lindeman, Michael W.
2012 Households and Production at Honey Bee Village, AZ BB:9:88 (ASM). In *Life in the Valley of Gold: Archaeological Investigations at Honey Bee Village, a Prehistoric Hohokam Ballcourt Village, Part 2*, edited by Henry D. Wallace, pp. 767–781. Anthropological Papers No. 48. Archaeology Southwest, Tucson.

Lombard, James P.
1987 Provenance of Sand Temper in Hohokam Ceramics, Arizona. *Geoarchaeology* 2:91–119.

Longacre, William A., Jingfeng Xia, and Tao Yang
2000 I Want to Buy a Black Pot. *Journal of Archaeological Method and Theory* 7:273–293.

Miksa, Elizabeth J.
2011 Half Million Points and Counting: Two Decades of Petrofacies Modeling in the Greater Tucson Basin and Avra Valley. In *Craft Specialization in the Southern Tucson Basin: Archaeological Excavations at the Julian Wash Site, AZ BB:13:17 (ASM): Part 2. Synthetic Studies*, edited by Henry D. Wallace, pp. 553–617. Anthropological Papers No. 40. Center for Desert Archaeology, Tucson.

Miksa, Elizabeth J., and James M. Heidke
2001 It All Comes Out in the Wash: Actualistic Petrofacies Modeling of Temper Provenance, Tonto Basin, Arizona, USA. *Geoarchaeology* 16:177–222.

Miksa, Elizabeth J., Mary F. Ownby, and Carlos P. Lavayen
2012 Petrographic Analysis of Pottery from Honey Bee Village. In *Life in the Valley of Gold: Archaeological Investigations at Honey Bee Village, a Prehistoric Hohokam Ballcourt Village, Part 1*, edited by Henry D. Wallace, pp. 123–181. Anthropological Papers No. 48. Archaeology Southwest, Tucson.

Miller, Daniel
1985 *Artifacts as Categories: A Study of Ceramic Variability in Central India*. Cambridge University Press, Cambridge.

Mindling, Eric
2010 *Barro y fuego: El arte de la alfarería en Oaxaca*. Editorial Arte Oaxaca, Oaxaca, Mexico.

Müller, Johannes
2009 Dating the Neolithic: Methodological Premises and Absolute Chronology. *Radiocarbon* 51:721–736.

Nováček, Karel
2009 Moulded Pottery from Istakhr. In *My Things Changed Things: Social Development and Cultural Exchange in Prehistory, Antiquity, and the Middle Ages*, edited by Petra M. Vlčková, Jana Mynářová, and Martin Tomášek, pp. 118–126. Institute of Archaeology of the Academy of Sciences of the Czech Republic, Charles University, Prague.

Pavúk, Peter
2010 Pottery Processing at Troy—Typology, Stratigraphy and Correspondence Analysis: How do They Work Together? In *Analysing Pottery: Processing—Classification—Publication*, edited by Barbara Horejs, Reinhard Jung, and Peter Pavúk, pp. 73–98. Studia Archaeologica et Medievalia, Comenius University, Bratislava.

Peeples, Matthew A., and Gregson Schachner
2012 Refining Correspondence Analysis-Based Ceramic Seriation of Regional Data Sets. *Journal of Archaeological Science* 39: 2818–2827.

Pitts, Martin
2005 Pots and Pits: Drinking and Deposition in Late Iron Age South-east Britain. *Oxford Journal of Archaeology* 24(2):143–161.

Ringrose, Trevor J.
1992 Bootstrapping and Correspondence Analysis in Archaeology. *Journal of Archaeological Science* 19:615–629.

Robertson, Ian G.
1999 Spatial and Multivariate Analysis, Random Sampling Error, and Analytical Noise: Empirical Bayesian Methods at Teotihuacan, Mexico. *American Antiquity* 64:137–152.

Russell, Frank
1980 *The Pima Indians*. Reprinted: University of Arizona Press, Tucson. Originally published 1908, 26th Annual Report of the Bureau of American Ethnology, 1904–1905, Smithsonian Institution. U.S. Government Printing Office, Washington, D.C.

Shennan, Stephen
1990 *Quantifying Archaeology*. Academic Press, New York.

Shepard, Anna O.
1936 The Technology of Pecos Pottery. In *The Pottery of Pecos*, vol. 2, by Alfred V. Kidder and Anna O. Shepard, pp. 389–587. Papers of the Phillips Academy Southwestern Expedition 7. Yale University Press, New Haven.
1942 Rio Grande Glaze Paint Ware: A Study Illustrating the Place of Ceramic Technological Analyses in Archaeological Research. In *Contributions to American Anthropology and History No. 39*, pp. 129–262. Carnegie Institution, Washington, D.C.
1963 *Beginnings of Ceramic Industrialization: An Example from the Oaxaca Valley*. Notes from a Ceramic Laboratory No. 2. Carnegie Institution, Washington, D.C.
1965 Rio Grande Glaze-Paint Pottery: A Test of Petrographic Analysis. In *Ceramics and Man*, edited by Frederick R. Matson, pp. 62–87. Viking Fund Publications in Anthropology No. 41. Aldine, Chicago.
1995 *Ceramics for the Archaeologist*. 5th ed. Reprinted: Braun-Brumfield, Ann Arbor, Michigan. Originally published in 1956 as Volume 609, Carnegie Institution, Washington, D.C.

Smith, Karen Y., and Fraser D. Neiman
2007 Frequency Seriation, Correspondence Analysis, and Woodland Period Ceramic Assemblage Variation in the Deep South. *Southeastern Archaeology* 26:47–72.

Stark, Miriam T.
1991 Ceramic Change in Ethnoarchaeological Perspective: A Kalinga Case Study. *Asian Perspectives* 30:193–216.

Wallace, Henry D.
2012a Honey Bee Village and the Valley of Gold in the Scheme of Things Hohokam. In *Life in the Valley of Gold: Archaeological Investigations at Honey Bee Village, a Prehistoric Hohokam Ballcourt Village, Part 2*, edited by Henry D. Wallace, pp. 783–827. Anthropological Papers No. 48. Archaeology Southwest, Tucson.
2012b Introduction. In *Life in the Valley of Gold: Archaeological Investigations at Honey Bee Village, a Prehistoric Hohokam Ballcourt Village, Part 1*, edited by Henry D. Wallace, pp. 1–28. Anthropological Papers No. 48. Archaeology Southwest, Tucson.

Weller, Susan C., and A. Kimball Romney
1990 *Metric Scaling: Correspondence Analysis*. Sage University Papers No. 07-075. Sage Publications, Newbury Park, California.

Wilcox, David R.
1979 The Hohokam Regional System. In *An Archaeological Test of Sites in the Gila Butte–Santan Region, South-Central Arizona*, edited by Glen E. Rice, David Wilcox, Kevin Rafferty, and James Schoenwetter, pp. 77–116. Anthropological Research Papers No. 18. Arizona State University, Tempe.

CHAPTER 7

The Contribution of Petrography to Understanding the Production and Consumption of Early Helladic Ceramics from Nemea, Mainland Greece

CLARE BURKE, PETER M. DAY, *and* DANIEL J. PULLEN

The Aegean Early Bronze Age (EBA ca. 3100–2100 BC) has been a focus of scholarly attention as the prelude to the early state organizations of the Middle and Late Bronze Ages (ca. 2000–1600 BC and 1600–1000 BC). A key aspect of this research has been to account for the emergence of societal complexity in different areas of the Aegean. In particular, such work has focused on changes in craft technology as an indicator of complexity, focusing on concepts such as craft specialization and its relation to sociopolitical organization (Pullen 2008; Renfrew et al. 1965; Renfrew 1967, 1972). While this approach has provided invaluable insight into many aspects of the Middle Bronze and Late Bronze Ages, it has not furnished us with a coherent picture of the character of Early Bronze Age society, its crafting practices, and motivations for consumption.

More recently, the approach has shifted, with scholars seeking to emphasize the social embeddedness of technological and consumption practices (Day and Wilson 1997, 2002, 2004; Day et al. 2010; compare Dobres 2010; Lemonnier 2002; Pfaffenberger 1992 for a general discussion of social aspects of technology). Focusing on the organization of production, networks of exchange, and evidence for consumption behavior, work undertaken on Crete and in the Cyclades has investigated the role of these activities in the negotiation of identity and cultural contact (Day et al. 1998; Day et al. 2012; Hilditch in press; Nodarou 2011). Such research has revealed a complex picture of early island society and its production and consumption of ceramic vessels. In contrast, the EBA ceramics of mainland Greece have continued to be examined from a primarily typological perspective, resulting in only a preliminary understanding of craft technology and consumption behavior for this area of the Aegean (for notable exceptions see Betancourt and Myer 2000; Hilditch et al. 2008).

The pioneering research by neutron activation analysis (NAA) undertaken by Michael Attas (1982; Attas et al. 1987) has highlighted the potential for ceramic analysis to go further. In this early work Attas investigated production and movement of Early Helladic (EH ca. 3000–2100 BC) ceramics from sites in the northeastern Peloponnese. His primary goal was to understand the distribution of ceramic production and gauge the level of interaction among sites during this formative period. Through his carefully considered analysis, Attas was able to suggest that ceramics were produced within close proximity to (if not actually at) all the sites within his study and identified small-scale movement of more common forms between sites and from areas outside his study.

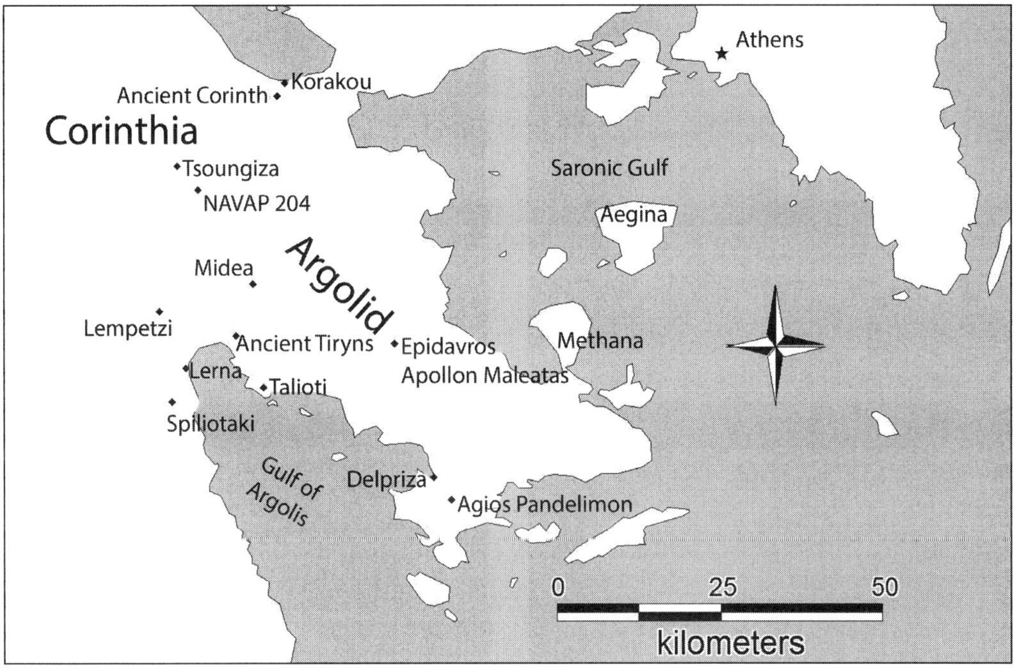

FIGURE 7.1. Location of the Early Helladic sites currently under study as part of this project (prepared by C. Burke).

This challenged early ideas about the movement of ceramic vessels, demonstrating the widespread production of both common vessels and wares and those previously assumed to be specialist products. In addition, Attas was able to document the exchange of common vessel types previously considered to have had very limited or no circulation. While this work was innovative, however, the reliance solely on chemical data resulted in limitations on the conclusions that could be reached, most significantly:

- The inability to source chemical groups that did not relate to reference groups;
- Difficulties refining groupings to differentiate between specific paste recipes within the same site, which meant that Attas was unable to discriminate between spatial and technological causes for the variability that he identified;
- An inability to comment on the technological aspects of the pottery analyzed;
- The chemical data distorted by the inclusion of coarsewares could not be used to characterize much of the EH ceramic repertoire (Attas et al. 1987:81).

Both Attas's early chemical results and those from petrographic studies of EBA ceramics elsewhere in the Aegean noted above have highlighted the need to look in more detail at the ceramics of the mainland. In the light of such research, it is timely to consider the technological behavior and interaction of the small-scale communities that most likely characterize this period. Discussion of this has been regularly neglected in favor of a focus on monumental sites, most prominently Lerna (Shriner 1999; Shriner and Murray 2001; Weingarten 1997).

Forming part of a broader program of multitechnique analytical research examining ceramic production and consumption, this chapter discusses the results of macroscopic and petrographic examination of EH vessels from two sites in the Nemea Valley, the excavated settlement of Tsoungiza (Pullen 2011) and Site 204 of the Nemea Valley Archaeological Project (NVAP) (Wright et al. 1990) (fig. 7.1 and table 7.1).

While the approach is diachronic, the emphasis is on an understanding of technology and tradition rather than specialization and the organization of production per se. By using

TABLE 7.1. Chronology and number of samples for each epoch of the Early Helladic period at NVAP 204 and Tsoungiza

Chronological phases	NVAP 204	Tsoungiza
EH I ca. 3100–2700	14	42
EH II ca. 2700–2450	28	
EHII Initial (Tsoungiza only)		36
EHII Developed (Tsoungiza only)		41
ca. 2700–2450		
EH III ca. 2300–2000	0	35
Total number of samples	42	154

Source: dates from Pullen 2011 and Wiencke 2000.

petrography not only to suggest provenance but also to show how raw materials are selected and manipulated, the intention is to look at every day practices of production and ceramic consumption, to explore the choices humans make in their interactions with ceramic material culture. Such a bottom-up approach aims to produce a representative picture of ceramic production and consumption *before* moving to questions of diachronic social change. The expectation should be for a period of research when everyday practices are illuminated, to understand the Early Bronze Age for itself, revealing choices and identities, rather than feeling bound to use these new understandings immediately to return to models of social complexity.

Site Location and Environmental Setting

The site of Tsoungiza is located on a small hill, directly west of the modern village of Ancient Nemea, in the southern portion of the Nemea Valley. NVAP 204 lies on a small bluff, approximately 5 km southeast of Tsoungiza. Located about 30 km west-southwest of Corinth, the subbasin of the Nemea Valley is one of several basins in the northeastern Peloponnese and has a long history of occupation dating back to the Neolithic. It is bordered by a number of hill ranges, including those of the Psyli Rachi massif to the southeast, which divides the Corinthia from the Argolid, and the Kyllini range to the west. These ranges are intersected by passes that connect the Nemea Valley to the surrounding area, including the Longopotamos Valley to the east, the location of the EH settlement site of Zygouries. The valley is drained to the northeast by the Asopos River, which is fed by a number of springs and small tributaries forming the plains of Phliasia to the west.

Geologically, the Nemea Basin is dominated by Pliocene-Pleistocene marls, sandstones, conglomerates, and limestones accompanied by Quaternary sediments. The thickness of these deposits varies throughout the basin, with the Upper Pliocene conglomerate deposits reaching a thickness of over 300 m in some places. The Neogene deposits include marl, sandstones, and sands (Dounias et al. 2006). The river plain contains fluvial deposits, including sands, silts, and clays derived from weathering and transportation of the Pliocene-Pleistocene deposits that form the valley.

Fieldwork Background

The EH remains of Tsoungiza have a long history of exploration, with systematic excavations by Carl Blegen beginning in 1924 and expanded by James Penrose Harland until 1927. In the 1970s and 1980s rescue excavations were conducted by the University of California, Berkeley, and the Greek Archaeological Service. Most recently, full-scale excavation was undertaken as part of the Nemea Valley Archaeological Project between 1984 and 1986, from which the material presented here was recovered (Pullen 2011).

Tsoungiza shows a long period of EBA occupation, with evidence for settlement dating from EHI through to what Pullen has termed the EHII Developed period (EHI ca. 3100–2700, EHII ca. 2700–2450) and then again in EHIII (ca. 2300–2000). EHII Developed is a phase of the EHII period identified by Pullen and applies to the site of Tsoungiza based on typological and architectural phasing present at the site (see Pullen 2011 for full discussion of the phasing and chronology at Tsoungiza). The excavation of this site has been most valuable in its contribution to understanding the stratigraphy and material culture of the EHI and early EHII periods in particular (Pullen 2011:37–39).

Our previous understanding of these periods has been limited due to incomplete excavation and/or publication of prehistoric sites across the mainland. This has resulted in a reliance on a small number of published sites, the majority of which have lacked the full sequence between EHI and EHII. For example, the site of Lerna has only limited evidence for EHI occupation but extensive EHII material (Rutter 1995; Wiencke 2000). The lack of full sequences has meant that scholars have been forced to use comparable material from sites that may often lie outside their particular region of interest, resulting in an inability to account for geographically specific trends. For example, we might cite the reliance on comparison with Eutresis for an understanding of EHI of the Peloponnese, despite its location in Boeotia, which lies much further to the north (Caskey and Caskey 1960; Goldman 1931; also noted by Pullen 2011:37, 56–57).

NVAP 204 was identified as part of the Nemea Valley survey led by James Wright under the auspices of the American School of Classical Studies at Athens. The site was recognized in 1984 when deep ploughing uncovered approximately 2,000 sherds of EHI and early EHII pottery, spread over one hectare. While the site was never formally excavated, the high concentration of well-preserved material bears strong similarities to those found in other small prehistoric settlements in the area (Wright et al. 1990:611), including Tsoungiza, and suggests the presence of a settlement in proximity to the area. The ability to compare two such coherent assemblages directly provided a good opportunity to investigate production and consumption of ceramics within the Nemea Valley.

Methodology

Macroscopic Examination

During fieldwork, approximately 2,000 sherds were recovered from NVAP 204 and over 80,000 sherds from Tsoungiza. The team at each site sorted the pottery, counted the sherds, and weighed and recorded details such as surface treatment and macroscopic fabric (work undertaken by colleagues for NVAP 204 and Daniel Pullen and colleagues at Tsoungiza; see Pullen 2011:10–14 and Wright et al. 1990 for discussion of recovery and recording procedures). For the purposes of sampling for petrographic analysis, all the available material from each site was laid out by context/collection unit and date to ensure that samples chosen characterized the technological variation present in the assemblage as a whole, including the variety of forms, decorative techniques, and macroscopic fabrics observed for each of the EH periods present (table 7.1 shows the number of samples chosen at each site).

Petrography

Thin sections were prepared at the Fitch Laboratory, Athens, using standard techniques. These were then examined petrographically with a polarizing light microscope. Each section was characterized and grouped qualitatively, utilizing standard visual charts, and examined using four primary criteria:

1. aplastic inclusion types;
2. size, shape, frequency, and distribution of aplastic inclusions;
3. textural concentration features such as clay pellets;
4. appearance of the micromass, including color, optical activity, and the presence and character of voids.

The groups were then described based upon these criteria, using the system proposed by Whitbread (1989, 1995).

The examination of these characteristics can tell us much about the possible production location of vessels based upon the identification of rock and clay types and relating them to their geological source. Importantly, it can also inform us about technological choices and practices involved in the manipulation of these raw materials to form a vessel, through the identification of construction methods such as remnant coils, particular paste recipe traditions, and the processing of raw materials (Shepard 1963; Sillar and Tite 2000; Whitbread 1995).

The petrographic groups identified were compared to both the initial macroscopic fabrics

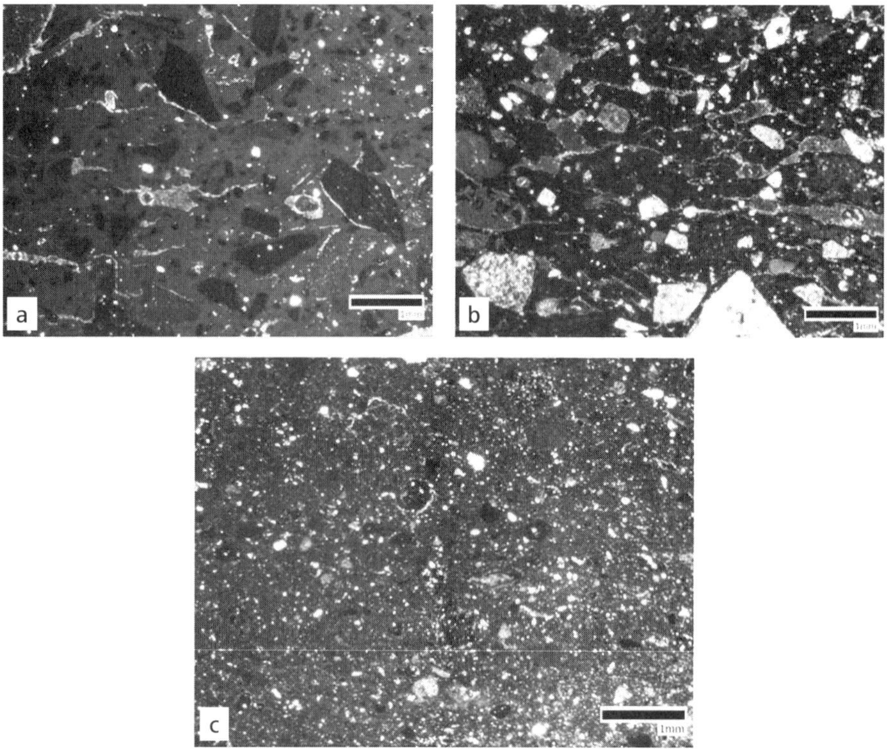

FIGURE 7.2. Petrographic images of ceramic fabrics identified: (*a*) high fired siltstone-tempered fabric; (*b*) chert, siltstone, and vegetal tempered fabric; (*c*) subrounded textural concentration features indicative of clay mixing (cross-polarized light).

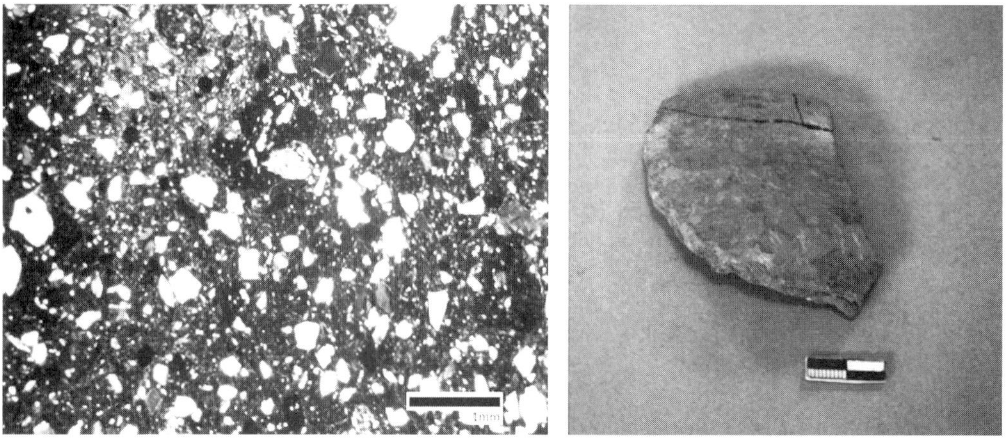

FIGURE 7.3. Andesite fabric (cross-polarized light) (*left*) and deep bowl displaying high-quality burnished surface (*right*).

noted by the field team and later observations made by the authors. It is important to undertake such comparison to test the degree to which fabric variation can be identified at the macroscopic level in the field. An examination of the degree of correlation among form, ware, and fabric was carried out to assess technological variation in the assemblage. This provides information on the potential repertoire of a production center/potter and on whether certain producers were making vessels with specific forms, fabrics, and surface treatment. Finally, the fabrics were compared to other material examined and analyzed as part of the broader project on Early Helladic ceramics of the Corinthia and Argolid to characterize the distribution of fabrics and interaction among different communities.

Results and Discussion

EHI-EHII Initial

The assemblages from Tsoungiza and NVAP 204 broadly share the same petrographic fabric groups (for detailed discussion of fabric classes and sample numbers see Burke Davies n.d.), suggesting that many of the ceramic vessels consumed at these sites were obtained from the same production locations. Tsoungiza contained more fabric groups, however, a reflection of the more extensive repertoire of forms and longer chronological period sampled. The fabrics of EHI and the early part of EHII from these two sites are notable for the significant degree of variability in raw material sources, representing diverse lithologies and paste recipes, suggestive of different potters working in a variety of geographical areas. It should be noted that in the context of this chapter the term "local" applies to those fabrics whose raw materials are consistent with the geology of the Nemea Valley and "nonlocal" refers to those fabrics whose raw materials are inconsistent with this geology.

At both sites the fabrics considered to be local are broadly characterized by the dominant presence of siltstones, mudstones, cherts, and calcite, all of which occur within a 2 km radius, based on assessment of resources from geological maps and prospection undertaken in the area. They also display characteristic variation in the choice and processing of these raw materials, however, most clearly evident when comparing just three of the local fabrics (fig. 7.2).

The first group (fig 7.2a) illustrates the technique of tempering a fine calcareous clay with coarse siltstones and mudstones. The second (fig. 7.2b) shows the use of crushed chert, mudstones/siltstones, and vegetal temper. Mixing is often evident from the remnants of clay pellets and striations within the matrix. Finally, the third group (fig. 7.2c) contains a mix of a terra rossa clay with a calcareous clay, illustrated by the presence of dark rounded sedimentary inclusions within a lighter calcareous matrix. While each of these fabrics contains raw materials available in the Nemea Valley, they represent distinct choices and ideas of the appropriate raw materials to be used to produce a vessel.

There does not appear to be a correlation among particular local fabrics, vessel forms, or ware types, with most of the EHI and early EHII repertoire present in each fabric. This evidence suggests that the fabric variation represents different potters with specific technological practices, utilizing sources geologically compatible with those surrounding NVAP 204 and Tsoungiza, rather than a single producer using particular paste recipes for functional or aesthetic purposes (such as the use of a particular fabric suitable for burnishing).

While there is a clear core of potentially local fabrics, a number of EHI groups are inconsistent with the geology of the area. These nonlocal groups consist of sedimentary, igneous, and metamorphic derived fabrics, suggesting consumption of vessels from a variety of producers located outside the Nemea Valley. Prominent among these nonlocal fabrics has been an andesite group found at Tsoungiza but not at NVAP 204 (fig. 7.3). This distinct fabric composition has been well documented for EHII and later across the mainland. Its geological composition matches that of the island of Aegina (Gauss and Kiriatzi 2011), but it is not yet well known for sites occupied in EHI.

This fabric group consists of small hemispherical bowls, notable during macroscopic examination for both the presence of angular black inclusions and the high quality of their burnished finish. In general the hemispherical

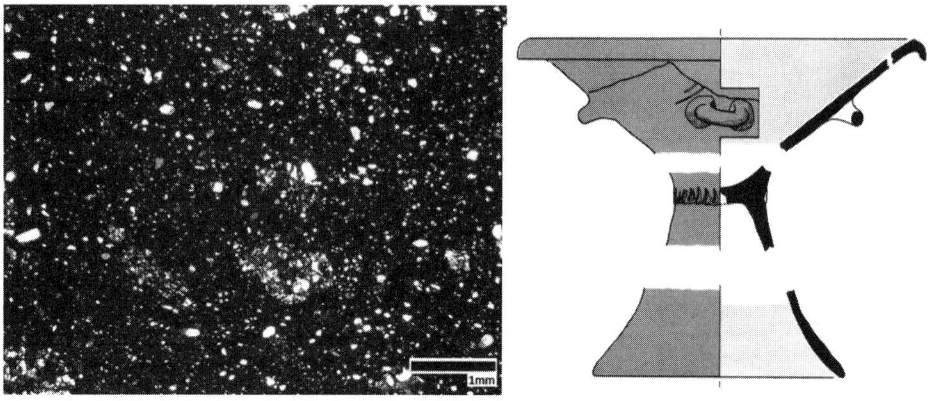

FIGURE 7.4. Sandstone to low-grade metamorphic fabric (cross-polarized light) (*left*) and the fruitstand form (*right*, after Pullen 2011:66).

bowl form is dominant in the repertoire of the EHI period and is present in most of the fabric groups at Tsoungiza, both local and nonlocal. The importation of such a widely produced and readily available item from Aegina, located approximately 60 km to the east, raises questions about the motivation for such long-distance movement and the social nature of such exchange. While the value of these vessels may have derived partially from their high-quality finish (the high shine of the burnish is striking, as can be seen in fig. 7.3), the association with such a distant place may also have been a factor. As Cyprian Broodbank (1993:315) has discussed, the movement of objects can be about more than just the accumulation of wealth. Their value can derive from association with their place of origin and ownership can act as a demonstration of knowledge of the world and of other communities, as Broodbank has shown.

This characteristic andesitic fabric was noted in small numbers during macroscopic evaluation of the Tsoungiza material (Pullen 2011:63) and does not appear to be present after the EHI period at the site, suggestive of low-level, small-scale consumption of Aeginitan vessels at Tsoungiza. The presence of such early imports, and their absence after EHI, is unexpected, as we have been able to identify EHII Aeginitan imports at many other sites currently being analyzed as part of our wider study.

A second characteristic fabric believed to originate outside the Pliocene geology of the Nemean sites is the low-grade metamorphic group. This fabric is characterized by a medium-fine clay matrix with subangular sandstone grading to low-grade metamorphic rock fragments (fig. 7.4). The nearest possible source of such metamorphic raw materials is likely to be the Argolid (IGSR 1970a and 1970b). We have found that this fabric dominates our Argolid assemblages and diminishes with distance from the area, which would add further support to the suggestion of an origin in the region.

Macroscopically, this fabric was distinctive for its hardness, the common presence of a gray core, and white and sometimes sparkling inclusions. It matches Pullen's macroscopic Class 1, which is largely confined to the fruitstand form at Tsoungiza (Pullen 2011:60). The fruitstand is a large pedestal bowl characterizing the vessel typology found at Talioti in the Argolid but present at many sites across both the regions of the Argolid and Corinthia (see fig. 7.4; compare Dousougli 1987). This form is believed to be part of dining sets, along with ladles and small bowls. Arguably, the other large bowl form also found in this fabric group could have served a similar function as a communal vessel from which individual portions were taken. We do not appear to have this fabric in the EHII period at Tsoungiza, with the only other example of the fabric belonging to an EHIII tankard. The fabric is present in EHII bowl and askos (vessel for pouring) forms at NVAP 204, indicating different consumption choices at the two sites.

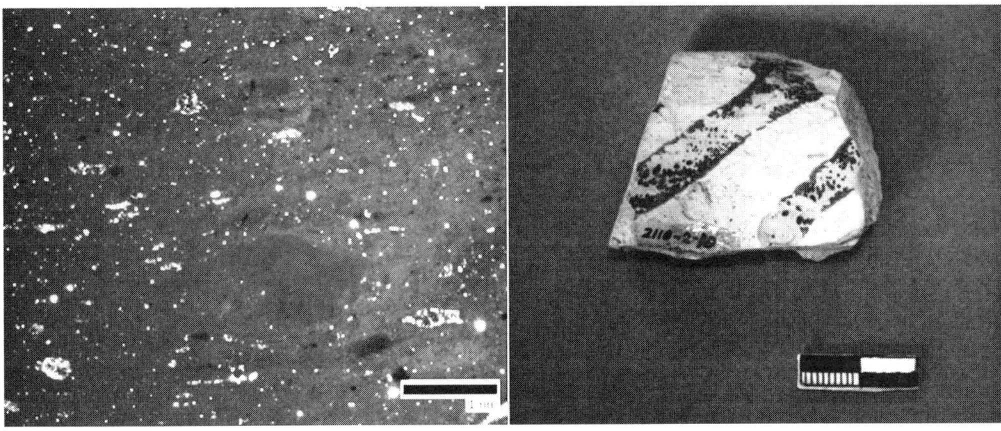

FIGURE 7.5. Fine green-firing fabric (cross-polarized light) (*left*) and the black pattern painted fruitstand sherd (*right*).

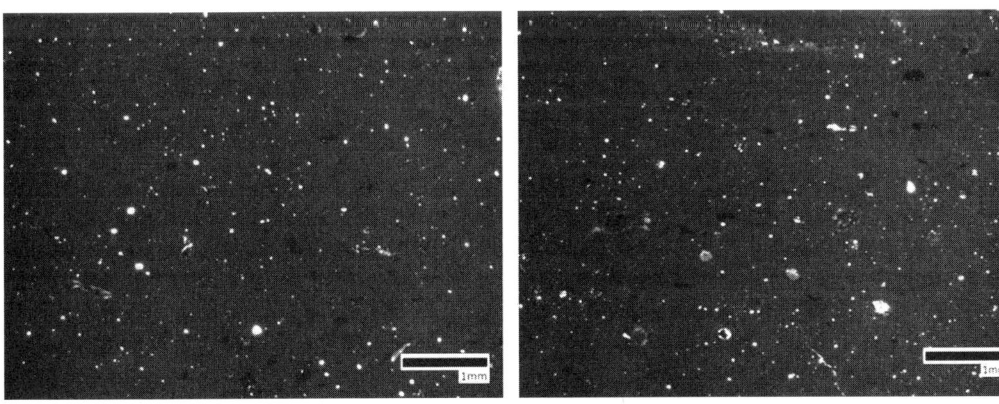

FIGURE 7.6. EHIII fine green-firing fabric from Tsoungiza (*left*) and EHII green-firing fabric from Keramidaki, Ancient Corinth (*right*) (cross-polarized light).

Two other fruitstands sampled from Tsoungiza belong to the third suspected imported group, a fine green-firing calcareous fabric (fig. 7.5). Macroscopically, this fabric is distinctive for its fine nature and light buff color. Unlike the other fruitstands from Tsoungiza, which are red painted with incised and/or applied decoration, the fruitstands in this fabric have a black slip/black painted decoration.

Petrographically, this fabric contains few aplastic inclusions (usually only containing clay pellets and quartz), making provenance ascription difficult. It does, however, compare directly with other material sampled from Corinthian sites as part of this project, which is believed to be local to Corinth (fig. 7.6). The chemical analysis (NAA by Peter Day, Kerasia Douni, Anno Hein, and Vasilis Kilikoglou) of this fabric identified at EHII Koropi in Attica has suggested that it is of Corinthian origin. Though further chemical and petrographic work is currently being undertaken to confirm this hypothesis, based on these initial comparative results this fabric has tentatively been assigned as Corinthian, indicating that all the fruitstands sampled in Nemea were imports. Such a suggestion is particularly striking because the fruitstand is taken to characterize EHI assemblages across the region, being a ubiquitous item in assemblages of this date.

The nonlocal origin of the fruitstand may explain the high proportion of repairs noted for these vessels (Pullen 2011:67, 93). If fruitstands were not widely available but were utilized by most households (see Pullen 2011:65) then it may have been necessary to repair them, while other vessel forms could have been more readily sourced from local producers.

The wide variety of fabrics present, most notably the Aeginitan, possible Corinthian, and Argive fabrics in these early periods, indicates that the communities of these two sites in Nemea consumed the vessels of a variety of producers, based both within the area of the Nemea Valley and beyond. The position of these two sites within the Nemea Valley, which acts as a natural route through this mountainous landscape, may have facilitated access to the pottery of such widely distributed producers (see also Wright et al. 1990:587). It is thus particularly striking that in the later EHII and by EHIII Tsoungiza appears to have become reliant on a smaller number of producing areas, suggesting a new motivation for the consumption behavior identified during these later periods.

EHII "Developed" and EHIII

The ceramics discussed next belong to what Pullen has termed the EHII "Developed" period (only applicable to Tsoungiza and equivalent to Lerna III phases late A–early C; see Pullen 2011:14–16 and see table 7.1) and the EHIII period and are present only at Tsoungiza. There appears to have been a short period of abandonment at Tsoungiza between EHII and EHIII, so we do not have the full EBA sequence.

In contrast to the fabrics of EHI and early EHII, those from later periods are remarkable in their consistency, with many of the fabrics familiar in earlier periods no longer present. In addition there appears to have been a change to more standardized pastes, with the dominance of fine fabrics, particularly the use of green-firing calcareous clay (see fig. 7.6).

While the dominance of finewares makes determining provenance by using petrography difficult, it is clear that many of the fabrics present in the EHI and early EHII periods, particularly the imports, do not continue. Those fabrics that do continue are primarily mudstone/siltstone tempered and calcite tempered, most often in coarsewares. Their compositions are consistent with the geology of the Nemea Valley, suggesting the continuation of local production.

The community at Tsoungiza appears to have consumed a high proportion of vessels belonging to the fine green-firing fabric that is dominant by the EHIII period. This fabric appears to be the same as the fine green-firing fabric of the EHI and early EHII periods, which we suggest is Corinthian (compare figs. 7.5 and 7.6). If indeed this is the case, it would suggest that the consumption of fabrics from Corinth became more prevalent at this site over time.

Macroscopically the shift to fine calcareous clays was also noted by the dominance of fine light buff fabrics. This is part of a wider regional diachronic shift from orange fabrics with red-orange decoration in the EHI and early EHII periods toward buff fabrics with contrasting black-brown Urfirnis slips (lustrous slipping type) during EHII and patterns painted dark on light decoration by EHIII. The possible dominance of Corinthian fabrics by later in EHII and in EHIII may be linked to the increased demand for such high-quality fine buff firing fabrics, which Corinth has a well-documented history of producing. Such a trend highlights the importance of these specific ware types within dining practices of these periods and is part of wider changes in the ceramic repertoire. During the later part of EHII and certainly by EHIII increasingly varied forms were produced and consumed, suggestive of more complex dining practices with an emphasis on individualized dining within commensal settings. This is indicated by a move from simple bowl forms for each individual to a range of vessels including kantharoi (the two-handled tankards that could be passed from one individual to another) and the "ouzo cup," which in contrast was evidently meant for use by one person only.

Conclusions

The petrographic and macroscopic examination of this material has enabled us to provide detailed information about the technological practices and areas of production during the Early Helladic period in the area in and around the Nemea Valley.

Through the identification of imports, most likely from the Argolid, Corinth, and the island of Aegina, it has been possible to document in detail the early movement of vessels from a wide number of producers in the EHI and early EHII period over considerable distances. It is clear that Nemea was linked to a wide number of producing areas that all made a varied repertoire of vessels, with the exception of fruitstands, which appear to have been more restricted in their production, primarily at a center located in the Argolid.

The identification of more homogeneous paste recipes in EHII "Developed" and EHIII at Tsoungiza may suggest more standardized production practices and/or perhaps the consumption of vessels from much smaller numbers of producing areas with limited geographic distribution. The rise in dominance of particular techniques or centers of production may have been linked to the increased popularity of buff fabrics with Urfirnis and pattern-painted decoration during the EHII and EHIII periods, with those centers able to produce such wares becoming dominant.

The ability to study the entire ceramic repertoire, including coarsewares, has enabled us to add a higher degree of resolution to Attas's initial suggestions about the distribution of ceramic production in the EH period and to identify different technological paste recipe traditions. Through detailing these traditions, their diachronic changes, and spheres of interaction with the surrounding region, the continuation and expansion of this research to other sites throughout the Corinthia and the Argolid will enable us to build a coherent picture of EH ceramic production and consumption choices in the northeast Peloponnese.

Acknowledgments

This work would not have been possible without the support of a number of organizations and individuals. The Institute for Aegean Prehistory has been generous in its funding of analysis of EH pottery from the Corinthia, while funds for attendance at this conference were provided by a Petrie Watson Scholarship from the University of Sheffield (Clare Burke) and the NARNIA ITN (New Archaeological Research Network for Integrating Approaches to Ancient Material Studies Innovative Training Network) of the EU Seventh Framework through the Marie Curie Actions (People Project) (Peter Day). The Nemea Valley Archaeological Project was supported by grants from the National Endowment for the Humanities, the National Geographic Society, Bryn Mawr College, and private donors and directed by James Wright in collaboration with John Cherry, Jack Davis, and Eleni Mantzourani. We are very grateful for their generous support and advice. For permission to sample pottery for analysis, we are grateful to the 37th Ephorate of Prehistoric and Classical Antiquities at Corinth, the British School at Athens, and the guards of the Nemea Museum. Thin sections of samples discussed here and associated chemical data are housed at the Laboratory of Archaeological Materials, National Centre of Scientific Research "Demokritos," Greece. Finally, our thanks go to Mary Ownby and Sophia Kelly for organizing the Society for American Archaeology session in Memphis.

References Cited

Attas, Michael

1982 Regional Ceramic Trade in Early Bronze Age Greece: Evidence from Neutron Activation Analysis of Early Helladic Pottery from Argolis and Korinthia. Ph.D. dissertation, McGill University, Montreal.

Attas, Michael, John M. Fossey, and Leo Yaffe

1987 An Archaeometric Study of EBA Pottery Production and Exchange in Argolis and Korinthia (Corinthia), Greece. *Journal of Field Archaeology* 14:77–90.

Betancourt, Philip P., and George H. Myer

2000 Appendix III. Petrographic Examination of Pottery Samples from Lerna III. In *Lerna IV: The Architecture, Stratification, and Pottery of Lerna III*, edited by Martha H. Wiencke, pp. 680–703. American School of Classical Studies at Athens, Princeton, New Jersey.

Broodbank, Cyprian

1993 Ulysses without Sails: Trade, Distance, Knowledge and Power in the Early Cyclades. *World Archaeology* 24(3):315–331.

Burke Davies, Clare
n.d. Crafting Continuity and Change: Ceramic Technology of the Early Helladic Peloponnese, Greece. Ph.D. dissertation, Department of Archaeology, University of Sheffield. Forthcoming.

Caskey, John L., and Elizabeth G. Caskey
1960 The Earliest Settlements at Eutresis: Supplementary Excavations, 1958. *Hesperia* 29:126–167.

Day, Peter M., Anno Hein, Lousie Joyner, Vasilis Kilikoglou, Evangelia Kiriatzi, Alexandra Tsolakidou, and David E. Wilson
2012 Petrographic and Chemical Analysis of the Pottery. In *The Hagia Photia Cemetery* II: *The Pottery*, edited by Costis Davaras and Philip P. Betancourt, pp. 115–138. Prehistory Monographs 34. Academic Press, Philadelphia.

Day, Peter M., Maria Relaki, and Simona Todaro
2010 Living from Pots: Ceramic Perspectives on the Economies of Prepalatial Crete. In *Political Economies of the Aegean Bronze Age: Papers from the Langford Conference, Florida State University, Tallahassee, 22–24 February 2007*, edited by Daniel Pullen, pp. 205–229. Oxford Books, Oxford.

Day, Peter M., and David E. Wilson
1997 Reassessing Specialization in Prepalatial Cretan Ceramic Production. In *TEXNH: Craftsmen, Craftswomen and Craftsmanship in the Aegean Bronze Age*, edited by Robert Laffineur and Philip P. Betancourt, pp. 275–289. Aegaeum 16. Université de Liège, Liège.
2002 Landscapes of Memory, Craft and Power in Prepalatial and Protopalatial Knossos. In *Labyrinth Revisited: Rethinking Minoan Archaeology*, edited by Yannis Hamilakis, pp. 143–166. Oxbow, Oxford.
2004 Consuming Power: Kamares Ware in Protopalatial Knossos. *Antiquity* 72:350–358.

Day, Peter M., David E. Wilson, and Evangelia Kiriatzi
1998 Pots, Labels and People: Burying Ethnicity in the EMI Cemetery at Aghia Photia, Siteias. In *Cemetery and Society in the Aegean Bronze Age*, edited by Keith Branigan, pp.133–149. Sheffield Studies in Aegean Archaeology 1. Sheffield Academic Press, Sheffield.

Dobres, Marcia-Anne
2010 Archaeologies of Technology. *Journal of Economics* 34:103–114.

Dounias, Georgios, Ioannis Fikiris, Anastasio Kollios, Mihailis Kavvadas, and Pavlos Marinos
2006 Η Κατολίσθηση της Νεμέας. Μηχανισμός Αστοχίας—Μέτρα Σταθεροποίησης (The Nemea Landslide: Trigger Mechanism-Remedial Measures). 5th National Conference of Geotechnical and Geoenvironmental Engineering, the Technical Chamber of Greece, Xanthi, Greece, May 31 to June 2, 2006.

Dousougli, Angelika
1987 Makrovouni—Kefalari Magoula—Talioti: Bemerkungen zu den Stufen FH I and II in der Argolis. *Prähistorische Zeitschrift* 62:164–220.

Gauss, Walter, and Evangelia Kiriatzi
2011 *Pottery Production and Supply at Bronze Age Kolonna, Aegina: An Integrated Archaeological and Scientific Study of a Ceramic Landscape*. Ägina-Kolonna. Forschungen und Ergebnisse, Band 5. Verlag der Österreichischen Akademie der Wissenschaften, Vienna.

Gosselain, Olivier P.

Goldman, Hetty
1931 *Excavations at Eutresis in Boeotia*. Harvard University Press, Cambridge, Massachusetts.

Hilditch, Jill
in press Ceramic Analysis under the Microscope: The Development of Petrographic Approaches within the Bronze Age Cyclades. In *Analytical Approaches to Prehistoric Ceramics: Technologies and Exchange in the Aegean*, edited by Peter M. Day. Sheffield Studies in Aegean Archaeology. Oxbow, Oxford.

Hilditch, Jill, Evangelia Kiriatzi, Kyriaki Psaraki, and Vassilis Aravantinos
2008 Early Helladic II Pottery from Thebes: an Integrated Typological, Technological and Provenance Study. In *Proceedings of the 4th Symposium of the Hellenic Society for Archaeometry, National Hellenic Research Foundation, Athens, 28–31 May, 2003*, edited by Yorgos Facorellis, Nikos Zacharias, and Kiki Polikreti, pp. 263–268. BAR-IS 1746. Archaeopress, Oxford.

IGSR (Institute for Geology and Subsurface Research)
1970a *Geological Map of Greece: Argos Sheet 1:50,000*. Institute for Geology and Subsurface Research, Athens.

1970b *Geological Map of Greece: Nemea Sheet 1:50,000.* Institute for Geology and Subsurface Research, Athens.

Lemonnier, Pierre

2002 *Technological Choices: Transformation in Material Culture since the Neolithic.* Routledge, London.

Nodarou, Eleni

2011 *Pottery Production, Distribution and Consumption in Early Minoan West Crete: An Analytical Perspective.* BAR International Series 2210. Archaeopress, Oxford.

Pfaffenberger, Brian

1992 Social Anthropology of Technology. *Annual Review of Anthropology* 21:491–516.

Pullen, Daniel J.

2008 The Early Bronze Age in Greece. In *The Cambridge Companion to the Aegean Bronze Age*, edited by Cynthia W. Shelmerdine, pp. 19–46. Cambridge University Press, Cambridge.

2011 *The Early Bronze Age Village on Tsoungiza Hill.* Nemea Valley Archaeological Project 1. American School of Classical Studies at Athens, Princeton, New Jersey.

Renfrew, Colin

1967 Cycladic Metallurgy and the Aegean Early Bronze Age. *American Journal of Archaeology* 71:1–20.

1972 *The Emergence of Civilization: The Cyclades and the Aegean in the Third Millennium B.C.* Methuen, London.

Renfrew, Colin, Joseph R. Cann, and John E. Dixon

1965 Obsidian in the Aegean. *Annual of the British School at Athens* 63:225–247.

Rutter, Jeremy B.

1995 *Lerna III: The Pottery of Lerna IV.* American School of Classical Studies at Athens, Princeton, New Jersey.

Shepard, Anna O.

1963 *Ceramics for the Archaeologist.* Volume 609, Carnegie Institution, Washington, D.C.

Shriner, Christine M.

1999 Ceramic Technology at Lerna, Greece in the Third Millennium B.C.: Economic and Social Implication. Ph.D. dissertation. Department of Anthropology, Indiana University, Bloomington.

Shriner, Christine M., and Haydn H. Murray

2001 Explaining Sudden Ceramic Change at Early Helladic Lerna: A Technological Paradigm. In *Archaeology and Clays*, edited by Isabelle C. Druc, 1–16. BAR-International Series 942. British Archaeological Reports, Oxford.

Sillar, Bill, and Michael S. Tite

2000 The Challenge of "Technological Choices" for Materials Science Approaches in Archaeology. *Archaeometry* 42(1):2–20.

Weingarten, Judith

1997 Another Look at Lerna: An EHIIB Trading Post? *Oxford Journal of Archaeology* 16: 147–166.

Whitbread, Ian K.

1989 A Proposal for the Systematic Description of Thin Sections towards the Study of Ancient Ceramic Technology. In *Archaeometry: Proceedings of the 25th International Symposium*, edited by Yannis Maniatis, pp. 127–138. Elsevier, Amsterdam.

1995 *Greek Transport Amphorae: A Petrological and Archaeological Study.* Fitch Laboratory Occasional Paper 4. British School at Athens, Oxford.

Wiencke, Martha H.

2000 *Lerna IV: The Architecture, Stratification, and Pottery of Lerna III.* American School of Classical Studies at Athens, Princeton, New Jersey.

Wright, James C., John F. Cherry, Jack L. Davis, Eleni Mantzourani, Susan B. Sutton, and Robert F. Sutton

1990 The Nemea Valley Archaeological Project: A Preliminary Report. *Hesperia* 59(4):579–659.

CHAPTER 8

The Use of Loess in Pottery Manufacture

A Comparative Analysis of Pottery from Yinxu in North China and Linearbandkeramik Sites in Belgium

James B. Stoltman

This chapter focuses upon the use of loess in the manufacture of prehistoric pottery. Loess is "windblown dust," composed predominantly of silt-size particles of quartz, "that covers wide areas in northern Europe, eastern China and the Mississippi Valley" (Bates and Jackson 1984:301). This chapter describes, compares, and contrasts the results of petrographic analysis of pottery thin sections from archaeological sites located in two of these major loess-covered landscapes, the North European Plain (Charlesworth 1966:figure 101) and the loess plateau of north China (Liu 1988). The goal of the study is to use the physical compositional data derived from the petrographic analyses to assess the ways in which loessic sediments were used to make pottery vessels through comparing and contrasting the ceramics from these two loess-rich regions.

Because the regions are completely disconnected from one another, this study in a sense can be considered a controlled experiment investigating the physical properties of loess as a medium for use in ceramic manufacture. When parallelisms occur, it is reasonable to attribute them to independent adaptations to the properties of local materials. In both regions loess was used primarily for the manufacture of finewares/noncooking vessels, whereas alluvium was the paste of choice when coarsewares/cooking vessels were made. Coarsewares/cooking vessels were invariably tempered in both regions, although the tempers used were different: grog in Belgium and sand or grit in China. Most Belgian finewares/noncooking vessels were also tempered, but in China such vessels, with rare exceptions, were made from untempered loess. In exposing these differential patterns in resource utilization by people in two widely separated parts of the world, this study reveals what appears to be an inherent property of loess: its unsuitability, even if tempered, for the manufacture of pottery containers used for cooking.

Methodology

Since the methodology used here has been described in several earlier publications (Stoltman 1989, 1991, 2001) only a brief outline is presented here. Using a 1 mm counting interval over the entire area of each slide, four quantitative indices were compiled for each thin section, two of them volumetric estimates (body and paste) and two of them size measures (temper size and natural sand size).

The main components of the two volumetric indices are three size grades of mineral inclusions as follows: matrix/clay (< 0.002 mm), silt (0.002 mm–0.0624 mm), and sand (0.0625 mm–2.00 mm). Once the temper was identified for tempered vessels, those counts were temporarily set aside so that paste (the mix of natural materials from which the vessel was made) could be recorded as the relative percentages of matrix/clay, silt, and sand. For untempered vessels and "raw" sediments, the paste index is the equivalent of bulk composition. When the temper counts are reconsidered, the body index (the bulk composition of the tempered vessel) was recorded as the relative percentages of matrix/clay (with silt now recorded here), sand, and temper.

The two size indices involve assigning a value in the following ordinal scale to each sand and temper grain encountered in point counting: (1) 0.0625–0.249 mm (very fine to fine); (2) 0.25–0.499 mm (medium); (3) 0.50–0.99 mm (coarse); (4) 1.00–1.99 mm (very coarse); and (5) 2.00 mm+ (gravel). The size values for each of the grains were then summed and divided by the total number of grains counted, thus providing a mean size index for each thin section of between 1 and 5 for both temper and natural sand.

All thin sections were prepared by Brian Hess, proprietor of Hess Petrographic of Madison, Wisconsin. All thin sections and associated records are housed in the Laboratory of Archaeology of the Department of Anthropology at the University of Wisconsin–Madison.

Yinxu

Yinxu (pronounced *yin-shoo*) is the Chinese name for the last capital of the Shang Dynasty, which is located adjacent to the modern city of Anyang in Henan province about 450 km south of Beijing (Stoltman et al. 2009:183). In the English literature (for example, Chang 1980; Li 1977) the site is commonly referred to as Anyang. It flourished for a little over a century from ca. 1200 BC to 1047 BC, when it fell to Zhou conquerors.

Local Sediments Used for Pottery Manufacture at Yinxu

A critical step needed for the evaluation of the paste properties of the locally made pottery at Yinxu was to gather samples of clay-rich sediments suitable for pottery manufacture from in and around the site. Three such sediments were identified in the field: alluvium, paleosols, and loess. Several samples of each were collected and prepared into test tiles that were then dried, fired, and mounted on thin section slides. Bulk compositions for 13 of these local sediments are recorded in table 8.1 and presented graphically in figure 8.1. The compositions of the paleosols are similar to those of the alluvium, so it appears that the three paleosols formed on top of alluvial sediments. Accordingly, we treat these 7 samples (the 4 alluviums and the 3 paleosols) as a single group for comparative purposes in marked contrast to the much siltier loessic sediments.

The alluvial sediments differ from the loess in possessing a higher incidence of clay and a lower incidence of silt (see table 8.1). The reason is that alluvial deposition normally occurs during high-water (flood) episodes when current flow is retarded and suspended sediments are allowed to settle out. During such episodes of alluvial deposition, the resulting sediments normally have a fining-upward texture because the coarser particles (sands and silts) will settle out sooner than finer particles (fine silts and clays). For potters, the clay-rich upper levels of alluvial deposits are excellent resources for their purposes and thus are commonly sought.

The Local Pottery of Yinxu

The preponderance of pottery from Yinxu (90 percent) is generally (and aptly) characterized as grayware (Cheng 1960:145–148; Li 1977:202–204). At present 49 grayware vessels identified by Anyang-station archaeologists as local manufactures on stylistic grounds constitute the sample of thin-sectioned vessels included in this study. (In an earlier study, only 23 such vessels were analyzed: Stoltman et al. 2009.) Yinxu grayware occurs in a great variety of forms (for example, Chang 1980:107–108; Cheng 1960:149). In terms of use, two general categories are readily discernible: those used for cooking such as *li* tripods (Cheng 1960:149, nos. 18 and 21) and steaming vessels (Cheng 1960:149, nos. 1 and 15; see also Thorp 2006:154–155) and noncooking vessels in a variety of bowl, basin, and jar forms (for example, Thorp 2006:154–155). Unlike the cooking vessels, these were not subject to recurrent heating and cooling associated with food preparation.

Turning first to grayware vessels used for noncooking tasks such as storage, serving, transport, and ritual, the sample of thin-sectioned vessels totals 32. Bulk compositions are plotted

TABLE 8.1. Bulk compositions for local sediments in and around Yinxu

Sediment type	Thin section #	% Matrix	% Silt	% Sand	Sand size index
Alluvium	HBOO-21	88	11	1	1.00
Alluvium	AYOO-34	94	5	1	1.50
Alluvium	AYOO-35	86	12	2	1.20
Alluvium	ALNO9-B7	94	5	1	1.00
Alluvium mean ± 1 std dev	$n = 4$	90.5±4.1	8.3±3.8	1.2±0.5	1.18±0.24
Paleosol: "Huanbei soil"	AYO9-3	84	8	8	1.65
Paleosol	ALNO9-B1	91	8	1	1.00
Paleosol	ALNO9-A1	89	11	0	—
Paleosol mean ± 1 std dev	$n = 3$	88.0±3.6	9.0±1.7	3.0±4.4	1.32±0.46
Alluvium + Paleosols mean ± 1 std dev	$n = 7$	89.4±3.8	8.6±2.9	2.0±2.7	1.23±0.29
Loess	HBOO-20	77	22	1	1.00
Loess	ALNO9-A4	78	21	1	1.00
Loess	ALNO9-A6	80	17	3	1.00
Loess	ALNO9-A8	83	16	1	1.00
Loess	ALNO9-B4	82	17	1	1.00
Loess	ALNO9-B6	72	24	4	1.00
Loess mean ± 1 std dev	$n = 6$	78.7±4.0	19.5±3.3	1.8±1.3	1.00±0.0

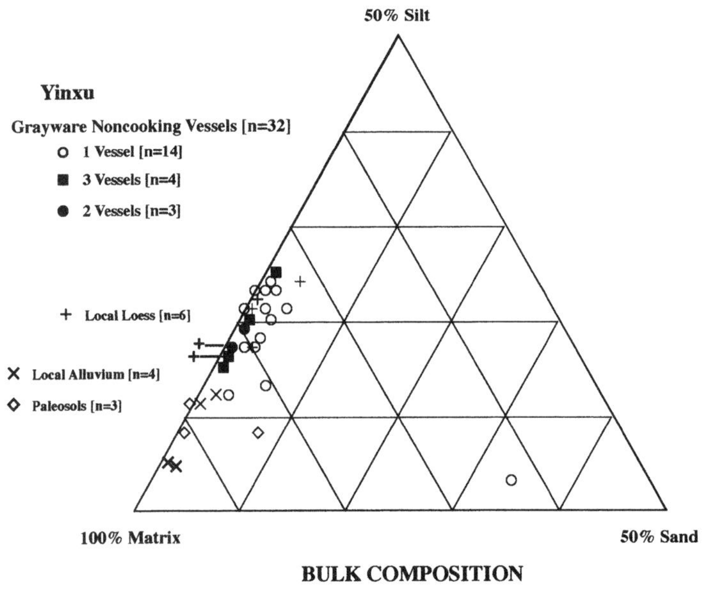

FIGURE 8.1. Ternary plot showing the bulk compositions of 32 grayware noncooking vessels versus 13 local sediments from Yinxu (note 4 instances of 3 vessels having the same value and 3 instances of 2 vessels having the same value).

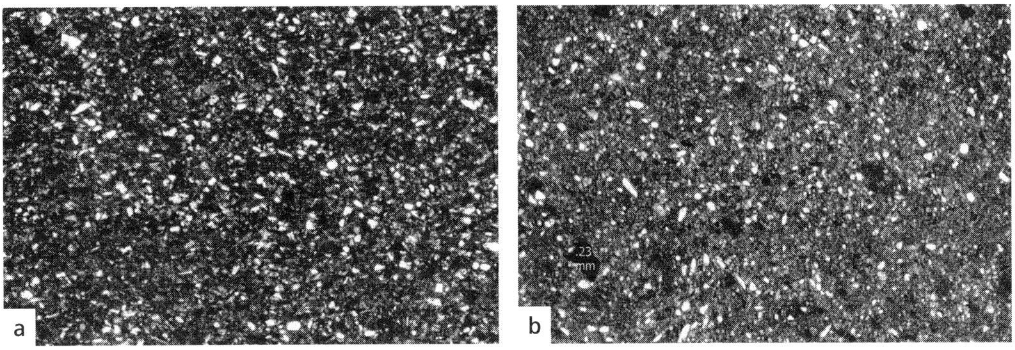

FIGURE 8.2. Photomicrographs (10× under crossed polars): (*a*) loess (ALN09-B6) with 24 percent silt, (*b*) an untempered Gui (AY00-3b) with 17 percent silt. Scale: large void near SW corner of (b) = 0.23 mm.

graphically for each of these vessels in figure 8.1 along with the bulk compositions for the 13 local clay-rich sediments. With the exception of one extraordinarily sandy jar (34 percent sand), none of these vessels is tempered. As figure 8.1 shows, 29 of the 32 noncooking, grayware vessels have compositional values that fall within the range of variation of the local loess. The critical variable in viewing these data is the silt. The range for the 6 loess samples is 16–24 percent, with a mean of 19.5 percent (see table 8.1). By contrast, the silt range for the 7 soil and alluvial samples is 5–12 percent with a mean of 8.6 percent. The two ranges do not overlap. The mean silt value for the 31 untempered noncooking grayware vessels is 19.2 percent, which is virtually identical to the 19.5 percent mean for the 6 loess samples. The compositional evidence thus demonstrates that Yinxu potters, with few exceptions, preferentially made their noncooking grayware vessels from untempered loess (fig. 8.2).

Turning now to the cooking vessels, a distinctly different compositional makeup is evident. Two types of cooking vessels are generally recognized within late Shang society, li tripods and steaming vessels (*yan* and *zeng* are the current Chinese terms for the latter: Cheng 1960: 149, nos. 1 and 15, respectively; see also Thorp 2006:155). Figure 8.3 shows that 13 of the 17 grayware cooking vessels have 15 percent or more sand-sized inclusions and thus are presumed to be tempered (fig. 8.4a).

Three temper types were identified among the tempered vessels: gneiss, granite, and sand. Three vessels with grit temper (two with gneiss and one with granite: see fig. 8.4b) were included in the study because they were adjudged to be stylistically "local," although there are no bedrock outcrops closer to Yinxu than the Taihang mountains 20 km to the west. In considering the issue of what was imported, tempers or pots, the case for the former is the more likely when the vessels with exotic tempers are stylistically local, not exotic.

Grog temper (made with recycled potsherds) is totally lacking among the current sample of "local" grayware vessels. A single grog-tempered li tripod was reported in the earlier study (Stoltman et al. 2009:197), but subsequent evaluation of the vessel by Anyang-station archaeologists was unanimous in considering the vessel to be stylistically exotic. Thus it is excluded from further consideration in the present study.

Four of the cooking vessels (the three li tripods and one steamer) are unusual in that they lack any evidence of temper (see fig. 8.3). Three of them were made from untempered loess, while the fourth (vessel AY06-11, a li) is unique (note the unusually high silt content) in having been made from processed loess using a recipe that characterizes bronze piece molds (Stoltman et al. 2009:191). All four of these vessels, especially AY06-11, could be regarded as aberrant, but it seems more prudent to view them as reflections of the sophistication of the Yinxu potters, who were willing and able to design the

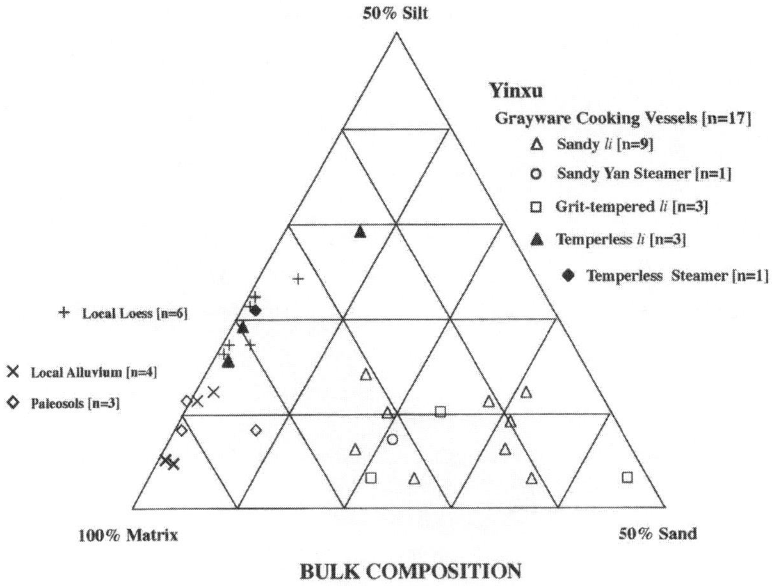

FIGURE 8.3. Ternary plot showing the bulk compositions of 17 grayware cooking vessels versus 13 local sediments from Yinxu.

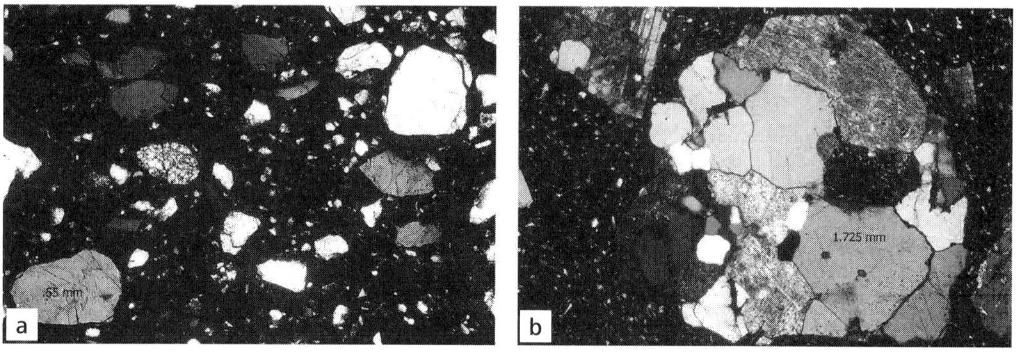

FIGURE 8.4. Photomicrographs (10× under crossed polars) of two li tripods from Yinxu made from alluvium: (*a*) sand-tempered vessel AY00-5 (note varied composition of rounded grains that are mostly monomineralic in composition, 6 percent silt); (*b*) gneiss-tempered vessel AY05-6 (note large, polymineralic grains typical of grit temper, 10 percent silt).

composition of their vessels in various ways to perform special or unique tasks.

One final compositional property of the tempered cooking vessels is that the majority (10 of 13) were made from alluvium, not loess. This inference is based on acceptance of the mean sand value for alluvium and loess (see table 8.1) as a reasonable cutoff criterion for identifying those sand grains in excess of 2 percent as humanly added temper.

Concluding Comments on Yinxu

The site of Yinxu is situated upon a loess-covered landscape devoid of outcrops of bedrock of any kind. Thus it is no surprise that loess was utilized in the manufacture of pottery. But the ways in which the loess was used are surprising in many respects. With some exceptions, two basic compositional recipes were used by Yinxu potters: (1) vessels not used for cooking purposes (various bowl, basin, and jar forms) were mostly made from untempered, "raw" loess; and (2) vessels used for cooking purposes (li tripods and steamers) were typically made from tempered alluvium.

Let us now turn to northern Belgium to analyze how Linearbandkeramik (LBK) peoples adapted their ceramic-making practices to a loess-covered landscape similar to that of north China.

Linearbandkeramik Pottery Making in North Belgium

The bases of the present study are thin sections of 5 loess samples and 35 pottery vessels from two LBK sites in the loess-covered Hesbaye region of northeast Belgium (Keeley and Cahen 1989: 159). The two sites, Longchamps and Darion, are located 3 km apart along the north bank of the Geer River. Excavations were conducted at these sites between 1981 and 1989. Both sites, which date to the interval ca. 4300–4000 BC, are characterized by classic LBK longhouses situated within ditched enclosures.

Two stylistic classes are generally recognized within LBK pottery: fineware and coarseware (Bogucki 1995:91; Keeley and Cahen 1989:163, 167). The former is thinner, regarded as "rarely tempered" (Bogucki 1995:91), and typically has stab-and-drag and/or incised decoration. Coarseware is rarely decorated except for fingernail impressions, has thicker vessel walls, and "frequently contains inclusions of sand grains." Bogucki (1995:92) regards both wares as being made from local clays, "probably from the loess soil found in the immediate environs of the settlements." While both wares may have served multiple functions, only the coarseware was used for cooking purposes.

I first analyzed 11 thin sections of Longchamps pottery and reported the findings at the first annual meeting of the Midwest European Archaeological Diaspora held at the University of Wisconsin on October 21, 1995. The main purpose of this study was to provide foundational data, a sort of feasibility study for the formulation of a doctoral dissertation proposal by a former student of Keeley's who had recently entered graduate school at the University of Wisconsin–Madison. As it turned out, the student withdrew from graduate school, and the data languished for several years. Eventually, thinking that these data had interesting implications, I began to dabble in the analysis of the unstudied thin sections (29 in number, including 15 from Darion) that were among the original 40 that had been prepared. The opportunity to prepare the present chapter provided the impetus to conclude the analysis of the last 5 thin sections that had remained unstudied.

The Composition of Longchamps Loess

Included among the thin sections were 5 prepared from fired test tiles derived from the B-horizons of loessic sediments at Longchamps. The bulk compositions of these 5 sediment samples are recorded in table 8.2. Unfortunately, no samples of alluvium were collected, but the loess samples provide a robust compositional signature for the local loess of the region for comparison with the ceramic compositions. Based on these data, 15 percent silt appears to be a reasonable minimal value for distinguishing loessic from alluvial sediments.

LBK Fineware Vessels

Of the 35 thin-sectioned vessels, 13 are fineware. Unlike the Yinxu potters, potters at the LBK sites were more sparing in their use of loess for the manufacture of fineware/noncooking vessels. Contrary to the suggestion of Bogucki (1995: 92), based on previous petrographic study, only 7 such LBK vessels have compositional values consistent with loess as the source of their pastes (compare tables 8.2 and 8.3 and fig. 8.5). Of these vessels, 3 are untempered (fig. 8.6a). Interestingly, the other 4 LBK fineware vessels made from loess (Nos. WLP-4, WLP-6, DA-26, and DA-29, table 8.3) were apparently sand tempered (fig. 8.6b). These 4 vessels, along with 2 that were apparently made from alluvium (Nos. WLP-7 and DA-27, table 8.3), have 12 percent or more sand. Without alluvial sediments for comparison, the possibility that naturally sandy sediments were used for these vessels cannot be ruled out. But, at least for those with loessic pastes, it is likely that some sand grains were intentionally

TABLE 8.2. Bulk compositions for Longchamps loess and untempered fineware from Longchamps and Darion

Site + Type	Thin section #	% Matrix	% Silt	% Sand	Sand size index
Longchamps					
Loess	WLP-12	77	20	3	1.00
Loess	WLP-13	71	26	3	1.00
Loess	WLP-14	72	25	3	1.00
Loess	WLP-15	77	20	3	1.00
Loess	"No. 5"	77	19	4	1.00
Mean ± 1 std dev	[*n* = 5]	74.4±2.8	22.4±3.0	3.2±.04	1.00±0.0
Untempered fineware	WLP-8	79	18*	3	1.00
Darion					
Untempered fineware	DA-30	80	17*	3	1.00
Untempered fineware	DA-40	67	27*	6	1.17

Note: Bulk compositions are equivalent to paste indices for tempered vessels.
* Denotes loessic silt values for pottery vessels.

TABLE 8.3. Bulk compositions for sand-tempered fineware from Longchamps and Darion

Site + Type	Thin section #	% Matrix	% Silt	% Sand	Sand size index
Longchamps					
Fineware	WLP-4	67	16*	17	1.08
Fineware	WLP-6	63	21*	16	1.08
Fineware	WLP-7	73	12	15	1.15
Mean ± 1 std dev	[*n* = 3]	67.7±5.0	16.3±4.5	16.0±1.0	1.10±0.04
Darion					
Fineware	DA-26	67	21*	12	1.04
Fineware	DA-27	73	6	21	1.22
Fineware	DA-29	62	25*	13	1.26
Mean ± 1 std dev	[*n* = 3]	67.4±5.5	17.3±10.0	15.3±4.9	1.17±.12

* Denotes loessic silt values for pottery vessels.

added by the potters. By contrast, there is only a single incidence of a sand-tempered vessel made from loess at Yinxu.

Each of 4 additional fineware vessels is tempered, 1 with bone and 3 with grog, 1 of which has far more glauconite than grog (table 8.4). Glauconite is a mineral associated with sedimentary rocks, almost certainly derived from marine sediments lying just north of the loess belt on the North European Plain. Accepting that the 6 fineware vessels with 12 percent or more sand can be considered tempered, 10 of the 13 fineware vessels in the current study are tempered (again contrary to Bogucki's view [1995:91] that finewares were "rarely tempered"). Recalling that 6 of the 13 were made from alluvium, not loess, the compositional diversity of LBK fineware/noncooking vessels contrasts markedly with the nearly universal practice by Yinxu potters of using untempered loess for such vessels.

LBK Coarseware Vessels

Two major compositional properties characterize the 22 coarseware vessels from Longchamps and Darion: (1) all are tempered, and (2) all but 1 were made from alluvium (tables 8.5 and 8.6).

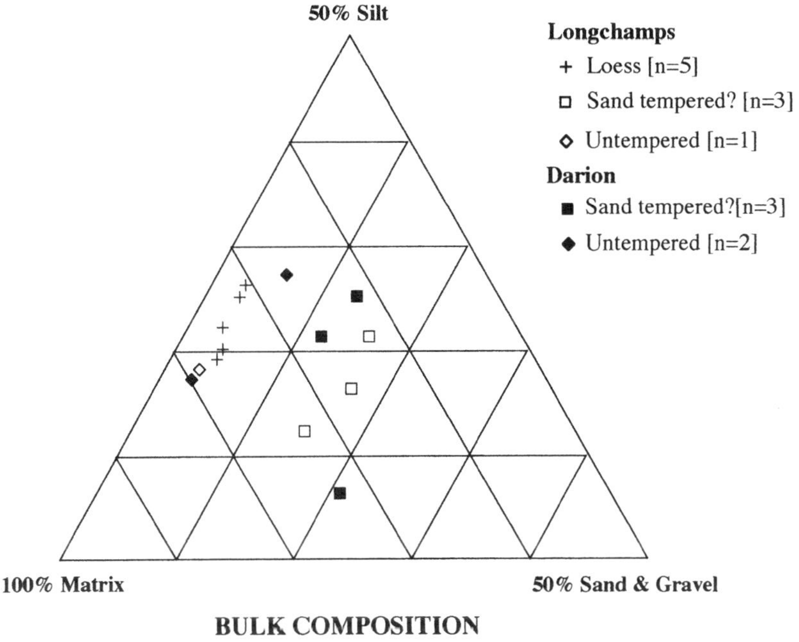

FIGURE 8.5. Ternary plot showing the bulk compositions for 5 loess samples from Longchamps and 9 fineware vessels, 4 from Longchamps and 5 from Darion.

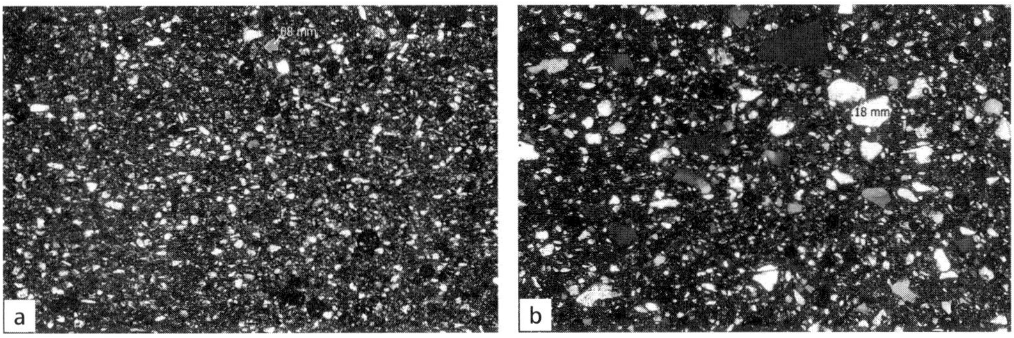

FIGURE 8.6. Photomicrographs (10× under crossed polars) of two fineware vessels from Longchamps: (*a*) untempered vessel WLP-8 with loessic paste (18 percent silt); (*b*) sand-tempered vessel WLP-6 with loessic paste (21 percent silt and 16 percent sand).

TABLE 8.4. Body and paste values for tempered fineware from Longchamps and Darion

		← BODY				→	← PASTE			→
Type	Thin section #	% Matrix	% Sand	% Temper	Temper type	Temper size index	% Matrix	% Silt	% Sand	Sand size index
Longchamps										
Fineware	WLP-5	71	4	23 2	glauconite/ grog	1.48 5.00	83	12	5	1.17
Fineware	WLP-9	84	5	11	bone	2.18	92	2	6	2.00
Darion										
Fineware	DA-28	90	2	6	grog	2.92	90	8	2	1.00
Fineware	DA-39	90	3	7	grog	2.73	89	7	4	1.00
Mean/std dev	[*n* = 2]	91.0 ± 1.4	2.5 ± 0.7	6.5 ± 0.7		2.82 ± .13	89.5 ± 0.7	7.5 ± 0.7	3.0 ± 1.4	1.00 ± 0

TABLE 8.5. Body and paste values for coarseware at Longchamps

		BODY					PASTE			
Type	Thin section #	% Matrix	% Sand	% Temper	Temper type	Temper size index	% Matrix	% Silt	% Sand	Sand size index
Coarseware	WLP-1	85	6	9	grog	4.44	86	7	7	1.33
Coarseware	WLP-2	88	5	7	grog	3.82	86	8	5	1.14
Coarseware	WLP-3	83	10	7	grog	3.79	82	7	11	1.20
Coarseware	WLP-16	91	6	3	grog	4.25	87	7	6	1.42
Coarseware	WLP-18	77	7	16	grog	3.62	88	4	8	1.18
Coarseware	WLP-19	82	6	12	grog	3.52	89	4	7	1.47
Coarseware	WLP-20	74	7	19	grog	3.64	88	4	8	1.12
Coarseware	WLP-21	73	4	23	grog	3.76	91	4	5	1.40
Coarseware	WLP-22	69	8	23	grog	4.12	83	6	11	1.05
Coarseware	WLP-23	82	11	7	grog	3.88	81	7	12	1.20
Coarseware	WLP-24	69	7	24	grog	4.38	84	7	9	1.44
Coarseware	WLP-25	74	11	15	grog	3.76	83	4	13	1.20
Mean/std dev	[*n* = 12]	78.9 ± 7.3	7.3 ± 2.3	13.8 ± 7.3		3.92 ± 0.31	85.7 ± 3.1	5.8 ± 1.6	8.5 ± 2.7	1.26 ± .014
Coarseware	WLP-10	76	3	21	dacite	3.03	80	16	4	1.00
Coarseware	WLP-11	82	3	11	trachyte	3.50	96	1	3	1.50
				4	sandstone	2.80				

TABLE 8.6. Body and paste values for Darion coarseware

		BODY					PASTE			
Type	Thin section #	% Matrix	% Sand	% Temper	Temper type	Temper size index	% Matrix	% Silt	% Sand	Sand size index
Coarseware	DA-31	70	16	14	grog	4.00	78	4	18	1.51
Coarseware	DA-32	75	17	8	grog	3.67	69	12	19	1.32
Coarseware	DA-33	75	12	13	grog	4.00	79	8	13	1.52
Coarseware	DA-34	71	9	9	grog/	3.67	80	9	11	1.78
				11	glauconite	1.33				
Coarseware	DA-35	75	11	14	grog	3.85	79	9	12	1.16
Coarseware	DA-36	77	17	6	grog	4.33	80	2	18	1.08
Coarseware	DA-37	79	5	16	grog	4.07	81	13	6	1.50
Coarseware	DA-38	72	5	23	grog	3.67	79	12	9	1.33
Mean/std dev	[*n* = 8]	74.3 ± 3.1	11.5 ± 5.0	14.2 ± 5.6		3.91 ± 0.024	78.1 ± 3.8	8.6 ± 3.9	13.3 ± 4.7	1.40 ± 0.022

The sole coarseware vessel that was possibly made from loess (WLP-10) is unique in another way: it is tempered with dacite, a fine-grained, extrusive igneous rock that is clearly of nonlocal origin.

The temper of choice for the coarseware vessels from both sites (12 of 14 at Longchamps and 7 of 8 at Darion) was grog (fig. 8.7, see tables 8.5 and 8.6). Vessel DA-34 from Darion was also grog tempered but had such a high incidence of glauconite that it was also recorded (with some trepidation) as temper (see table 8.6). Of the vessels from Longchamps, 2 are departures from the general pattern of grog temper for LBK

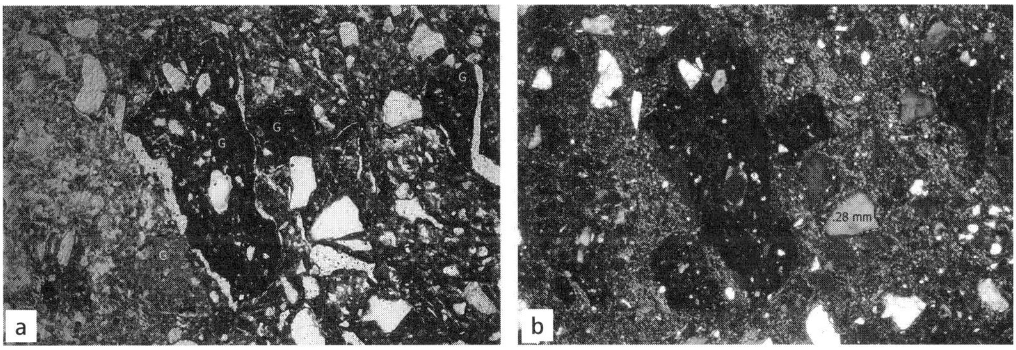

FIGURE 8.7. Photomicrographs (10×) of coarseware vessel WLP-2: (*a*) view under plane polars with "G" marking four grog grains; (*b*) same view as (*a*) under crossed polars. Silt content of paste = 8 percent, indicative of alluvium.

coarsewares (WLP-10 and WLP-11, see table 8.5). Both are characterized by fine-grained, extrusive igneous rocks, dacite for the former (as noted) and trachyte plus some sandstone for the latter. Both vessels are strong candidates for status as imports.

Summary and Conclusions

The petrographic analysis of thin sections of pottery vessels from two of the world's most prominent loess-covered regions (northeastern China and northeastern Belgium) reveals several interesting compositional properties, both shared and not shared, that offer valuable insights into how loess was used as a ceramic-making material in prehistory. By comparing and contrasting loess use by potters who were separated in space by several thousand miles and at least 3,000 years in time, it is reasonable to treat the resulting pottery of these artisans as expressions of independent adaptations to this resource. Observed compositional similarities can then be viewed as direct reflections of physical constraints that loessic sediments imparted to pottery manufacture in both regions.

Despite the dominance of loess as a surficial blanket over both regions, its use in ceramic manufacture was selective. Importantly, in both regions loess was almost exclusively reserved for finewares/noncooking vessels. At Yinxu most fineware/noncooking vessels (29 of 32 in this study) were made from untempered loess. At the Belgian sites most of the finewares were also made from loess, but only 3 of 13 fineware vessels from the Belgian sites were made from untempered loess. Of the remaining 10 vessels, 4 were made from sand-tempered loess and 6 were made from variously tempered (for example, grog, glauconite, and bone) alluvium.

The most prominent similarity in ceramic composition between the two regions involves coarsewares/cooking vessels. With relatively few exceptions, coarsewares/cooking vessels in both regions were made from tempered alluvium. Loess was rarely used in the manufacture of such vessels: 10 of 17 Yinxu cooking vessels have alluvial pastes, while 21 of the 22 LBK coarseware vessels were made from alluvium. The sole LBK exception, vessel WLP-10, is probably an import.

Despite the shared practice of using primarily tempered alluvium for coarsewares/cooking vessels, there were distinct differences in the kinds of tempers used in the two regions. Except for a few apparently imported vessels with igneous rock tempers, the majority of LBK coarseware vessels (20 of 22) had grog temper. By contrast, not a single locally made Yinxu vessel was grog tempered. Instead sand temper characterizes most cooking vessels at Yinxu, with a few such stylistically local vessels having igneous rock tempers.

Cooking vessels were nearly always tempered in both regions. This empirical observation has a likely explanation: efforts to counter the effects of thermal shock associated with recurrent heating and cooling. Several experimental studies provide evidence that coarse (tempered) wares serve better for cooking vessels than finewares because they are more resistant to the propagation of thermally induced cracks (for example, Braun 1983:123; Bronitsky and Hamer 1986:96; Steponaitis 1984:106). Among the very best tempers for controlling thermally induced cracking is grog (Rye 1976:115; see also Rice 1987:229), which was the main inclusion used by LBK potters for their cooking vessels. Why Shang potters never employed grog temper for their cooking vessels is unknown, but, as Prudence Rice (1987:230) has noted, evidence from the ethnographic record indicates that for some potters "their behaviors run counter to prediction, at least with regard to thermal properties."

It is interesting to speculate about why potters in both China and Belgium showed strong preferences for using alluvial clays and a decided avoidance of loessic pastes in the manufacture of coarsewares/cooking vessels. It seems likely that this pattern reflects inherent liabilities of loess when subjected to repeated heating and cooling. In discussing the use of loess by Neolithic potters of China, Rose Kerr and Nigel Wood (2004:102) describe as its main drawbacks "rather low refractoriness" and "a poor resistance to thermal shock in its fired state." These observations certainly ring true and would provide an explanation of why loessic materials were rarely used for cooking vessels. This also provides evidence of the sophistication of the potters in widely distant regions who reached similar technological decisions in adopting practices that take best advantage of the properties of the raw materials available to them.

References Cited

Bates, Robert L., and Julia A. Jackson (eds.)
1984 *Dictionary of Geological Terms*. 3rd ed. Anchor Press/Doubleday, Garden City, New York.

Bogucki, Peter
1995 The Linear Pottery Culture of Central Europe, Conservative Colonists? In *The Emergence of Pottery*, edited by William K. Barnett and John W. Hooper, pp. 89–97. Smithsonian Institution Press, Washington, D.C.

Braun, David P.
1983 Pots as Tools. In *Archaeological Hammers and Theories*, edited by James A. Moore and Arthur S. Keene, pp. 107–134. Academic Press, New York.

Bronitsky, Gordon, and Robert Hamer
1986 Experiments in Ceramic Technology: The Effects of Various Tempering Materials on Impact and Thermal-Shock Resistance. *American Antiquity* 51(1):89–101.

Chang, Kwang-Chih
1980 *Shang Civilization*. Yale University Press, New Haven.

Charlesworth, J. K.
1966 *The Quaternary Era*, Vol. 1. Edward Arnold Publishers, London.

Cheng, Te-K'un
1960 *Archaeology in China, Volume II, Shang China*. W. Heffer and Sons, Cambridge.

Keeley, Lawrence H., and Daniel Cahen
1989 Early Neolithic Forts and Villages in NE Belgium: A Preliminary Report. *Journal of Field Archaeology* 16(2):157–176.

Kerr, Rose, and Nigel Wood
2004 *Science and Civilization in China, Volume 5: Chemistry and Chemical Technology, Part XII: Ceramic Technology*. Cambridge University Press, Cambridge.

Li, Chi
1977 *Anyang*. University of Washington Press, Seattle.

Liu, Tungsheng
1988 *Loess in China*. 2nd ed. Springer-Verlag, Berlin.

Rice, Prudence M.
1987 *Pottery Analysis: A Sourcebook*. University of Chicago Press, Chicago.

Rye, Owen S.
1976 Keeping Your Temper under Control: Materials and Manufacture of Papuan Pottery. *Archaeology and Physical Anthropology in Oceania* 11:106–137.

Steponaitis, Vincas P.
1984 Technological Studies of Prehistoric Pottery from Alabama: Physical Properties and Vessel Function. In *The Many Dimensions of Pottery*, edited by Sander E. Van der Leeuw and Alison C. Pritchard, pp. 79–122. Universiteit van Amsterdam, Amsterdam.

Stoltman, James B.

1989 A Quantitative Approach to the Petrographic Analysis of Ceramic Thin Sections. *American Antiquity* 54(1):147–160.

1991 Ceramic Petrography as a Technique for Documenting Cultural Interaction: An Example from the Upper Mississippi Valley. *American Antiquity* 56(1):103–120.

2001 The Role of Petrography in the Study of Archaeological Ceramics. In *Earth Sciences and Archaeology*, edited by Paul Goldberg, Vance T. Holliday, and C. Reid Ferring, pp. 297–326. Kluwer Academic/Plenum Publishers, New York.

Stoltman, James B., Zhichun Jing, Jigen Tang, and George (Rip) Rapp

2009 Ceramic Production in Shang Societies of Anyang. *Asian Perspectives* 48(1):182–203.

Thorp, Robert L.

2006 *China in the Early Bronze Age*. University of Pennsylvania Press, Philadelphia.

CHAPTER 9

Clay Pellets in Hohokam Red-on-Buff Pottery

Shifts in Pottery Recipes and the Organization of Ceramic Production

Sophia E. Kelly

The Hohokam economy developed on a social and environmental landscape characterized by large, stable population centers and subsistence intensification in a desert ecosystem. The cultural developments in the Hohokam region are rooted in the tradition of deep sedentism in central and southern Arizona throughout prehistory (Clark and Gilman 2012; Fish 2006 [1989]; Fish and Fish 2012; Hill et al. 2004:689; White and Lekson 2001:99). Permanent villages were established in the Tucson Basin as early as 2100 BC (Mabry 1999, 2008). By AD 500–650 the material markers that archaeologists use to identify Hohokam culture, such as pottery, particular stylistic motifs, pithouse dwellings, and large-scale irrigation agriculture, appear at settlements throughout central and southern Arizona (Abbott 2000:27; Haury 1976; Wallace et al. 1995). Hohokam communities continued to develop over the next 1,000 years, often in the same areas, until the disintegration of the regional system after AD 1400 (Abbott 2003; Ackerly 1988; Dean 2007).

Hohokam decorated pottery is characterized by the overlay of painted designs using hematite pigment on light buff unslipped ceramic pastes. Painted designs on Hohokam decorated pottery were introduced in the late Pioneer period (ca. AD 650–750). Before this, decoration on vessels primarily consisted of incised designs. Initially, decorated wares did not differ dramatically in color from plainware pottery. Both were dark gray.

Through time Hohokam potters consistently lightened the color of their pastes. One likely explanation is that the light-colored surface provided a better canvas for the bright red designs. The regularity with which paste color became lighter through time indicates that potters were actively seeking methods to make their wares brighter.

In order to achieve light-colored pastes, Hohokam potters appear to have manipulated the chemistry of the ceramic pastes and carefully controlled firing conditions. The research of David Abbott (2007) suggests that potters selected calcium-rich, and possibly salt-rich, clays for buffware production. Nodules of calcium carbonate that form naturally in semiarid soils (caliche) were added to the clay to lighten the paste even further. Most importantly, the firing temperature was carefully controlled between 800°C and 900°C to induce a chemical reaction between the calcareous clay and caliche and to prevent spalling. This complex process was not developed quickly. Ample evidence of experimentation during the late Pioneer period suggests that potters were actively adjusting their methods to obtain light-colored pottery.

Argillaceous Inclusions in Hohokam Red-on-Buff Wares

This chapter explores the possibility that the technological process used to manufacture red-on-buff pottery may have been responsible for introducing red argillaceous inclusions into the pastes of Hohokam buffwares in the form of clay pellets or clay temper. Ceramic analysts have long noted the presence of reddish argillaceous inclusions within fired red-on-buff pottery and the absence of these inclusions in Hohokam plainware vessels (Kelly 2013; for raw clays, see Beck 2006 and Beck et al. 2012). In addition, analysts generally note an increase in the presence of these inclusions through time. Their presence in Hohokam Red-on-buff wares during a time when potters may have been experimenting with the chemistry of these wares to achieve light, nonspalling vessels suggests a relationship between these inclusions and the process of Hohokam Red-on-buff production.

If the argillaceous inclusions in decorated wares are clay pellets, their presence may indicate that Hohokam potters selected calcium-rich clays that had natural globules of salt and iron-rich clay that would increase the temperature at which lime spalling would occur (Brewer 1976). Alternatively, Hohokam potters may have achieved light-colored decorated pottery without lime spalling by mixing clays with different chemistries. For instance, the mixture of calcium-rich and salt-rich clays may have provided the appropriate chemical balance to prevent lime spalling while simultaneously creating a buff exterior. The combination of different clays may have resulted in unmixed portions of clays that were mixed together (clay temper). Therefore the positive identification of the argillaceous inclusions within Hohokam Red-on-buff wares as clay pellets or clay temper will indicate what technique Hohokam potters used to achieve a chemical balance in their ceramic pastes.

Identifying Types of Argillaceous Inclusion

The primary objective of this analysis is to determine what type of argillaceous inclusion is present in Hohokam Red-on-buff wares and then to propose why these inclusions are present. The analysis principally aims to determine whether the reddish inclusions present are natural clay pellets or added clay temper. To that end the classification of these argillaceous inclusions relies on Ian Whitbread's seminal work on the subject. Whitbread (1986) defines three different types of argillaceous inclusions: (1) argillaceous rock fragments, (2) grog (pottery fragments), and (3) clay pellets and clay temper. He distinguishes each type of inclusion on the basis of boundaries, roundness, shape, optical density, internal features, external features, constituents, color, and prominence.

The boundaries of clay pellets and clay temper tend to be less sharp than those associated with other argillaceous inclusions such as grog and argillaceous rock fragments (ARFs). These boundaries can range from sharp to merging with the ceramic paste. Clay pellets and clay temper are also rounder in shape than grog or ARFs, which most often appear as angular or subangular fragments. Similarly, clay pellets and clay temper most often appear as equant and ovoid shapes in contrast to the prolate forms that ARFs and grog assume.

The optical density of all types of argillaceous inclusions can vary from high to low depending on composition. In general clay pellets and clay temper can be distinguished from other varieties of argillaceous inclusions because they lack internal features. In contrast, ARFs and grog sometimes have internal features. ARFs can possess polygonal fractures, parallel alignment, banding, or graded bedding. Grog can have voids or parallel alignment. The external features of argillaceous inclusions tend to be classified as discordant depending on how the inclusions are oriented and/or if they are distorted.

The constituents of argillaceous inclusions can be used to distinguish them. In general ARFs can have inclusions, such as micas, opaques, and quartz, which differ from those in the surrounding clay matrix. Grog, as previously fired pottery, can have a consistency similar to the ceramic paste if the two ceramic types are similar. Differences in how the clay is prepared and fired, however, may generate differences in the constituents present in the grog and the ceramic paste. Clay pellets and clay temper can also have a very similar array of constituents similar to those of

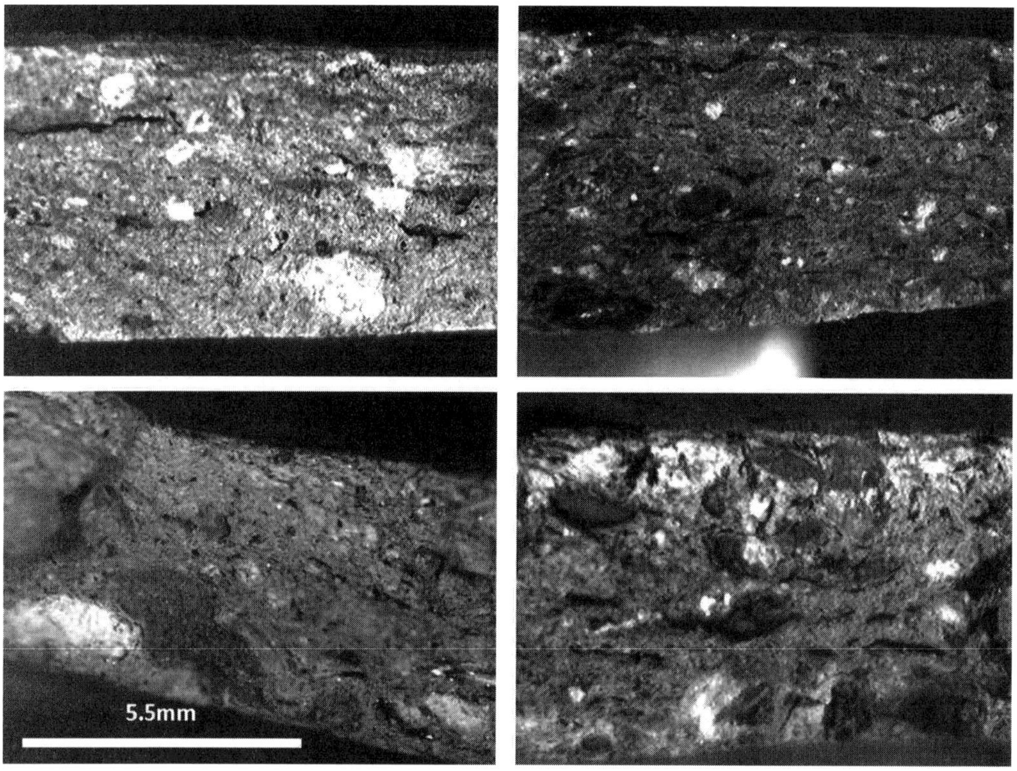

FIGURE 9.1. Cross sections of Hohokam Red-on-buff pottery with clay pellets. Clay pellets appear as rounded reddish globules or reddish streaks within the light-colored paste.

the ceramic paste, depending on whether the clays are similar. Whitbread (1986) notes that clay temper may diverge more dramatically from the clay paste than from clay pellets.

The color of clay pellets and clay temper is typically darker than the clay matrix. But, as noted above, very little color distinction may be noted between the inclusions and the paste if the clays are similar. Grog can also appear to have a color similar to the surrounding paste. ARFs can be similar or very different in color from the ceramic paste, depending on composition. The prominence of clay pellets and clay temper can vary widely from prominent to faint. Grog varies similarly. ARFs are typically classified as either prominent or distinct.

Characterizing Inclusions in Hohokam Red-on-Buff Wares

Observations made during macroscopic and petrographic analyses were used to determine that the argillaceous inclusions in Hohokam Red-on-buff pottery were predominantly natural clay pellets or clay temper instead of argillaceous rock fragments (Whitbread 1986). In addition, the analysis determined that the inclusions were not ground sherd temper, using the criteria of Cuomo di Caprio and Vaughn (1993) to distinguish between grog and argillaceous materials in ceramics.

Using Whitbread's (1986) criteria, argillaceous inclusions in Hohokam Red-on-buff pottery were identified and characterized through macroscopic and petrographic analyses. Macroscopically, the argillaceous inclusions are typically distinguished from the buff color of the surrounding paste by their reddish or orange color (fig. 9.1). This color difference is likely related to a higher iron and lower calcium content in the argillaceous inclusions in comparison to the surrounding clay paste. Macroscopic identification of argillaceous inclusions also relied on the texture of these inclusions in comparison to the surrounding clay matrix. Specifically, most

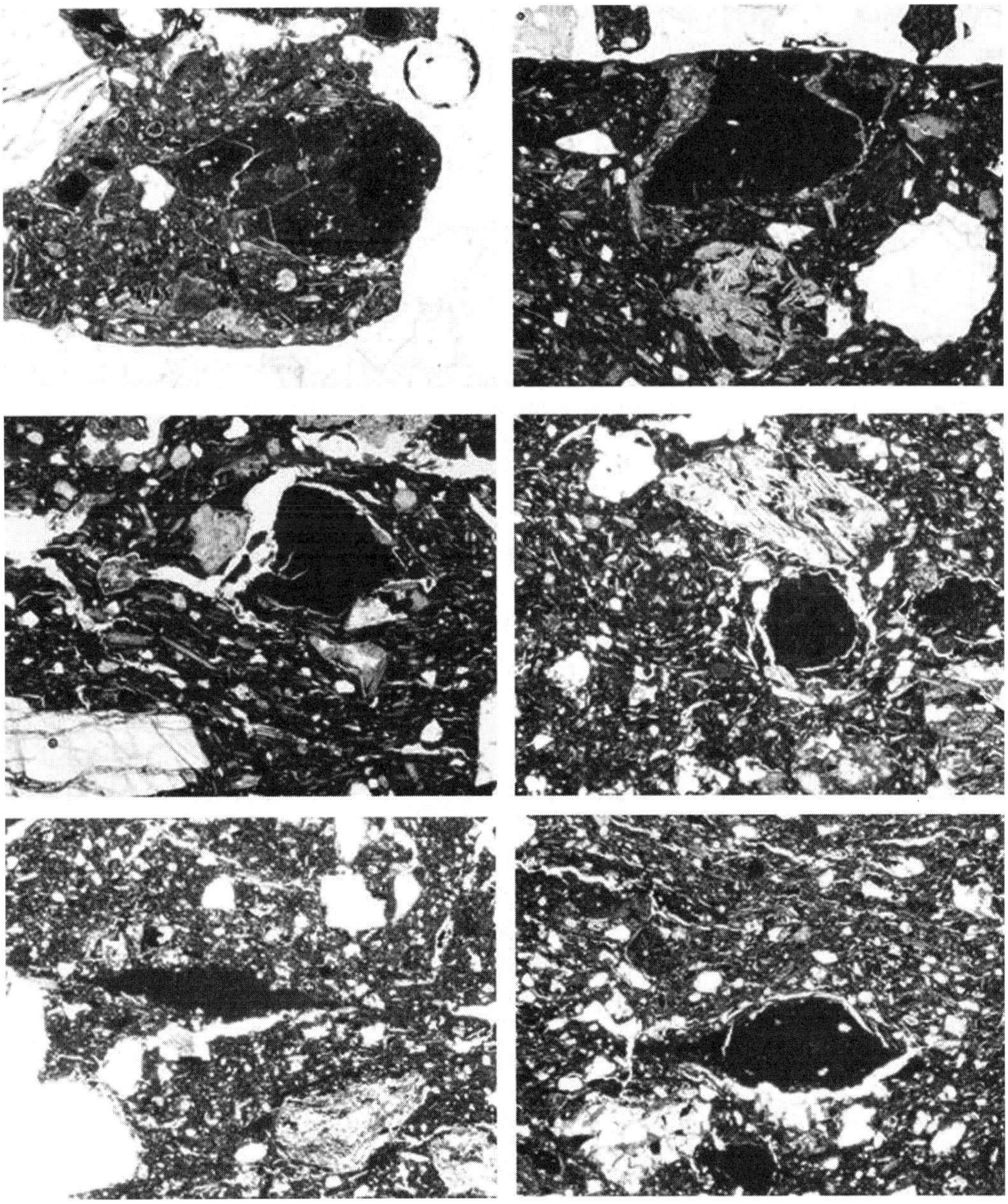

FIGURE 9.2. Microphotographs of clay pellets in Hohokam sherds (plane polarized light, 10×). Clay pellets appear as reddish areas with relatively few inclusions.

argillaceous inclusions generally have fewer natural aplastic inclusions than the rest of the clay paste.

Petrographically, the argillaceous inclusions are distinguished by similar characteristics noted in the macroscopic analysis, color and texture. The color of the argillaceous inclusions tends to be darker and redder than the surrounding paste (fig. 9.2). The inclusions are characteristically rounded or subrounded. Some inclusions appear as streaks that run through the clay paste. Streaks may represent less consolidated inclusions that were stretched as the potter kneaded the clay paste. The argillaceous inclusions typically ranged between 0.5 and 2 mm in diameter. The boundaries of the argillaceous inclusions ranged from sharp to merging. The shapes of many inclusions were equant and were concordant with the directionality of the clay matrix from the kneading of the clay by the potter. The optical density of these inclusions was generally neutral, and they typically lacked internal features such as aplastic inclusions. A few inclusions in some of the thin sections were more likely ARFs.

The macroscopic and petrographic analyses of the Hohokam Red-on-buff wares determined that the inclusions were not grog or ARFs but did not reveal whether or not the argillaceous inclusions were clay pellets or clay temper. Whitbread (1986:84) writes that it is "particularly difficult to identify clay temper positively and to differentiate it from plastic clay pellets." He notes that clay pellets result from particular depositional environments and that clay temper specifically refers to inclusions of unfired clay mixed within the clay matrix by the potter. Mixing different clays and poor wedging of these clays can result in visible clay temper inclusions.

What Can Argillaceous Inclusions Tell Us about Hohokam Ceramic Production?

The binocular and petrographic inspections of Hohokam Red-on-buff sherds indicate that the argillaceous inclusions are either clay temper or natural clay pellets. If clay pellets are present, Hohokam potters may have intentionally selected calcium-rich clays that had higher concentrations of these natural inclusions. The inclusions could have changed the chemistry of the base clay to lower the temperature at which vitrification occurs, thus preventing lime spalling (Rice 1987:98). If clay temper is present, Hohokam potters appear to have been mixing two different clays together to alter the chemistry of their finished wares. As noted above, it is possible that this mixing is an attempt to change the chemistry of the clay paste to achieve a light-colored exterior while also avoiding lime spalling. Mixing a calcium-rich clay with a salt-rich clay could have been one way that potters accomplished this result. An iron-rich clay would also have acted as a flux, lowering the vitrification point (Rice 1987:94).

The microscopic analyses could not positively identify the inclusions as clay pellets or clay temper, so this study relies on temporal and spatial information on Hohokam buffwares as an additional line of evidence to determine their origin. Specifically, the presence of argillaceous inclusions in Hohokam Red-on-buff pottery was examined to assess variation by time and/or production area. The analysis focuses on sherds collected from excavated features that have been dated by Wallace (2001, 2004) or Abbott (2009) using Wallace's refined red-on-buff typology (see also Lack in press). Thirteen sites located on the Salt and the Gila Rivers were analyzed: Chee Nee, El Caserio, Grewe, La Ciudad, La Lomita, La Villa, Las Colinas, Las Ruinitas, Los Hornos, Lower Santan, Sacaton Park, Snaketown, and Upper Santan (see fig. 9.3 for site locations). The sites were selected to obtain even spatial coverage across the Phoenix Basin.

Provenance Determination

Due to a varied and complex geologic terrain, raw materials analyses in the Phoenix Basin provide the basis for comprehensive ceramic provenance assignments (for example, Abbott 1994; Fournier 1989; Gladwin 1937; Hepburn 1984; Lombard 1987; Miksa 1995; Schaller 1994; Walsh-Anduze and Abbott 1994). Miksa and her colleagues (Miksa 2001a, 2001b; Miksa and Castro-Reino 2001; Miksa et al. 2004) developed a technique for sourcing sand temper in central Arizona through the characterization of unique and geographically isolated sand composition zones called petrofacies, which identifies the movement of pottery over distances as little as 5 km (see the discussion in the introduction).

The Phoenix Basin petrofacies were defined by collecting and point-counting 80 sand samples from the Salt River valley and 180 sand samples from the Gila River valley (Miksa 2001a, 2001b; Miksa and Castro-Reino 2001; Miksa et al. 2004). The point-count data from the raw sand samples were then examined using a series of correspondence analyses. These exploratory analyses delineated three major groups in the sand data: mineral-rich, lithic-volcanic, and lithic-metamorphic sands, which were geographically discrete. A series of nested discriminant models determined that these sand petro-

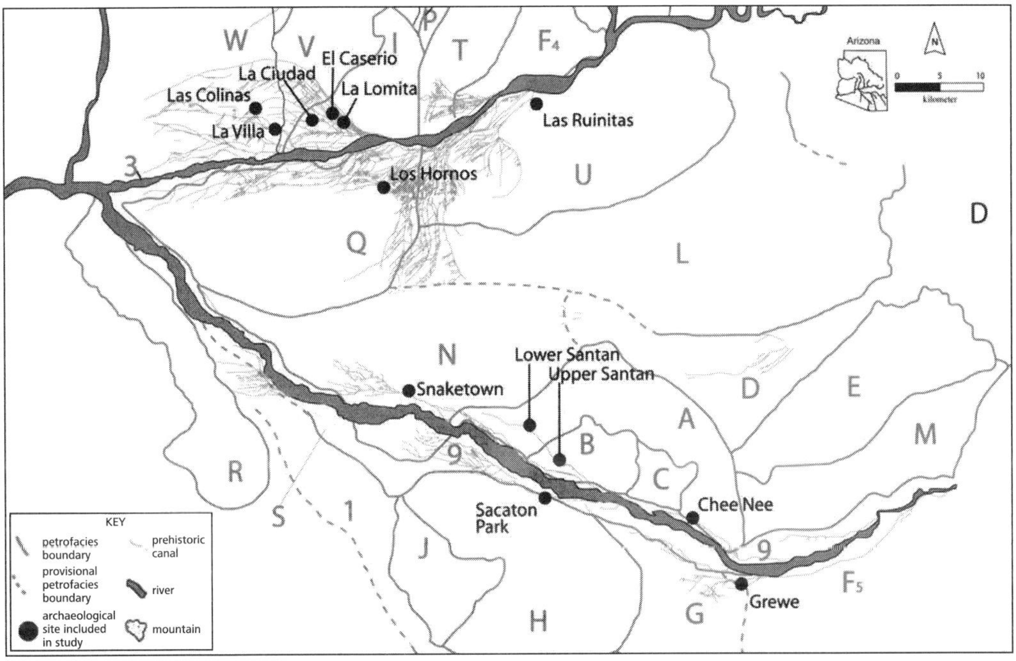

FIGURE 9.3. Map of the Phoenix Basin with sand petrofacies zones. Study sites included in this analysis are marked.

facies were statistically distinguishable from one another and the precise criteria by which the groups were separated. The results were used to establish boundaries for different petrofacies based on geologic maps of the region. Finally, Miksa developed detailed descriptions of the mineralic and lithic composition for each of the sand petrofacies. Using these descriptions and a detailed flowchart, analysts can now identify sands from different areas using a low-powered binocular microscope.

Binocular Microscope Examination

In order to relate provenance and the presence of argillaceous inclusions, the temper in 4,310 red-on-buff sherds was examined according to Miksa's classification scheme (table 9.1). A fresh cross section of each sherd was observed with a binocular microscope and a magnification of up to 30×. Approximately 82 percent of the sample was tempered with enough sand to source to specific petrofacies. The remaining 18 percent of the sample was predominantly tempered with coarse-grained mica schist that could not be sourced to a specific production locale (Kelly 2012; Neff and Dudgeon 2006; Walsh-Anduze 1993). For sand-tempered sherds the estimated percentages for each rock and mineral type in the sand temper provided a guide to navigate through a detailed and comprehensive flow chart for the sand petrofacies (Miksa et al. 2004:figure 2.12). Qualitative descriptions of the petrofacies were also used to make petrofacies determinations. Petrographic analyses determined that while some petrofacies were accurately identified in macroscopic analysis, such as Petrofacies N, others were not (Kelly 2013). These include Petrofacies A, B, C, and H. In the following analyses these petrofacies are lumped together as Petrofacies A/B/C/H.

Petrographic Analysis

The petrographic analysis of 182 thin sections of red-on-buff pottery pulled from the larger sample provided a check on petrofacies assignments made with the binocular microscope and an opportunity for further analysis of argillaceous inclusions. The selected sherds were pulled from collections from various sites and temporal phases (table 9.2), were at least 9 cm^2, and

TABLE 9.1. Sherd counts for macroscopic analyses by phase

Sites	Snaketown	Gila Butte	Santa Cruz	Early Sacaton	Total
Chee Nee	53				53
El Caserio			109	226	335
Grewe	127	121	131	174	553
La Ciudad	99	788	162		1,049
La Lomita				262	262
La Villa	58	339	131		528
Las Colinas		28			28
Las Ruinitas				104	104
Los Hornos	91	179		112	382
Lower Santan				105	105
Sacaton Park				20	20
Snaketown	302	145	86		533
Upper Santan	171	123	64		358
Total	901	1,723	683	1,003	4,310

were thought to contain enough sand temper to match to a specific petrofacies. Several sections perpendicular to the vessel wall were taken and then mounted on a single slide to increase the surface area available for identification. The sections were partially stained for feldspar identification. Similar to the low-powered binocular microscope analysis, the petrographic analysis relies on methods developed by Miksa and her colleagues for the identification of sand temper in Hohokam sherds (Miksa and Castro-Reino 2001; Miksa et al. 2004). In this study, however, ceramic samples were not point-counted as in Miksa's methodology. Temper particles were qualitatively assessed and compared to Miksa's classification scheme based on quantitative point-count data using a method developed and tested by Sophia Kelly (2013). A complete discussion of Miksa's methods is available in the introduction to this volume. Thin sections from this analysis are housed at the Pueblo Grande Museum, the Arizona State Museum, and Arizona State University museum collections.

Results

After the petrographic and binocular microscope analyses matched the sand temper in Hohokam Red-on-buff sherds with specific sand petrofacies across the Phoenix Basin, the study examined the argillaceous inclusions in each sample. Although some variation existed in the quantity of clay pellets in the clay fraction, the study only recorded whether or not pellets were present.

River System

The presence of clay pellets in the pastes of pottery manufactured on the Salt and Gila Rivers varies significantly. In particular, Gila River red-on-buff wares tended to have clay pellets, while Salt River red-on-buff wares rarely have pellets in their ceramic pastes (fig. 9.4). This result is likely related to the fact that Salt River potters used different clays for decorated pottery production than Gila River potters did. Margaret Beck et al. (2012) have argued that buffware production focused on iron-leached alluvial clays deposited in the middle Gila River valley 800–1,500 years ago. These clays only appear in particular depositional environments, such as certain middle Gila River riverine deposits. Some surviving remnants of buff-firing clays used by prehistoric potters on the Gila River have been identified (Beck et al. 2012; Eiselt and Woodson 2002; Waters and Ravesloot 2000). In a previous study Beck (2006) suggests that iron leaching from these deposits could have occurred in situ or, alternatively, in another depositional environment from which the sediments were redeposited in a new location. Similar sedi-

Table 9.2. Petrographic thin sections by site, time, and petrofacies determination

Sites in each phase	A	B	D	F5	G	H	I	N	Q	U	V	X	Schist only	Total
Snaketown														
Chee Nee	3				1	1							1	6
Grewe	1			2				1						4
La Ciudad	2					4		1					2	9
La Villa	1												1	2
Upper Santan	2					7		1					2	12
Gila Butte														
Grewe	1				1	1								3
La Ciudad	8	2	2			3		7	1				5	28
La Villa	4		1			1	5	8				2	1	22
Los Hornos		1	1								1			3
Upper Santan	5	1				2		2					1	11
Gila Butte–Santa Cruz														
La Ciudad	3													3
La Villa	1										1			2
Las Colinas	2		4					2			1			9
Los Hornos			1					1						2
Santa Cruz														
La Ciudad	1					2		1						4
La Villa			3			4	1	9						17
Upper Santan		1	2			1		5						9
Early Sacaton														
Grewe	1	1	1		1	1		8						13
Los Hornos	3	1	2			2		7		2			2	19
Upper Santan	1													1
No date														
Los Hornos			1			2								3
Total	39	7	18	2	3	31	6	53	1	2	3	2	15	182

ments have not been identified in the Salt River valley.

Temporal Change

The presence of clay pellets in pottery manufactured by middle Gila River potters also rose significantly through time (fig. 9.5). The largest increase occurred from the Gila Butte to the Santa Cruz phase (ca. AD 750–950) when the presence of clay pellets in pottery increased from 17 to 40 percent. This sharp increase implies that Hohokam potters changed their red-on-buff production methods, either by shifting clay sources or through clay mixing, which resulted in changes to the ceramic pastes of their wares.

Major Production Areas

Finally, the two major decorated pottery production locales in the Phoenix Basin, located in Petrofacies N and Petrofacies A/B/C/H, indicate similar trends in the presence of clay pellets through time (fig. 9.5). Petrofacies A/B/C/H

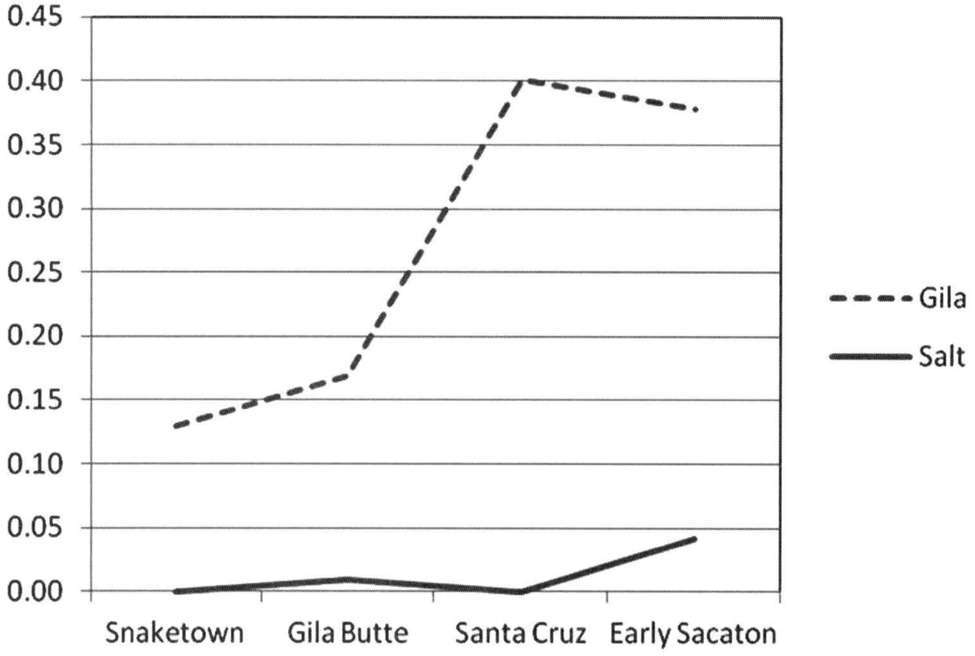

FIGURE 9.4. Proportion of red-on-buff pottery with clay pellets by river system through time. Phase dates: Snaketown (AD 700–750), Gila Butte (AD 750–850/900), Santa Cruz (AD 850/900–950), and early Sedentary (AD 950–1020).

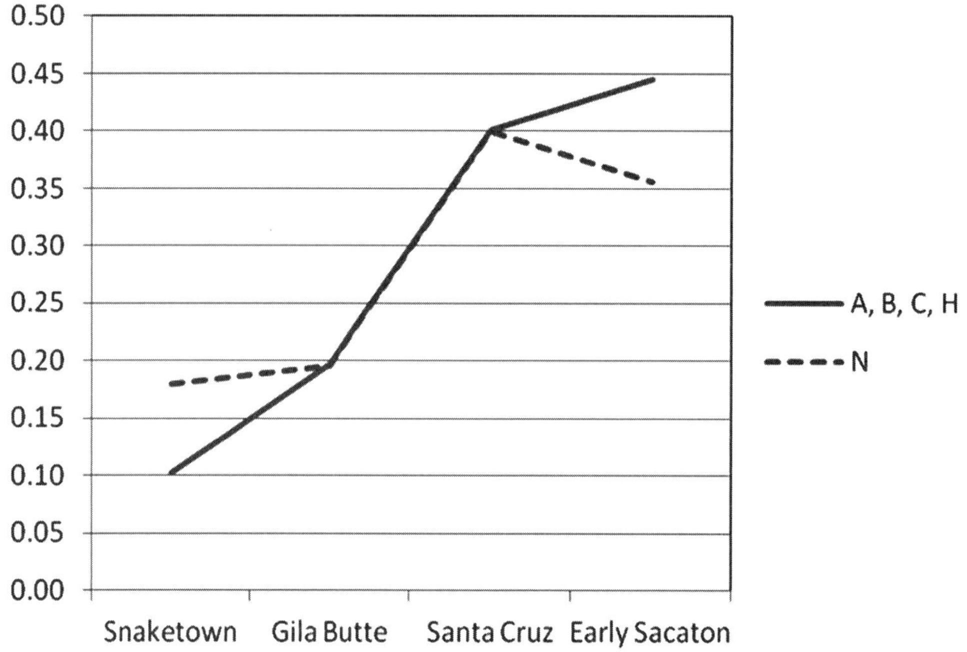

FIGURE 9.5. Proportion of red-on-buff pottery with clay pellets in the clay paste from the two major production locales on the Gila River: Petrofacies N and Petrofacies A/B/C/H. Phase dates: Snaketown (AD 700–750), Gila Butte (AD 750–850/900), Santa Cruz (AD 850/900–950), and early Sedentary (AD 950–1020).

were grouped together because they are difficult to separate using the low-powered binocular microscope. But current research now indicates that most specialized ceramic production within these petrofacies boundaries was concentrated near Gila Butte in Petrofacies A (Kelly 2013). These data suggest that potters working in both areas of the middle Gila River valley were influenced by similar changes in ceramic production methods. The increase in clay pellets in Hohokam Red-on-buff pottery manufactured on the middle Gila River valley directly coincides with increases in the concentration and supply from specialized producers located in this area. Specifically, potters working on the Snaketown canal system in Petrofacies N appear to have expanded their production and distribution between the Gila Butte and Santa Cruz phases (Kelly 2013). In particular, these specialists began supplying consumers in both the Salt and the Gila River valleys with most of their decorated pottery. As the proportion of red-on-buff wares in household assemblages increased, the percentage of pottery obtained from specialists on the Snaketown canal system increased in tandem.

Discussion

The alluvial clays that Hohokam potters used on the Gila River may have contained natural, unweathered clay pellets that analysts now see in Hohokam pottery pastes. Numerous authors have described the natural occurrence of consolidated, unweathered clay pellets within natural clays (Brewer 1976; Bullock et al. 1985; Kemp 1985; Whitbread 1986:83–84). Depending on the clays selected by Hohokam potters, these clay pellets may have been an innate part of the raw materials used in ceramic manufacture. Petrographic examination of the fired raw clay samples collected by Beck et al. (2012) enabled an assessment of the presence of clay pellets in the Middle Gila River clays. All of the raw clay samples were fired to 850°C, which is the estimated firing temperature for prehistoric Hohokam pottery.

The petrographic appearance of the buff-firing clay suggests that these raw clay deposits share a similar texture with Hohokam Red-on-buff pottery. In particular, thin sections of the Casa Blanca clay indicate that natural clay pellets are present within the clay fabric (figs. 9.6c and 9.6d). Although these clay pellets are not identical to the form and shape of clay pellets in Hohokam Red-on-buff sherds, further sampling of the Casa Blanca clay could reveal that other portions of the clay bed contain these pellets. This clay is chemically separate from other alluvial clays collected along the Gila River. It is also possible that the clays that prehistoric Hohokam potters used were not the same as the extant buff-firing clays available in the valley. Beck et al. (2012:329) note that the movement of the Gila River channel may have obscured clay beds that potters used in the past.

In contrast, another alluvial clay that fired buff in oxidation tests performed by Beck (2006; Beck et al. 2012) does not have clay pellets (figs. 9.6a and 9.6b). This clay (Komatki Weco) appears to exhibit internal compositional variation, because some of the samples were buff-firing and some were brown-firing in Beck's oxidation tests. The Komatki Weco samples were placed in the same compositional group in Beck's chemical analysis. The Komatki Weco clays are currently used by Pee Posh potters (Stoeppelmann and Fernald 1995) but may not have been used by prehistoric Hohokam potters. The Komatki Weco clay that fired to a buff color in oxidation tests also is similar in appearance and texture to clays that fired to a brown color (fig. 9.7). Neither clay pellets nor argillaceous rock fragments were observed in high frequencies in the selected samples of natural brown-firing clays collected along the middle Gila River valley.

If clay pellets in Hohokam Red-on-buff pottery pastes were a natural artifact of the clays used in ceramic production, variation in the incidences of clay pellets in Hohokam pottery would be linked to variation in the clay beds used by potters. For instance, an increase in the presence of clay pellets through time may be explained by a shift to using clays that have more natural clay pellets, such as the Casa Blanca clays. The coincidence of clay pellets with intensification in red-on-buff ceramic manufacture

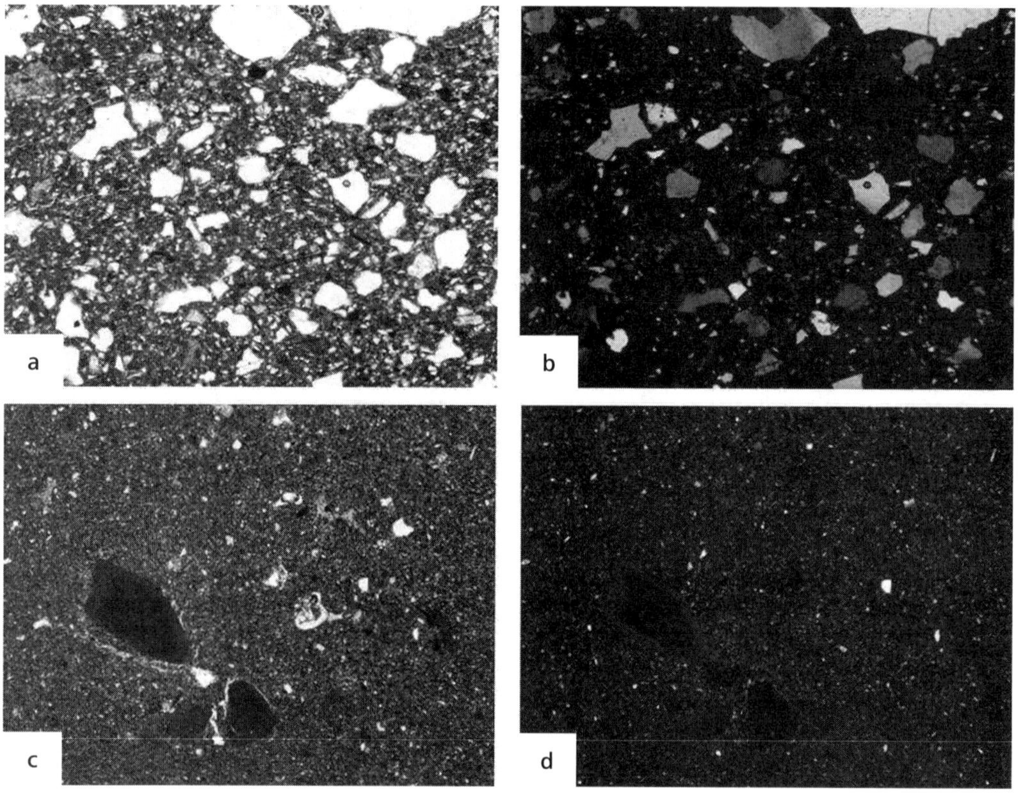

FIGURE 9.6. Micrographs of buff-firing clay samples. Komatki-Weco Pit 4, Location 1 clay: (*a*) PPL 4×; (*b*) XPL 4×; Casa Blanca clay: (*c*) PPL 4×; (*d*) XPL 4×.

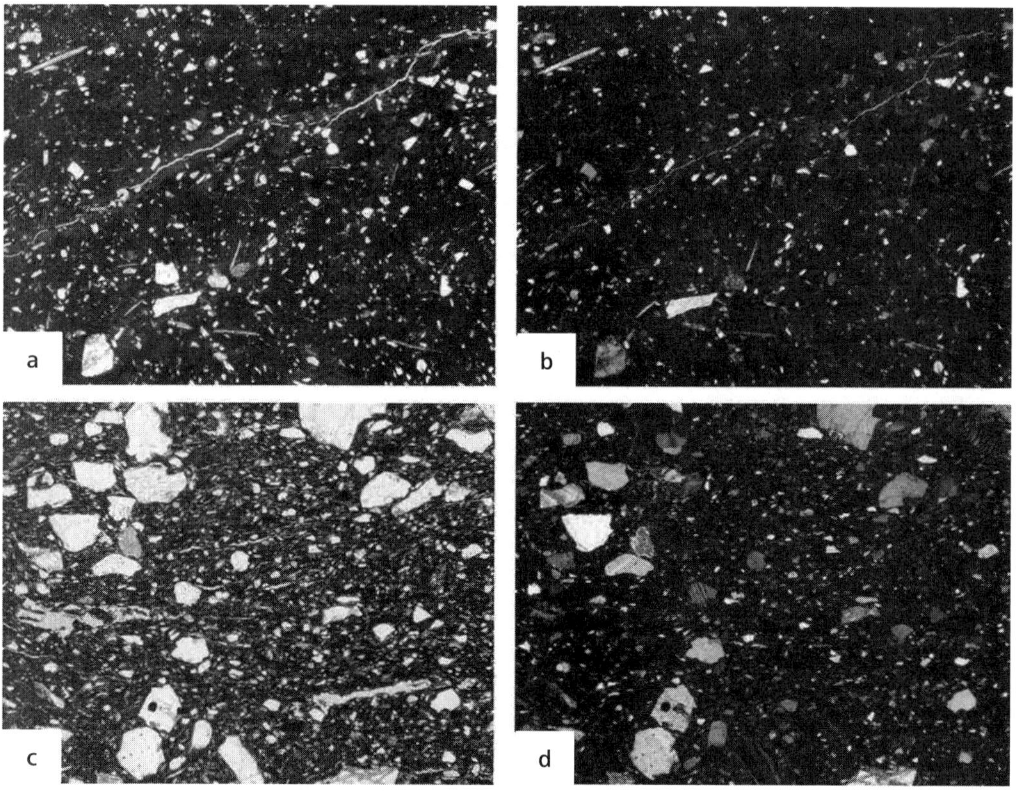

FIGURE 9.7. Micrographs of brown-firing clay samples. Pima Butte #1 clay: (*a*) PPL 4×; (*b*) XPL 4×; Komatki Weco Pit 1, Location 1 clay: (*c*) PPL 4×; (*d*) XPL 4×.

may indicate that potters were mining greater volumes of clay and may have been scraping into older clay beds. Using alluvial clays in close contact with established clay deposits may introduce more clay pellets.

In contrast to the theory that natural clay pellets in buff-firing clays along the Gila River account for the argillaceous inclusions in Hohokam Red-on-buff pottery, these inclusions may be an artifact of clay mixing (clay temper). Although the Casa Blanca buff-firing clays do have natural clay pellets, these pellets are not as pronounced as those that appear in prehistoric Hohokam sherds. Thin sections of brown-firing clays, which were also fired to 850°C, share the appearance and texture of some of the clay pellets in Hohokam pottery (see fig. 9.7). It is possible that buff-firing clays were mixed with smaller amounts of brown-firing clays, which served as clay temper and lowered the temperature at which vitrification occurred, preventing spalling. The mixing of two different clays may result in chunks (pellets) of unweathered clay material that differs in texture and color from the mixed ceramic paste. Whitbread (1986:85) notes the presence of clay pellets as a result of clay mixing in his analysis of Corinthian amphorae. In addition, his analysis of argillaceous inclusions in pottery suggests that clay temper, or clay mixing, may result in unweathered pellets of clay within the ceramic paste (Whitbread 1986).

Temporal variation in clay pellets within Hohokam Red-on-buff pottery suggests that potters shifted their production methods through time. In particular, as the output and concentration of pottery production on the middle Gila River valley increased, the incidences of clay pellets in red-on-buffwares increased as well.

One explanation for the increase in clay pellets in wares manufactured by concentrated specialists on the Gila River was that potters were scraping the bottom of secondary calcareous clay deposits to obtain more clay. As the number of pots manufactured increased, potters may have needed additional clay and used extant clay beds more intensively. It is possible that potters who expanded the areas where they collected clay and gathered clay in greater volumes were scraping clays adjacent to the original clay deposits. These clays may have had greater amounts of unweathered, argillaceous pellets. It is also possible, though, that the introduction of inclusions from scraping of additional clay deposits might not produce the consistent results observed in the pottery sherds.

An alternative theory for an increase in the presence of clay pellets in Hohokam pottery wares through time may be the mixing of two different clays. It seems reasonable that Hohokam potters, who were intimately familiar with local materials and experimented with them frequently, may have used clay mixing as one technique to create light-colored pottery from calcium-rich clay while avoiding the issues caused by lime spalling. Calcareous clays and clays that have been mixed with calcium-rich materials such as caliche are unstable at temperatures above 800–900°C. Abbott (2007) notes that by adding salt to clay mixtures Hohokam potters could have lowered the vitrification point so that lime spalling did not occur. Therefore mixing clay that is high in sodium might have served as a means by which the potters introduced salt into ceramic recipes, which is also necessary for the buff color. Chemical testing is required to evaluate this hypothesis, however. An iron-rich clay would lower the temperature of vitrification but not affect the surface color. Mixing clays may also have increased the workability of pottery pastes. For instance, combining stiff and plastic clays together creates a composite that might be easier to shape and form than either clay on its own.

Conclusion

Hohokam potters in central Arizona, over the course of more than 700 years of pottery production, experimented and refined decorated pottery recipes. The technological skill necessary to produce Hohokam Red-on-buff pottery suggests significant investment in the aesthetic qualities of the pots. Recent research has highlighted the complex recipe of buff ware pastes that contributes to its light color when fired. Archaeologists believe that ancient potters worked to lighten the pottery paste so that it would provide a contrastive background to the red painted designs.

Variation in the presence of clay pellets indicates that the clays and techniques that specialized potters used to manufacture Hohokam Red-on-buff pottery changed through time. In particular, the increase of clay pellets with an increase in the output and concentration of pottery production indicates that these pellets are linked to shifts in the organization of pottery production. The differentiation in the presence of clay pellets in Salt and Gila River pots suggests differences in the raw materials available in each river valley and the ceramic production methods used by potters in these areas. Two possibilities for these patterns may be scraping the bottom of existing clay pits to obtain more clay and mixing two different clays together.

Future research will explore the origin of the clay pellets in the ceramic samples through chemical analyses of clay pellets in Salt and Gila River pottery, the base clays of Salt and Gila River ceramics, argillaceous inclusions in raw clays, and a sample of raw clays available in the study area. Similarities in the chemical profiles of clay pellets in ceramics to either the argillaceous inclusions in raw clays or the base raw clays themselves may indicate whether the clay pellets in the ceramics were a product of natural inclusions or resulted from clay mixing. These analyses may ultimately illuminate specific clay processing and selection techniques that Salt and Gila River potters used and can further understanding of specialized ceramic manufacture in the Phoenix Basin.

Acknowledgments

This analysis utilized Pima-Maricopa Irrigation Project (P-MIP) ceramic and raw clay samples gathered by the Cultural Resource Management Program (CRMP) of the Gila River Indian Community (GRIC). P-MIP is federally funded by the U.S. Department of the Interior, Bureau of Reclamation, under the Tribal Self-Governance Act of 1994 (PL 103-413). The raw clay samples thin-sectioned and photographed for this chapter were collected and processed by Linda Morgan and Lorie Sinclair for the Cultural Resource Management Program for the Gila River Indian Community. Sinclair and Morgan provided access to these samples. All work done by me with P-MIP pottery collections and raw clay samples was conducted under a Research Collaborator Agreement with P-MIP and GRIC, which is on file at the CRMP offices at GRIC and with me.

All sands, grain boxes, and initial training for petrofacies assignments were generously provided by Elizabeth Miksa, formerly at Desert Archaeology, Inc. The sand samples and grain boxes are housed in the research collections for the Laboratory for Sonoran Ceramic Research at Arizona State University (ASU). The ASU Anthropology Collections, the Pueblo Grande Museum (PGM), and the Arizona State Museum (ASM) provided Salt River pottery collections used in this project. In particular, I am grateful to Holly Young (PGM), Lindsey Vogel (PGM), Arleyn Simon (ASU), Dolma Roder (ASU), and Arthur Vokes (ASM) for assisting with the collections. Funding for thin-sectioning the clay samples was provided by Kostalena Michelaki (ASU). The clay samples were analyzed and photographed using equipment in the Laboratory for Sonoran Ceramic Research and the Ceramic Technology Microscopy Laboratory at ASU. The clay samples were fired with the assistance of Blake Rogers (ASU). This chapter is based on a conference paper presented at the 2011 Arizona Archaeological Council meeting in Tucson. I am grateful to Scott Thompson, David Abbott, and Andrew Lack for comments on earlier drafts of this chapter.

References Cited

Abbott, David R.

1994 Appendix B: Categories and Criteria for Temper Classification during the Detailed Analyses of Whole Vessels and Sherds. In *The Pueblo Grande Project, Volume 3: Ceramics and the Production and Exchange of Pottery in the Central Phoenix Basin*, edited by David R. Abbott, pp. 582–585. Soil Systems Publications in Archaeology No. 20, Phoenix.

2000 *Ceramics and Community Organization among the Hohokam*. University of Arizona Press, Tucson.

2003 *Centuries of Decline during the Hohokam Classic Period at Pueblo Grande*. University of Arizona Press, Tucson.

2007 The Process, Location, and History of

Hohokam Buff Ware Production: Some Experimental and Analytical Results. *Journal of Archaeological Science* 35(2):388–397.
2009 Extensive and Long-Term Specialization: Hohokam Ceramic Production in the Phoenix Basin, Arizona. *American Antiquity* 74(3):531–557.

Abbott, David R., Alexa M. Smith, and Emiliano Gallaga
2007 Ballcourts and Ceramics: The Case for Hohokam Marketplaces in the Arizona Desert. *American Antiquity* 72(3):461–484.

Ackerly, Neal W.
1988 False Causality in the Hohokam Collapse. *Kiva* 53(4):305–319.

Beck, Margaret E.
2006 Linking Finished Ceramics to Raw Materials: Oxidized Color Groups for Lowland Desert Clays. *Kiva* 72(1):93–118.

Beck, Margaret E., Jill Onken, B. Sunday Eiselt, J. Andrew Darling, and Jeffrey R. Ferguson
2012 Geomorphological Setting and Native American Acquisition of Buff-Firing Ceramic Clays in the Lower and Middle Gila River Valley, Arizona. *Journal of Archaeological Science* 39:321–331.

Brewer, Roy
1976 *Fabric and Mineral Analysis of Soils.* Krieger, New York.

Bullock, Peter, Nicholas Fedoroff, A. Jongerius, Georges Stoops, and Tatiana Tursina
1985 *Handbook for Soil Thin Section Description.* Waine Research Publications, Wolverhampton, UK.

Clark, Jeffrey J., and Pat Gilman
2012 Persistent and Permanent Pithouse places in the Basin and Range Province of Southeastern Arizona. In *Southwestern Pithouse Communities, AD 200–900*, edited by Lisa C. Young and Sarah A. Herr, pp. 61–77. University of Arizona Press, Tucson.

Cuomo di Caprio, Ninina, and Sarah J. Vaughan
1993 An Experimental Study in Distinguishing Grog (Chamotte) from Argillaceous Inclusions in Ceramic Thin Sections. *Archaeomaterials* 7(1):21–40.

Dean, Rebecca M.
2007 Hunting Intensification and the Hohokam "Collapse." *Journal of Anthropological Archaeology* 26(1):109–132.

Eiselt, B. Sunday, and M. Kyle Woodson
2002 Changing Patterns of Pottery Production in the Casa Blanca Area of the Middle Gila River Valley. Paper presented in the symposium entitled "Visible Archaeology on the Gila River Indian Reservation" at the 67th Annual Meeting of the Society for American Archaeology, Denver, Colorado.

Fish, Paul R.
2006 [1989] The Hohokam: 1,000 Years of Prehistory in the Sonoran Desert. In *Dynamics of Southwest Prehistory*, edited by Linda S. Cordell and George J. Gumerman, pp. 19–63. University of Alabama Press, Tuscaloosa.

Fish, Suzanne K., and Paul R. Fish
2012 Hohokam Society and Water Management. In *The Oxford Handbook of North American Archaeology*, edited by Timothy R. Pauketat, pp. 571–584. Oxford University Press, New York.

Fournier, Dale M.
1989 Appendix A: The Results of the Petrographic Analysis of 86 Prehistoric Sherds from Sites AZ T:12:14 (ASU) and AZ T:12:16 (ASU). In *Archaeological Investigations at the Grand Canal Ruins: A Classic Period Site in Phoenix, Arizona, Volume 1*, edited by Douglas R. Mitchell, pp. 965–975. Soil Systems, Phoenix.

Gladwin, Nora
1937 Petrography of Snaketown Pottery. In *Excavations at Snaketown: Material Culture*, edited by Harold S. Gladwin, Emil W. Haury, Edwin B. Sayles, and Nora Gladwin, pp. 230–232. Medallion Papers No. 25. Gila Pueblo, Globe, Arizona.

Haury, Emil W.
1976 *The Hohokam, Desert Farmers and Craftsmen: Excavations at Snaketown, 1964–1965.* University of Arizona Press, Tucson.

Hepburn, Judith R.
1984 Appendix B: Ceramic Petrographic Analysis. In *Hohokam Archaeology along the Salt-Gila Aqueduct, Central Arizona Project, Vol VIII: Material Culture*, edited by Lynn S. Teague and Patricia L. Crown, pp. 341–352. Arizona State Museum Archaeological Series 150. University of Arizona, Tucson.

Hill, J. Brett, Jeffrey J. Clark, William H. Doelle, and Patrick D. Lyons
2004 Prehistoric Demography in the Southwest: Migration, Coalescence, and Hohokam Population Decline. *American Antiquity* 69(4):689–716.

Kelly, Sophia E.
2012 *A Refined Methodology to Source Hohokam Schist-Tempered Pottery.* Paper presented in the symposium "Petrography's Continued Role in Ceramic Studies: New Advances and Debates" at the 77th Annual Meeting of the Society for American Archaeology, Memphis, Tennessee.
2013 A Multi-Factor Analysis of the Emergence of a Specialist-Based Economy among the Phoenix Basin Hohokam. Ph.D. dissertation, Arizona State University, Tempe.

Kemp, Rob A.
1985 *Soil Micromorphology and the Quaternary.* Quaternary Research Association Technical Guide 2. Quaternary Research Association, Cambridge.

Lack, Andrew D.
In press A Stylistic Analysis of Hohokam Red-on-Buff Ceramics from the Lower Santan Site (AZ U:13:6 ASM). In *The Archaeology of the Pima-Maricopa Irrigation Project, in the Santan Area of the Gila River Indian Community, Volume 2: Material Culture (Part 1: Ceramic Studies at Lower Santan Village)*, edited by Christopher P. Garraty. P-MIP Technical Report No. 2008-01. Cultural Resource Management Program, Gila River Indian Community, Sacaton, Arizona.

Lombard, James P.
1987 Provenance of Sand Temper in Hohokam Ceramics, Arizona. *Geoarchaeology* 2:91–119.

Mabry, Jonathan B.
1999 Las Capas and Early Irrigation Farming. *Archaeology Southwest* 13(1):14.
2008 *Las Capas: Early Irrigation and Sedentism in a Southwestern Floodplain.* Anthropological Papers No. 28. Center for Desert Archaeology, Tucson.

Miksa, Elizabeth J.
1995 Petrographic Analysis of Pottery from Pueblo Blanco. In *Archaeological Investigations at Pueblo Blanco: The MCDOT Alma School Road Project*, edited by David E. Doyel, Andrew T. Black and Barbara S. Macnider, pp. 169–186. Archaeological Resources Report No. 90. Archaeological Consulting Services, Tempe, Arizona.
2001a *Provenance Characterization of Schist-Tempered Buff Ware Pottery.* Petrographic Report No. 99-6. Center for Desert Archaeology, Tucson.
2001b Temper Provenance Studies. In *The Grewe Archaeological Research Project, Vol. 2: Material Culture, Part I: Ceramic Studies*, edited by David R. Abbott, pp. 7–45. Anthropological Papers No. 99-1. Northland Research, Tempe, Arizona.

Miksa, Elizabeth J., and Sergio F. Castro-Reino
2001 *An Updated Sand Petrofacies Model for the Middle Gila River Basin.* Petrographic Report No. 2001-02. Desert Archaeology, Tucson.

Miksa, Elizabeth J., Sergio F. Castro-Reino, and Carlos Lavayen
2004 A Combined Petrofacies Model for the Middle Gila and Phoenix Basins, with Applications to Pottery from the Sky Harbor Site. In *Hohokam Farming on the Salt River Floodplain: Refining Models and Analytical Methods*, edited by T. Kathleen Henderson, pp. 7–44. Anthropological Papers No. 43, Center for Desert Archaeology, Tucson, Arizona.

Neff, Hector A., and John V. Dudgeon
2006 LA-ICP-MS of Ceramics and Ceramic Raw Materials from the Gila River Indian Community, Arizona. Gila River Indian Community, Cultural Resource Management Program, Sacaton, Arizona.

Rice, Prudence M.
1987 *Pottery Analysis: A Sourcebook.* University of Chicago Press, Chicago.

Schaller, David M.
1994 Geographic Sources of Phoenix Basin Hohokam Plainware Based on Petrographic Analysis. In *Pueblo Grande Project, Volume 3: Ceramics and the Production and Exchange of Pottery in the Central Phoenix Basin*, edited by David R. Abbott, pp. 17–90. Soil Systems Publications in Archaeology No. 20, Phoenix.

Stoeppelmann, Janet, and Mary L. Fernald
1995 *Dirt for Making Things: An Apprenticeship in Maricopa Pottery.* Northland Publishing, Flagstaff, Arizona.

Wallace, Henry D.
2001 Time Seriation and Typological Refinement of the Middle Gila Buffware Sequence: Snaketown through Soho Phases. In *The Grewe Archaeological Research Project, Volume 2: Material Culture, Part 1: Ceramic Studies*, edited by David R. Abbott, pp. 177–262. Northland Research, Tempe, Arizona.

2004 Update to the Middle Gila Buff Ware Ceramic Sequence. In *Hohokam Farming on the Salt River Floodplain: Refining Models and Analytical Methods*, edited by T. Kathleen Henderson, pp. 45–124. Anthropological Papers No. 43, Center for Desert Archaeology, Tucson, Arizona.

Wallace, Henry D., James M. Heidke, and William H. Doelle
1995 Hohokam origins. *Kiva* 60(4):575–618.

Walsh-Anduze, Mary-Ellen
1993 The Sourcing of Hohokam Red-on-Buff Ceramics Using Inductively Coupled Plasma Spectroscopy: "Schist Happens." M.A. thesis, Department of Anthropology, Northern Arizona University, Flagstaff.

Walsh-Anduze, Mary-Ellen, and David R. Abbott
1994 Pueblo Grande Whole Vessel Study: An Examination of Production and Short Distance Exchange. In *The Pueblo Grande Project, Volume 3: Ceramics and the Production and Exchange of Pottery in the Central Phoenix Basin*, edited by David R. Abbott, pp. 149–259. Soil Systems Publications in Archaeology No. 20, Phoenix.

Waters, M. R., and John C. Ravesloot
2000 Late Quaternary Geology of the Middle Gila River, Gila River Indian Reservation, Arizona. *Quaternary Research* 54:49–57.

Whitbread, Ian K.
1986 The Characterisation of Argillaceous Inclusions in Ceramic Thin Sections. *Archaeometry* 28(1):79–88.

White, Devon, and Stephen H. Lekson
2001 Early Hohokam. In *Encyclopedia of Prehistory: Volume 6, North America*, edited by Peter N. Peregrine and Melvin Ember, pp. 98–114. Springer, New York.

CHAPTER 10

Looking for the Right Outcrop

Ceramic Petrography in the Peruvian Andes

Isabelle C. Druc, Kinya Inokuchi, Victor Carlotto, *and* Pedro Navarro

In 2010 a series of compositional analysis was initiated to study the production and provenance of the ceramics found at the archaeological site of Kuntur Wasi, in the department of Cajamarca, Peru. Kuntur Wasi is a ceremonial site of the Formative Period occupied during most of the first millennium BC. The site is located at 2,300 m above sea level, 35 km west of the city of Cajamarca in the north-central Peruvian highlands (fig. 10.1). Excavations were carried out from 1988 to 2002 by the Kuntur Wasi archaeological project and a Japanese-led team. The site acquired much fame when eight intact tombs were excavated, bearing many gold objects, crowns, pendants, beads, and beautiful ceramics (Onuki et al. 1995; Onuki and Inokuchi 2011). In addition to the burials, the four occupation levels (spanning some 900 years) yielded a very large quantity of ceramic fragments. The major architectural and stylistic changes at the beginning of the second archaeological phase point to a strong foreign influence or a nonlocal origin of some of the people responsible for many of the changes observed (Onuki et al. 1995; Inokuchi 2010). Stylistic analysis also suggests the presence of local styles as well as foreign styles related to the coast and to the northern highlands.

Paste analysis involved petrography, X-ray diffraction (XRD), and scanning electron microscopy (SEM) of archaeological and comparative samples, in conjunction with the ethnoarchaeological study of the modern local ceramic tradition. The first step was to investigate the availability of potting materials around the archaeological site, the local geology, and the type of clay, temper, and practices of traditional potters in the region (Druc 2011; Druc et al. 2013). The analysis of the archaeological ceramics showed the existence of several paste groups, one of which was characterized by the presence of intrusive materials. The absence of intrusive bodies at the local level prompted the search for the provenance of the intrusive temper seen in the Kuntur Wasi ceramics. This search for the right outcrop is also an inquiry into understanding whether the wares manufactured with intrusive temper were produced on site with foreign material or brought to the site. This query was dictated by questions of interregional sociocultural or economic relationships in view of the architectural and stylistic changes observed at Kuntur Wasi.

We first briefly present the archaeological and ethnographic background of the research and then describe the local and regional geology. Later we turn to the compositional analysis of the archaeological and comparative samples related to this aspect of the study. The analysis results are discussed in light of the recent studies of the archaeological ceramics. We conclude with a tentative interpretation of ceramic production at or for the ceremonial site of Kuntur Wasi.

Research Background

The Archaeological Data

The ceremonial center of Kuntur Wasi (fig. 10.2) was occupied during the first millennium BC (Inokuchi 2010). The study of the construction phases, tombs, ceramics, and other cultural materials suggested a local initial occupation period (Idolo phase, 950–800 BC), followed by an important transformation in the ceremonial center and a population input of possible coastal provenance (Kuntur Wasi phase, 800–550 BC). New constructions and expansion of the site mark the third major occupation phase (Copa phase, 550–250 BC), while the last phase is characterized by drastic construction changes and quasi abandonment of the site as a major ceremonial center (Sotera phase, 250–50 BC). Apart from these four phases is the "Sangal complex" within the Kuntur Wasi phase, identified on the basis of a set of ceramics displaying similar stylistic characteristics (Inokuchi 2010; Onuki et al. 1995).

In terms of ceramic production, so far no evidence was found suggesting on-site production. But excavations have targeted the ceremonial area, where ceramic production was probably not taking place. Sixty-one different ceramic stylistic types have been identified (Inokuchi 2010). Several stylistic ceramic types display local characteristics, with a paste matching resources available within a 7–9 km range, as petrographic analysis showed.

Albeit fitting chronologically within the stylistic and cultural Chavín Horizon, Kuntur Wasi does not appear to have received direct impact from the ceremonial center of Chavín de Huántar in the north-central Andes (Inokuchi 1998). Rather, stylistic analysis suggests influences from the northern coast of Peru and northern highlands.

The Ethnographic Data

No clay is mined near Kuntur Wasi in present times. Modern ceramic production is only carried out in two villages several kilometers from Kuntur Wasi. Mangallpa, the closest production center in the San Pablo area, is 8 km straight north of the archaeological site (see fig. 10.1). This represents a three-hour walk for the potters

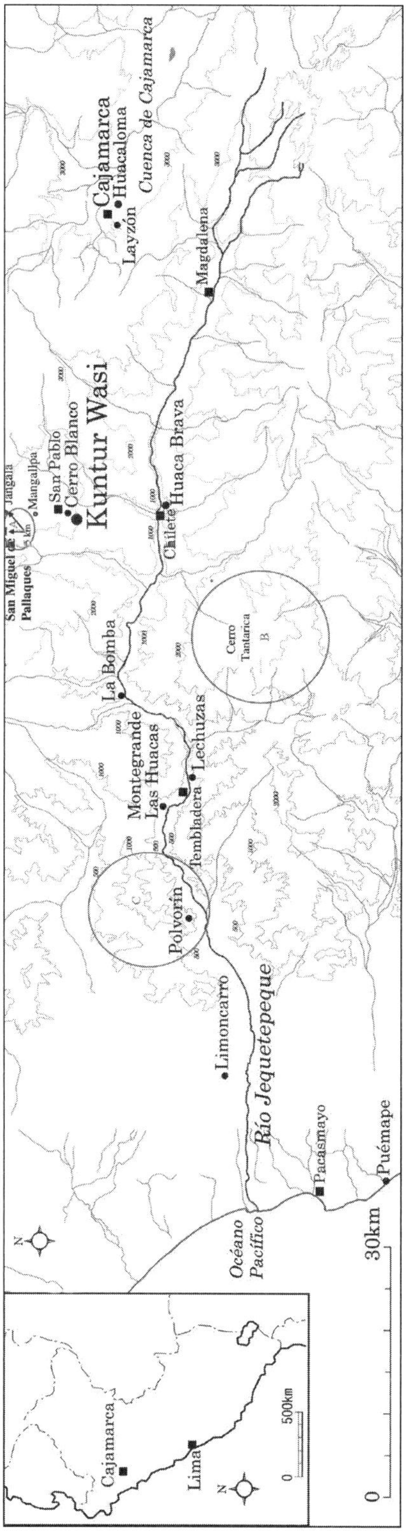

FIGURE 10.1. Location of Kuntur Wasi, the Jequetepeque Valley, and main sites mentioned in the text. Adapted from Inokuchi 2010:figure 1.

FIGURE 10.2. Ceremonial center of Kuntur Wasi, with the town of San Pablo in the background.

who come down to San Pablo to sell their wares at the Sunday market. The other area of production is around Jangala, a small village east of San Miguel de Pallaques, 16 km north of San Pablo in a straight line (22 km by road). The resource areas mined by the potters are generally located less than 3 km from the production areas, on the western slope of Cerro Cuscuden for Mangallpa and north of the same mountain for Jangala. Cerro Cuscuden is covered by extensive volcanic pyroclastic deposits, which are mined by potters of both villages. Additional local sources are exploited in the village of Jangala. The raw materials used are coarse clays and volcanic sand as temper, with quartz, plagioclase, biotite, and pyroclastic materials with glass shards, pumice (welded—where the glass vesicles are flattened and compacted in the same direction—and nonwelded), and volcanic tuff (Druc 2011). Both clay and temper are found close to the potter's workshop in fields or quarries ten to thirty minutes away by foot. The mix is one measure of clay for one to two of sand, according to the size of the pot to produce. The wares produced are cooking pots, jars, and cooking pans for domestic consumption (fig. 10.3). The Mangallpa potters follow a tradition using the paddle-and-anvil technique, for both the elaboration and the decoration of their pots, with plain and striated paddles and a round boulder as counterweight. They do not use a ceramic plate to support the wares. In Jangala the tradition is to use bivalve molds with interior incised decoration (see Druc 2011 for more details). Firing is done on the ground, in the patio or yard of the house of the potter, with cow dung at the bottom and straw on top. It lasts about one hour, then the pots are left to cool down. In both villages this is a masculine activity.

An important point in the ceramic production of Mangallpa is the itinerant character of the trade during one to three months of the year. In the past, between August and October, many potters would leave the village with their ma-

FIGURE 10.3. Example of traditional modern production of Mangallpa (photograph by I. Druc).

terials already crushed and sieved, offering their skills to villages close and far (up to the coastal valleys, the highlands, or the Amazon). This custom is still going on, on a reduced scale, in terms of both number of potters traveling and extension of their travels. The Mangallpa potters are providing the whole region with pots produced either at their homes or en route. The tradition is old and could well account for the production of ceramics in prehispanic times and should be considered a possible scenario when interpreting the ceramic data of archaeological sites in the north Andes.

The Geological Data

The regional geology (fig. 10.4) consists of a substratum of sedimentary sequences of pelitic, siliciclastic, and carbonate compositions of Upper Jurassic (Titonian) to Upper Cretaceous ages (Santonian). Different volcanic deposits from indistinct emission centers overlie the sedimentary layers (Navarro et al. 2008), dating from the Paleocene until the Miocene (56 to 10 Ma). This volcanism is represented by welded ignimbrites; rhyolites with abundant phenocrysts of quartz, plagioclase, and biotite; pumice with eutaxitic texture; and polymictic lithic fragments. These

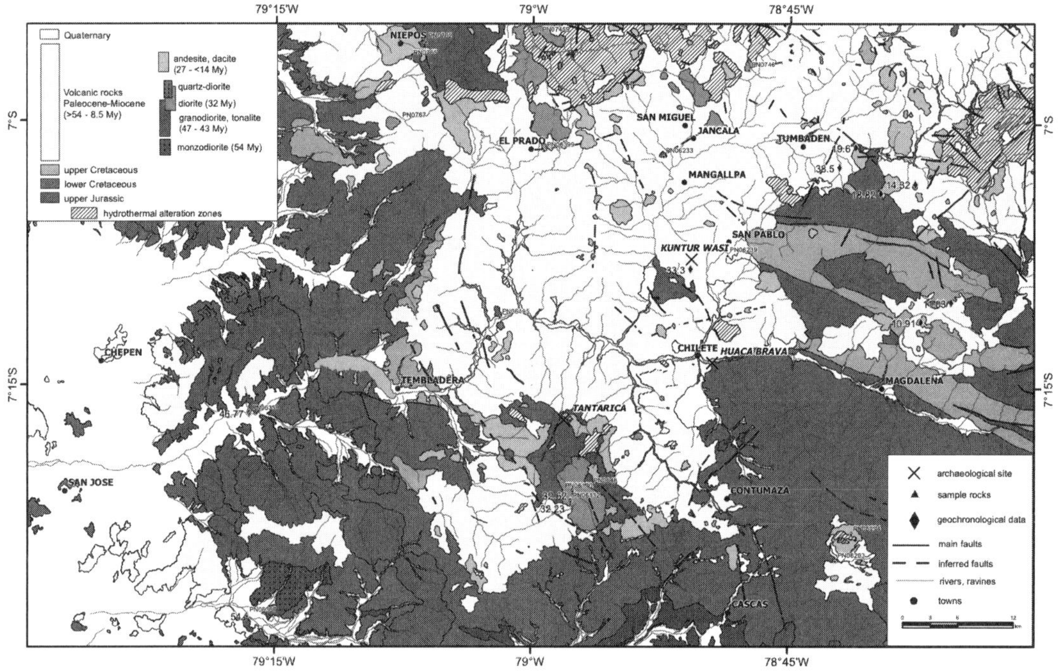

FIGURE 10.4. Geological map of the area of study.

are found from the west-southwest region of Kuntur Wasi to west of the town of Chepen. West of Tembladera, in the Jequetepeque Valley, thick andesitic lava layers alternate with flows of pyroclastic ash, intruded by granodiorites of 46.77 ± 0.16 Ma Ar-Ar age.

At the local level rhyolitic tuff surrounds the Kuntur Wasi site. A rhyolite cliff protrudes east-west just south of the site. Two andesite outcrops exist: a small one near Sangal south of the site and a very large one north of it, west of San Pablo (Yomoda 2002). Mixed layers of quartzite, limestone, and tuff are also found in a few places. The area has seen extensive landslides and much hydrothermal alteration according to Dr. Masaaki Shimizu (personal communication, February 2013) and Shinsuke Yomoda (2002). North-northeast of Kuntur Wasi are found pyroclasts of ash, lava, and welded ignimbrite with eutaxitic texture. Subvolcanic dacitic bodies intrude these pyroclastic layers with ages from 20 to 11 million years. A recent survey by Druc and Navarro also showed the presence of subvolcanic dacitic material with subround lithoclasts and many broken plagioclase crystals in a pyroclastic ash flow less than 6 km northeast of the archaeological site.

South of Kuntur Wasi, along the Jequetepeque Valley, are many flows of andesitic lava covered by diverse andesitic and dacitic pyroclastic deposits and a few ash layers with abundant accretional lapilli. One lava sample near Sangal (just south of the ceremonial center of Kuntur Wasi) yielded an age of 33.3 ± 0.8 Ma Ar-Ar (Navarro 2012) and 33.3 ± 0.21 Ma for an ignimbrite (Noble and Loayza 2004). These sequences underlie volcaniclastic deposits, purple gray conglomerates with subround to subangular volcanic fragments, layers of limestone of medium-grain size, and andesitic ash, suggesting a period of volcanic inactivity in the region. To the west of Contumazá (a town in the mountain range south of the Jequetepeque River), diverse subvolcanic and intrusive bodies cut interlayers of lava and ash pyroclasts. One diorite sample yielded dates of 32.63 ± 0.12 and 32.26 ± 0.13 Ma Ar-Ar (Navarro 2012). To the southeast intrusive bodies of granodiorite and tonalite composition intrude other lava and pyroclastic sequences.

The three intrusive bodies closest to Kuntur Wasi are presented below. The letters A, B, and C correspond to the respective areas in figure 10.1, while the given regional names are only valid for the present discussion (see fig. 10.4 for details; distances are geodesic). These bodies are:

(A) the San Miguel intrusives, 20 km north of Kuntur Wasi;
(B) the Contumazá/Cerro Tanta Rica intrusives, south of the Río Jequetepeque and Chilete (more than 22 km from Kuntur Wasi);
(C) the Tembladera intrusives, 25 km southwest of San Pablo (Navarro and Flores 2007; Navarro 2008, 2011).

These outcrops vary from granodiorite, diorite, to tonalite compositions. The San Miguel body (A) is more mafic than the Contumazá one (B), while the body west of Gallito Ciego (Tembladera region C) is much older (ca. 48 Ma) than the Contumazá body (30–35 Ma).

Geological survey of lithic materials (grinding stones and others) from Kuntur Wasi by Dr. Masaaki Shimizu (personal communication, April 2011) showed occasional use of granite porphyry with large-grained crystals, diorite porphyry, dolerite, tonalite-granodiorite, and diorite. These are intrusive rocks, which are not found surrounding the site. The building materials, however, are all of local origin, mainly quartzite and altered glassy tuff but also rhyolite and altered rhyolite tuff breccia (Yomoda 2002:111).

We now turn to the presentation of the analysis results, with a summary of previous studies conducted on the Kuntur Wasi ceramics and comparative raw materials. We then focus on the analysis of the rock samples and their comparison with the intrusive material in the ceramic thin sections.

Analysis Results

Summary of Previous Analysis of the Kuntur Wasi Ceramics and Comparative Samples

Petrographic and X-ray diffraction analyses have been conducted to study the archaeological and comparative samples of Kuntur Wasi and the surrounding region. The archaeological ceramic corpus for petrography consists of 110 thin sections from the different phases and styles observed at the site. Additional comparative materials include local clays, soils, sands, possible tempers, modern ceramics, and rock fragments to gain insight into the composition of possible potting materials around the archaeological site. The petrographic study was conducted on a Leitz petrographic microscope at 40× and 100×, with plain and crossed polarizers in the Department of Geosciences and the Department of Anthropology of the University of Wisconsin–Madison by I. Druc. The study included qualitative and point-counting analyses using the point-count method: identifying the grain directly under the crosshair at intervals of 0.5 or 1 mm according to granulometry in the thin section, for a total of 250 to 300 grains (see Stoltman 1989 for a description of the method). The photomicrographs were taken with a Nikon camera attached to the microscope. The thin sections are presently under the care of I. Druc.

Petrographic analysis allowed us to subdivide the corpus into five petrogroups of different mineral compositions. Petrogroup A (46.36 percent) and Petrogroup C (6.36 percent) present volcanic-rich compositions of different resource areas, and Petrogroup D (13.64 percent) has a mixed volcano-sedimentary composition. Most ceramics in Petrogroups A and D are of local production according to comparison with local geology, while Petrogroup C ceramics are probably nonlocal. Petrogroup E (23.64 percent) groups locally produced ceramics rich in subvolcanic material. The provenance and production area(s) of the ceramics with intrusive material in their paste (Petrogroup B, 10 percent) are considered nonlocal, as no intrusive outcrop exists near the site.

The study of the comparative samples shows that the composition of the volcanic-tempered ceramics matches the temper used in the Mangallpa area, obtained from sources 7 to 9 km away from Kuntur Wasi. But closer sources for volcanic material might have been used in the past. Internal variability within the petrogroups in terms of amount of inclusions, granulometry, and minor components suggests the existence of

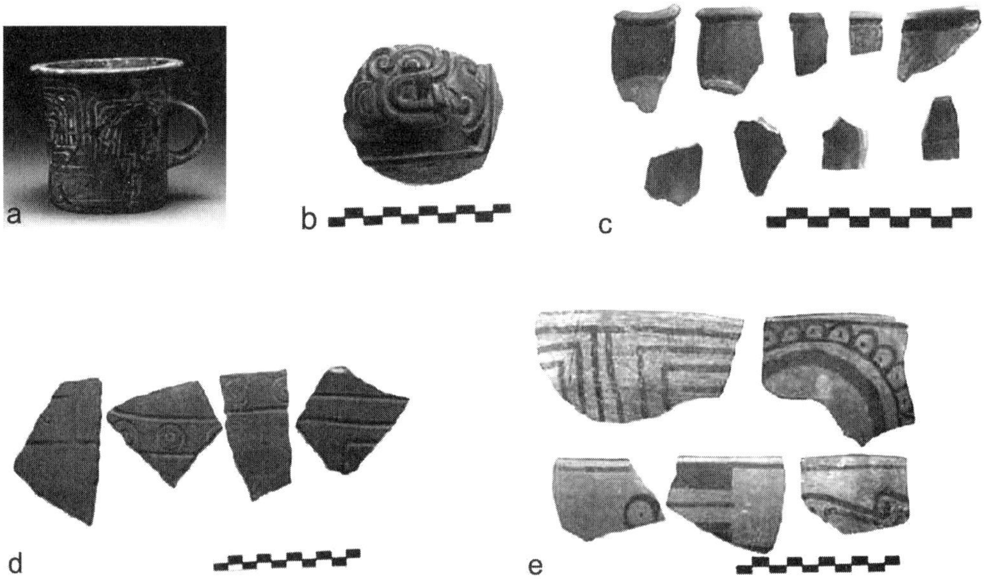

FIGURE 10.5. Samples of ceramics from Kuntur Wasi with intrusive temper stylistic types (Kuntur Wasi phase): (*a*) Negro Fino cup (11 cm high, 13 cm in diameter), (*b*) Negro Grafitado bottle fragment, (*c*) Rojo Fino, (*d*) Rojo Grafitado, (*e*) Rojo sobre Anaranjado (photographs by Kinya Inokuchi, montage by I. Druc).

multiple producers. Also, different recipes and trends can be observed in the course of the site's development.

The Kuntur Wasi ceramics tempered with intrusive material (Petrogroup B) display a variety of textures and compositions ranging from granitic to granodiorite and tonalite as well as varied stylistic, formal and temporal attributions (see figs. 10.5 and 10.6). Some thin sections show a heavy hornblende component. No one type of intrusive paste is reserved for the production of a certain ware style type, or form. In terms of ceramic production, interesting trends and technological traditions can be recognized throughout the development of the site.

During the first archaeological phase (Idolo) production is local, with volcanic temper, and only rare imports are seen (two with intrusive paste). The second phase (Kuntur Wasi) shows evidence of the introduction of ceramics tempered with intrusive-rich material on a regular basis, at a time when the site sees dramatic changes and an important input of influences from the coast. The third archaeological phase (Copa), in contrast, witnesses a strong decrease in nonlocal ceramics and the use of a new local source rich in subvolcanic material, in parallel with the local volcanic-temper tradition. During the fourth phase (Sotera), ceramic production declines in both quantity and quality. The different compositional groups presented above and internal variability suggest a diversity of production locations or resources used.

The XRD analysis focused on eight raw clays from local deposits and used by Mangallpa potters, as well as one low-fired clay sample from Sangal from a stream close to the archaeological site. Bulk and clay-suspension analysis were conducted, using a Scintag Pad V Diffractometer. The first analysis analyzed randomly oriented grains and only required a basic powder preparation; the second one involved a preparation by suspension of the sample to allow for the analysis of oriented inclusions (Druc et al. 2013). The results of the XRD study shows that the Sangal clay found around the site, albeit white and plastic, is not kaolin but an illite. Potter Sara Moller showed that this clay can be used for ceramic production, even without the addition of tempering material. It is thus possible that in earlier times this clay could have been used for ceramic production, as it is found in large amounts around the archaeological site. The other collected local clays were found not to be clays of

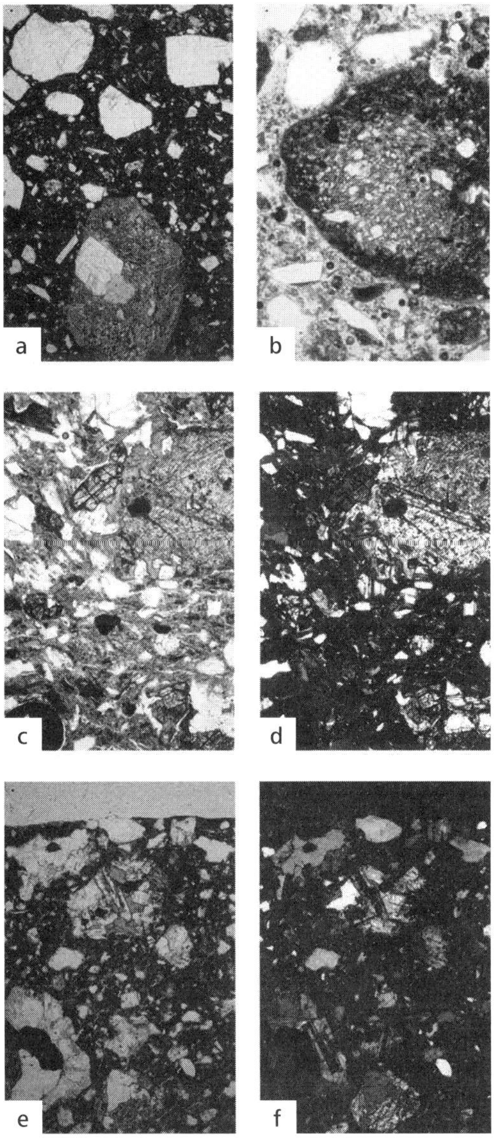

FIGURE 10.6. Photomicrographs of ceramic pastes from Kuntur Wasi: (*a*) Petrogroup A, (*c–f*) Petrogroup B; and (*b*) traditional ceramic from Mangallpa: (*a*) CP59, (*b*) MA26, (*c*) ID16, (*d*) ID16, (*e*) KW37, (*f*) KW37. Photomicrographs (*a*)–(*c*) and (*e*) taken in plain polarized light, (*d*) and (*f*) with crossed polarizers, all at 10× (4× objective and 2.5× camera magnification).

ceramic production range. The XRD analysis also suggests that the Mangallpa clay deposits mined by the actual potters are homogeneous, to the extent of the samples collected. The presence of manganese was also tested to account for the thoroughly black color of the paste in several of the archaeological ceramics, but the analysis failed to detect any significant amount of that mineral. Reduction firing is considered responsible for the color of the paste. A small spectroscopy analysis of four clay samples from the Middle Jequetepeque Valley was conducted using a TerraSpec spectrometer (Druc et al. 2013). All showed the presence of smectite and kaolin, which distinguish these clays from the clay samples from the region of Kuntur Wasi and San Pablo.

Looking for the Right Outcrop

In order to investigate the possible resources and production locations for the intrusive material found in the archaeological ceramics, we first sampled material from the intrusive bodies north and south of Kuntur Wasi (table 10.1, fig. 10.7). Sampling locations were dictated by ease of access by car or by foot. Ancient potters (as well as modern ones) might have used other quarries than those exposed now and followed more direct paths to get their material.

The petrographic analysis of the intrusive rock fragments sampled in the region of San Miguel de Pallaques (north of Kuntur Wasi and south of Chilete, in the region of Contumazá) and their comparison with the intrusive fragments in the Kuntur Wasi ceramics allow the following observations:

1. The crystals in the samples of intrusive rocks and conglomerates R7, R10, R13, and R14 are much bigger than the phenocrysts observed in the rock fragments of the ceramic thin sections. It does not seem probable that these rocks were used as temper for the production of the ceramics found at Kuntur Wasi, even crushed. In addition, most intrusive fragments in the ceramics are rounded and not angular.
2. The volcanic conglomerate samples R8 and R9 from the San Miguel region and the diorite sample R11 from El Rupe south of Chilete have smaller crystals (altered) and round to subround grains, which could correspond in their aspect to the type of temper observed in several ceramic thin sections from Kuntur Wasi. The volcanic conglomerate R8 with

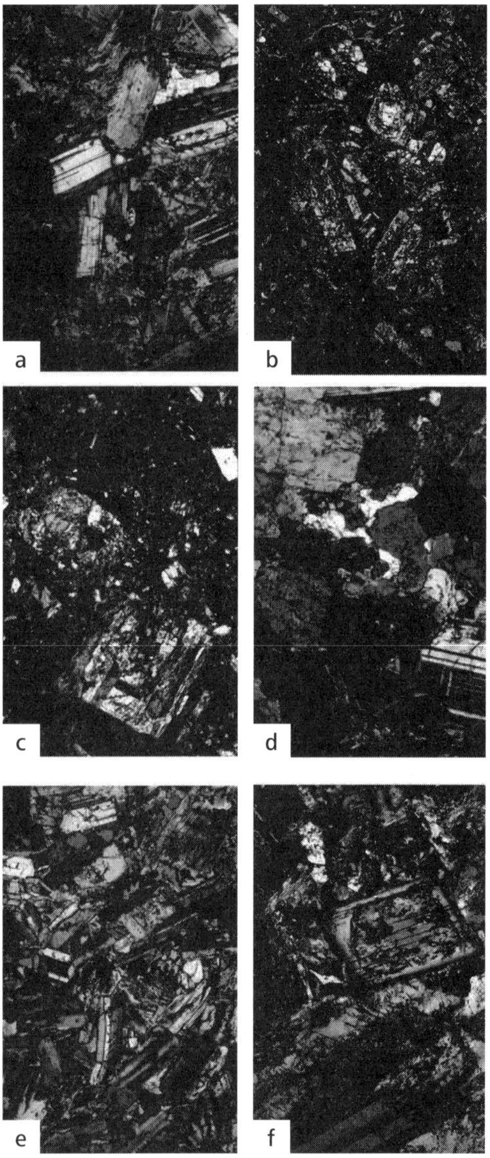

FIGURE 10.7. Photomicrographs of comparative rock fragments: (*a*) R7, (*b*) R8, (*c*) R9, (*d*) R10, (*e*) R11, (*f*) R14 (see table 10.1 for provenance and composition of rock samples). All images taken with crossed polarizers at 10× (4× objective and 2.5× camera magnification).

rounded lithics of intrusive origin, sampled 11.9 km south of San Miguel, is the best candidate for the material seen in the ceramics. Some ceramics, however, have altered intrusive fragments that correspond better to the crystal types seen in R11 (El Rupe) or R14 (Contumazá), south of Chilete.

These conclusions led us to sample sands and sediments during subsequent fieldwork, in particular from the Jequetepeque area. A few samples from ravines and riverbeds in the Tembladera area point to the Middle Jequetepeque as possible resource locations for the intrusive material found in the Kuntur Wasi ceramics. Also, in 2012 we happened to sample a potters' quarry in Jancos Alto, close to the potting center of Mangallpa, which we had already visited several times. The sample revealed the presence of some lithoclasts of subvolcanic composition (possibly dacitic), which to a certain degree look like intrusive fragments of acid to intermediate composition (a granodiorite would be the coarse-grained equivalent of a dacite), of subangular to subrounded shape, and of medium to granule size. This deposit can be classified as eluvial and pyroclastic, with tuff, pumice, quartz, plagioclase, biotite, and andesite fragments. The soil crumbles easily and has a sandy texture. This source is about 9 km away from the Kuntur Wasi site. The amount of subvolcanic material is low, however, and cannot easily account for the inclusions seen in the Copa phase ceramics of Kuntur Wasi. The best match for such a material was found in 2013, while sampling a pyroclastic flow from Cerro El Monton, just north of the village of San Pablo, less than 6 km from the ceremonial site. This sediment revealed a high percentage of plagioclase and broken rounded subvolcanic lithoclasts of dacitic composition, very similar to the type of mineral inclusions seen in the Kuntur Wasi Copa phase ceramics. This resource may well be the one used by the Copa potters.

Discussion: Compositional Variability and Production Diversity

The compositional variability observed within Petrogroup B from one ceramic thin section to the other suggests multiple production places or producers using different sources, quarries, or paste recipes. The analysis results point to the use of derived intrusive material from regions north and south of Kuntur Wasi for ceramic

TABLE 10.1. Comparative rock samples and their composition

Samples	Coordinates	Location	Composition
R7	07' 00" 23 78' 49" 15	Km 3.5 on the road San Miguel to Llapa	Diorite with prismatic crystals of plagioclase, augite, amphiboles, micas, opaques and oxides, rare quartz (≤1%)
R8	07' 01" 38 78' 52" 09	Km 14 on the road San Miguel to Llapa	Conglomerate with altered crystals, volcanic fragments of trachytic texture, calcite and sedimentary fragments, and a few very altered intrusive fragments
R9	07' 01" 18 78' 52" 04	11.9 km directly south of San Miguel	Volcanic conglomerate with volcanic microcrystals and subrounded altered diorite fragments; no glass or welded tuff
R10	07' 18" 01 78' 48" 33	Km 10 El Rupe, on the road Chilete to Contumazá	Altered diorite fragment
R11	07' 18" 01 78' 48" 33	Km 10 El Rupe, on the road Chilete to Contumazá	Diorite similar to R14 but not altered, fresh plagioclase crystals, augite, altered clinopyroxene
R12	07' 18" 01 78' 48" 33	Km 10 El Rupe, on the road Chilete to Contumazá	Volcanic conglomerate with trachyte fragments, round large quartz, altered pyroxenes; no glass or welded tuff
R13	07' 22" 12 78' 53" 13	Km 13 on the road Contumazá to Cruz Grande	Volcanic fragment (andesine type) with altered plagioclase, k-feldspar, ortho- and clino pyroxenes, small amount of quartz, oxides, olivine
R14	07' 21" 30 78' 53" 34	Km 25, Cruz Grande	Diorite with altered plagioclase, many large altered augite, amphibole, white mica, less than 10% quartz

production, not the direct mining of intrusive bodies. Thus different sources must be sought, circumventing the intrusive outcrops (in ravines or riverbeds for example). It is possible that ceramic production took place in villages close to these areas, operating during the same periods or at different times.

Potters in the Mangallpa area where actual production is found would have been well positioned for accessing the northern San Miguel intrusive body, while still remaining within walking distance of the ceremonial center for trading their wares. The Cerro Cuscuden is also proposed as the resource area for the production of the volcanic-tempered ceramics found in Kuntur Wasi. The local source recently found north of San Pablo is an excellent candidate for the production of the Copa phase ceramics with subvolcanic-sand temper. Ceramics with intrusive-sand temper, on the contrary, represent a nonlocal technological tradition, mostly seen during the second archaeological (Kuntur Wasi) phase. Their presence in Kuntur Wasi calls for long-distance ware distribution or for production by itinerant potters. The origin of such a tradition is not yet identified, but the Jequetepeque Valley could be a candidate, as coastal influences in many ceramic styles of this period at Kuntur Wasi are observed. In terms of distances, although the northern San Miguel outcrop is 16 to 20 km away from Kuntur Wasi, the area where the volcanic conglomerate with diorite inclusions comes from (sample R9) is 7 km from Kuntur Wasi (geodesic distance). Topography, however, would imply a longer distance to reach that source. During the Copa phase a community of artisans, distinct from the local one using volcanic material, must have been established close to Kuntur Wasi. They produced in great quantity, mining the subvolcanic-rich sediments of Cerro El Monton.

Looking at the possible provenances of ceramics with derived intrusive material from the Jequetepeque Valley, the sample location for R10 is 10 km from Chilete, in a small ravine along a stream running down the mountain. A small granodiorite outcrop of late Paleogene age lies less than 1 km away and is probably related to the large outcrop west of Cruz Grande/Contumazá sloping down toward the Jequetepeque/Tembladera area. Similar intrusive-rich conglomerates and sediments are found along the valley, in particular in the Tembladera area, in the quebradas and riverbeds of the streams rushing down the mountains to the Jequetepeque River.

In summary, petrographic analysis allows us to offer a production scenario involving the use of intrusive material mined not directly from intrusive outcrops but from sandy deposits or sediments derived from intrusive outcrops. The Jequetepeque Valley is only 12 km from Kuntur Wasi in a straight line, but the Tembladera area is another 38 km away down valley (circa 25 km diagonally from Kuntur Wasi). Such distances are not regular walking distribution routes for potters unless they travel to offer their services, as itinerant potters of Mangallpa still do. For ceramic provenance from the Jequetepeque/Tembladera area in the Formative Period we must envision long-distance or interregional trade, which may be linked to other economic or sociocultural relationships with the Jequetepeque sites. The coincidence of the appearance of many intrusive-tempered ceramics in Kuntur Wasi when the site is also registering very important architectural changes and stylistic influences traced to the coast might be indicative of a greater connection between the coastal valleys and the highland at that time. For the ceramics, this could mean higher trade or the adoption of a different technological tradition by a community of potters (local or not), while the local, volcanic-tempered production still continued. The possibility of itinerant potters cannot be discarded either.

Ancient potters probably used sand from intrusive or volcanic origin as temper (and not crushed rock). No particular paste recipe in the ceramics found at Kuntur Wasi is observed to produce only a certain form or style of ware, which allows us to speak in favor of technological styles or traditions linked to different regions of origin: one using an intrusive-rich paste, the other volcanic temper.

Several researchers have shown that technological styles are indicative of social identities (for example, Arnold 2005; Cecil 2004, 2009; Cecil and Neff 2006; Druc 2009; Gosselain 2002; Roux and Courty 2005). If this can be applied to the Kuntur Wasi case, the different paste traditions observed could point to the existence of (or interactions with) different sociocultural groups, at least at the beginning of the "intrusive" trend.

Conclusion

Tracing intrusive sources for ceramic production became important when petrographic analysis revealed the presence of intrusive material in many ceramics from Kuntur Wasi, a material unavailable locally. A survey of intrusive bodies at the regional level and petrographic analysis of rock samples showed that two areas could be proposed as possible origins of the temper seen in the ceramics. Also, the compositional variability observed within the group of ceramics tempered with intrusive material called for multiple production places or producers. In addition, results based upon the comparison of the rock and temper compositions allow us to suggest that the temper material is not derived from fresh, crushed intrusive rock but more probably from intrusive-rich sediments, from sedimentary or volcanic sands, which bear intrusive inclusions of smaller grain size than the granodiorite and diorite porphyry rocks characterizing the intrusive outcrops considered.

Possible origins for this temper are located some 12 km north (in a straight line) from the ceremonial site or 16 to 25 km south of the site, in the region south of Chilete or in the Middle Jequetepeque Valley. The valley area would suggest interregional trade, probably involving other Formative sites. As for the origin of the subvolcanic material seen in many Copa phase ceramics, Cerro el Monton close to San Pablo and the ceremonial site seems to be a very probable resource area.

Aside from the ceramic tradition of using intrusive temper, which appears to be a nonlocal tradition, and the Copa phase production using local subvolcanic material, the main tradition observed in Kuntur Wasi throughout the existence of the site is characterized by the use of volcanic temper. In situ production is yet to be proven for Kuntur Wasi, but a production place near Sangal, on the southern outskirts of the site, is possible in view of the existing volcanic resources (Yomoda 2002). The existence of a large resource area for clay and volcanic material in Mangallpa-Cerro Cuscuden, 8 km away from the site, could also indicate production in the outskirts of the site.

Finally, considering Arnold's model (1985), if we regard a production area encompassing the actual resources areas up to 8 km away from the ceremonial site, the ceramics from Kuntur Wasi with volcanic and subvolcanic tempers could be called local, while the intrusive-tempered ceramics are not. The possibility of on-site ceramic production by itinerant potters cannot be discarded, however, in view of the long tradition of this working mode in the area.

Earlier fieldwork and analysis have determined what to look for and narrowed down the search area for the provenance of intrusive-tempered ceramics, so we need to correlate the geological data with an analysis of the technological traditions observed for contemporary archaeological sites in the region of the Middle Jequetepeque and adjacent coastal valleys. This should help us to trace back the origin of the ceramics with intrusive-rich sand temper seen during the first and second phases of the site. A scenario involving different contemporaneous production zones and distribution networks is possible. Underlying ceramic production is the larger picture of socioeconomic or political relationships that shape and give direction to the movement of goods and services. Ceramics bear testimony to these relationships, but we need to interpret the data correctly.

Acknowledgments

This study has been possible thanks to funding from the Japan Society for the Promotion of Science (grant 23401040) and Saitama University. This funding helped us conduct the 2010 and 2011 fieldwork, prepare thin sections, and conduct the petrographic analyses. The Kuntur Wasi project provided lodging and transport, in particular to sampling locations. Lisenia Chavez from INGEMMET (Instituto Geológico Minero y Metalúrgico) in Lima helped with some geological questions. At the University of Wisconsin–Madison, Dr. John Valley facilitated our work in the Department of Geoscience and in the petrographic laboratory, while Dr. Jim Stoltman, from the Anthropology Department at UW–Madison, generously let us use his petrographic microscope for microphotography.

References Cited

Arnold, Dean E.
1985 *Ceramic Theory and Cultural Process.* Cambridge University Press, Cambridge.
2005 Linking Society with the Compositional Analyses of Pottery: A Model from Comparative Ethnography. In *Pottery Manufacturing Processes: Reconstitution and Interpretation*, edited by Alexandre Livingstone Smith, Dominique Bosquet, and Rémi Martineau, pp. 15–21. BAR International Series 1349, Oxford.

Cecil, Leslie
2004 Inductively Coupled Plasma Emission Spectroscopy and Postclassic Petén Slipped Pottery: An Examination of Pottery Wares, Social Identity and Trade. *Archaeometry* 46:385–404.
2009 Technological Styles of Slipped Pottery and Kowoj Identity. In *The Kowoj: Identity, Migration, and Politics in Late Postclassic Petén, Guatemala*, edited by Prudence M. Rice and Don S. Rice, pp. 221–237. University Press of Colorado, Boulder.

Cecil, Leslie G., and Hector Neff
2006 Postclassic Maya Slips and Paints and Their Relationship to Socio-political Groups in El Petén, Guatemala. *Journal of Archaeological Science* 33:1482–1491.

Druc, Isabelle C.
2009 Tradiciones alfareras, identidad social y el concepto de etnias tardías en Conchucos, Ancash, Perú. *Bulletin de l'Institut Français d'Études Andines* 38(1):87–106.
2011 Tradiciones alfareras del valle de Cajamarca y cuenca alta del Jequetepeque, Perú. *Bulletin de l'Institut Français d'Études Andines* 40(2):307–331.

Druc, Isabelle C., Kinya Inokuchi, and Shen Zhizhang
2013 Análisis de arcillas y material comparativo para Kuntur Wasi, Cajamarca, Perú, por medio de difracción de rayos X y petrografía, *Arqueología y Sociedad* 26:91–110.

Gosselain, Olivier
2002 *Poteries du Cameroun méridional: Styles techniques et rapports à l'identité*. CRA 26 Monographies. CNRS Editions, Paris.

Inokuchi, Kinya
1998 La cerámica de Kuntur Wasi y el problema Chavín. *Boletín de Arqueología Pontificia Universidad Católica del Perú* 2:161–180.
2010 La arquitectura de Kuntur Wasi: Secuencia constructiva y cronología de un centro ceremonial del Periodo Formativo. *Boletín de Arqueología Pontificia Universidad Católica del Perú* 12:219–248.

Navarro, Pedro
2008 *Mapa geológico del cuadrángulo de Chota*, sheet 14-f-III, 1:50,000. INGEMMET (Instituto Geológico Minero y Metalúrgico), Lima.
2011 *Mapa geológico del cuadrángulo de Chepén*, sheet 15-e, 1:50,000. INGEMMET (Instituto Geológico Minero y Metalúrgico), Lima.
2012 *Nuevas dataciones en el segmento sur de Cajamarca*. XVI Congreso Peruano de Geología, Lima.

Navarro, Pedro, and Carmen Flores
2007 *Mapa geológico del cuadrángulo de Cajamarca*, sheet 15-f-IV, 1:50,000. INGEMMET (Instituto Geológico Minero y Metalúrgico), Lima.

Navarro, Pedro, Robert Monge, and Astrid Flores
2008 *Centros volcánicos Oligo-Miocénicos (Grupo Calipuy) en la zona sur de Cajamarca*. XIII Congreso Latinoamericano de Geología and XIV Congreso Peruano de Geología, Lima, Perú, Resúmenes Extendidos. Sociedad Geológica del Perú, Lima.

Noble, Donald, and Carlos Loayza
2004 *Field Trip: Volcanic Rocks and Paleogene Geological History in the Vicinity of Chilete*. Guía de Campo, XII Congreso Peruano de Geologia. Sociedad Geológica del Perú, Lima.

Onuki, Yoshio, and Kinya Inokuchi
2011 *Gemelos prístinos: El tesoro del templo de Kuntur Wasi*. Fondo Editorial del Congreso del Perú and Minera Yanacocha, Lima.

Onuki, Yoshio, Yasutake Kato, and Kinya Inokuchi
1995 La primera parte: Las excavaciones en Kuntur Wasi, la primera etapa, 1988–1990. In *Kuntur Wasi y Cerro Blanco*, edited by Yoshio Onuki, pp. 1–126. Hokusen-Sha, Tokyo.

Roux, Valentine, and Marie-Agnès Courty
2005 Identifying Social Entities at a Macro-Regional Level: Chalcolithic Ceramics of South Levant as a Case Study. In *Pottery Manufacturing Processes: Reconstitution and Interpretation*, edited by Alexandre Livingstone Smith, Dominique Bosquet, and Rémi Martineau, pp. 201–214. British Archaeological Reports International Series S1349, Oxford.

Stoltman, James
1989 A Quantitative Approach to the Petrographic Analysis of Ceramic Thin Sections. *American Antiquity* 54(1):147–160.

Yomoda, Shinsuke
2002 Application of Mineral Sciences to the Archaeological Samples of Pre-Andes Civilization. Master's thesis, Graduate School of Science and Engineering, University of Toyama, Japan (in Japanese).

CHAPTER 11

Petrography and Behavior When the Minerals Do Not Change

Textural Analysis of Disaster Impacts on Historic Hidatsa Potting Practices, North Dakota

Kacy L. Hollenback

In archaeology, petrography is used to address questions of provenance and technology (Orton et al. 1993:133; Rice 1987:372–382). This study focuses on technology and asks: "What can analyses of temper tell us about human behavior?" By combining petrographic techniques with analysis of historic and early ethnographic records, the impacts of eighteenth-century and nineteenth-century smallpox epidemics on historic Hidatsa potting practices are explored, with an emphasis on raw material selection, preparation, and decisions about labor investment.

Northern Plains ceramics underwent dramatic changes after contact (Ahler and Swenson 1993; Lehmer et al. 1978; Mitchell 2011). It is argued that vessels became cruder or "decadent," based on observed changes in rim forms and design motifs (Lehmer et al. 1978). Existing models explain regional ceramic change as resulting from the abandonment or senescence (*sensu* Hollenback and Schiffer 2010) of traditional technologies in favor of Euroamerican trade goods as labor-saving mechanisms, the perceived prestige/superiority of metal objects, or the death of craft specialists in the wake of introduced diseases (Krause 1972; Lehmer et al. 1978:183–185; Wedel 1957:91). While combinations of these processes were likely the ultimate drivers of technological transformations, systematic study of *persistence* and *change* during this period will lead to a better understanding of the variability in postcontact and postepidemic indigenous histories.

If archaeologists want to explore the complex interactions between demographic disasters such as epidemics and postcatastrophe behaviors and choices, we must understand how these processes affect the production, use, distribution, and meaning of objects—our focus of study. Catastrophic impacts on technological systems, however, are the least developed aspect of disaster research (Hollenback 2012). The current study explores temper-related technological changes and explicitly investigates the relationship between disaster and technology at the community level by identifying changes in the production of pottery among the Awatixa Hidatsa.

This study addresses two major questions. How did each epidemic affect Hidatsa pottery-making practices? And what social and cultural coping mechanisms were in place that could explain potters' ability to maintain or modify their craft in postepidemic conditions (for example, knowledge loss through the death of craftspeople)? The Hidatsa case study, with a focus on the Awatixa subgroup, is ideally suited for scrutinizing disaster impacts on technology because it is founded upon a rich but underutilized body of historical and anthropological knowledge of this group. Petrographic textural analysis of temper offers unique insight into the reallocation of women's postepidemic labor, while also highlighting continuities in practice previously overlooked.

The Hidatsa, Culture Contact, Epidemics, and Pottery

Historically, the Hidatsa were semisedentary earthlodge villagers who practiced a mixed subsistence economy based on horticulture and bison hunting (Ahler et al. 1991; Bowers 1965; Stewart 2001). Three autonomous subgroups, each with its own dialect and origin/migration stories, lived in independent communities at the mouth of the Knife River in present-day North Dakota: the Awatixa, Awaxawi, and Hidatsa-Proper (Matthews 1877; Wood 1986). Today the descendants of these groups recognize themselves as members of the Hidatsa tribe.

Contact with Euroamericans occurred relatively late on the Northern Plains. Trade goods first appeared in village contexts in the seventeenth century, arriving indirectly through indigenous trade networks (Wood 1980). But direct contact with Euroamerican explorers took place only in the eighteenth century, with interactions increasing exponentially with the fur trade (Thiessen 1993; Wood and Thiessen 1985).

With contact came the exchange of goods and ideas but also Old World epidemic diseases, such as measles, influenza, whooping cough, cholera, plague, and smallpox. The resultant "virgin soil" epidemics were catastrophic for indigenous populations in the Americas (Crosby 1976; Dobyns 1983, 1993; Ramenofsky 1987; Thornton 1987). Such epidemics occur when a pathogen is introduced to communities with no prior exposure and therefore no immunity. Consequently, broader portions of the population are infected (the young, fully mature, and elderly) and mortality rates tend to be higher than in epidemics of endemic pathogens.

Much historical and archaeological research on epidemics in the New World has focused on identifying when and where outbreaks occurred as well as how many people died (Thornton et al. 1991). Yet recent research has demonstrated that the timing, spread, and experience of epidemics varied (Thornton 1987, 1997). Examinations of these phenomena require historical understandings of disease outbreaks at community scales, consideration of coping strategies and the experiences of survivors, as well as explorations of the disasters' lasting legacies. This can be done through systematic study of technological change and continuity after such traumatic events, because these yield insights into the decisions and practices of survivors.

Smallpox and the Hidatsa

The Hidatsa experienced multiple outbreaks of different diseases in the eighteenth and nineteenth centuries, yet two outbreaks are still discussed on the reservation today, with accounts passed on through oral tradition and social memory (Hollenback 2012). The first occurred just over two decades before the Meriwether Lewis and William Clark expedition (ca. 1804–1806). Details were recounted to the American captains by neighboring tribes such as the Mandan and Arikara (Jenkinson 2003). This 1781–1782 Northern Plains smallpox outbreak (Wood 2010) was part of a larger pandemic that swept across the Americas (Fenn 2001). Fifty-five years later *Variola* sp. was once again carried to the region, this time aboard the steamship *St. Peters* (Chardon 1997; Robertson 2001; Trimble 1985, 1988, 1989, 1994). No known mortality rates exist for the earlier epidemic, but estimates for the later one range between 50 and 80 percent. Needless to say, this loss of life affected every aspect of Hidatsa culture, especially specialized skill sets such as ritual and crafting activities (Hollenback 2012).

Hidatsa Pottery Making

It is unknown how many Hidatsa potters lived in each village at any time in the past, but it is believed that pottery making was a form of part-time craft specialization (Wilson et al. 1977). General consensus in the 1930s was that only a few families produced ceramics, and historical and ethnographic sources indicate that Hidatsa women were the potters (for accounts of male participation in pottery production, see Bowers 1965:165–166).

Hidatsa pottery making was "a secret and sacred activity" (Bowers 1965:165; for a detailed overview, see Hollenback 2012). The mystery and secrecy of the art were maintained through restricted access to knowledge of production. Specifically, potters controlled who witnessed and learned practical and sacred knowledge by retreating from public realms when making pots and performing sacred rites (Bowers 1965:345, 373). The structure of Hidatsa potting communities mirrors Hélène Wallaert-Pêtre's (2001) concept of "closed" learning systems, in which the exchange of knowledge is highly structured and restricted to a few individuals. Closed learning frameworks are more suceptible to disruption by demographic disasters because the knowledge and practical skillsets are not widely shared (Hollenback 2012). The death of artisans and ritual specialists therefore represents a significant loss to a community.

Among the Hidatsa, knowledge and training related to pottery production was "purchased" from older skilled craftswomen—often by matrilineally related relatives (Bowers 1965; Weitzner 1979; Wilson et al. 1977). A novice wishing to learn approached an older woman with goods and food in exchange for instruction. If the offer was accepted, the older woman took the initiate to acquire raw materials for production and demonstrated the complete sequences of activities for making a pot, which included sacred rites, stories, and songs. The exchange of knowledge was based on demonstration and trial and error (Wilson 1912:24–25). If the novice could not successfully produce a pot on her own, which was often the case, she would return to the older woman for another lesson.

Pots were made by mixing local clays with crushed, fire-cracked granite (Curtis 1909:181; Witte and Gallagher 2012:158). Wilson (1910:283) reported:

> The fire pit by accounts was in old times the usual place where the stones were gotten. But they may have been burned specially. Stone must be burned until it "explodes" as the Indians express it: by which they seem to mean merely that it has to be burned until it cracks.... The sweat lodge fire pit however doubtless furnished most of the needed stone.

In the early twentieth century Wilson (1910: 285) estimated that it took the potter about twenty minutes to make temper by pounding burned granite on a flat rock with a stone hammer (fig. 11.1; Wilson 1910:274).

When forming vessels, potters sequestered themselves either away from the village or within their lodges, shutting the door and covering the smoke hole (Bowers 1965:165–166; Wilson 1910: 286–287). The techniques for making cooking and water storage vessels involved pinching or "mass modeling" the vessel into the desired shape then thinning and strengthening the walls with paddle and anvil technique (Krause 2007: 36; Wilson et al. 1977). Pots were dried for several days and then fired in open, bonfire features.

The basic details of Hidatsa pottery manufacture have been provided by historical and ethnographic sources. Yet these often male-biased documents provide little insight into the daily lives of women or indications of how traditional crafting practices, in a closed learning framework, managed to persist through several demographic disasters. Studying the objects themselves can help address these issues.

For a systematic exploration of how potters maintained or modified their practices after the 1781–1782 and 1837–1838 epidemics, a life history framework (*sensu* Schiffer 1972) was used to compare procurement, processing, manufacture, use, reuse/recycling, and discard behaviors over time (Hollenback 2012). An attribute analysis of technical choices (*sensu* Chilton 1996), which holds that artisans make a sequence of decisions during production processes and that object attributes reflect these choices, was also considered. This combined approach facilitates the examination of change and persistence in manufacturing behaviors as well as the ability to make inferences about the allocation of artisans' time, learning frameworks, and patterns of

FIGURE 11.1. Hides-and-Eats, an elderly Hidatsa woman, "pounding rotten granite for stone sand" to be mixed with clay in the production of a pot. Photograph taken on the Fort Berthold Indian Reservation, North Dakota, in 1910 (reproduced from the original in the Gilbert Wilson photos of the Minnesota State Archives with permission from the Minnesota Historical Society and American Museum of Natural History).

TABLE 11.1. Expectations of observations

Attribute	Expectation	Reasoning
Temper type	Greater variety of temper types	Loss of knowledge through the death of specialists, combined with a need for time and labor conservation could result in experimentation with other temper types (such as sand)
Temper percentage	Decrease in temper density	As an "invisible" attribute, potters may use less temper so that less time is spent grinding to conserve energy
Temper size	Increase in temper size	Dedicating less time to grinding temper would result in larger pieces of granite and lower quantities, with less even sorting
Temper sorting	Poorer sorting	Less time dedicated to grinding and sorting activities would conserve energy and result in larger pieces of granite with less even sorting

enculturation. The following sections examine decisions and behaviors of potters relating to temper acquisition and processing (for a detailed discussion of all life history phases of Hidatsa pots, see Hollenback 2012).

Hypotheses

Several hypotheses address the impacts of epidemics on temper in Hidatsa pottery making (table 11.1). Changes might occur in raw material selection and use after an epidemic. It is assumed that knowledge about the location of clays and tempers was closely guarded if pottery making was a secret and sacred activity. If many potters died after an epidemic, novice survivors might not know where to obtain or how to process raw materials. Additionally, new and pre-epidemic potters might have experimented with alternative clays and tempers (such as sand instead of crushed rock) to save time and labor. Collecting and grinding rock takes much energy, and survivors may have had more pressing labor demands with the death of household members and other relatives. Sand, which required no grinding, was readily available along the both the Knife and Missouri Rivers.

With increased demands on women's labor, both new and surviving potters could alter raw material collection and preparation activities in other ways that would conserve time and energy. Because temper is "invisible" to consumers, producers could spend less time grinding and sorting to produce a product that nonetheless had the same general appearance. Temper size, percentage, sorting, and density are proxies for labor and time investment in the preparation of crushed rock (see table 11.1).

Methods

Textural analysis is often used in regions with "monotonous" geologies over large areas, such as those dominated by quartz sand and/or lacking distinctive minerals (Middleton et al. 1985:64; for overviews of textural analysis, see Darvill and Timby 1982; Orton et al. 1993:141–144; Peacock 1971; Streeten 1982). Adapted from sedimentology, the approach considers differences in "texture" of inclusions or paste instead of mineral identification. The technique is useful in addressing questions of continuity and change in technology or how a ceramic was created (Orton et al. 1993:133; Rice 1987:373). This is especially true in instances where tempers, such as bone, shell, or rock, are prepared by grinding, as textural analysis provides a means to measure diachronic change in labor and time investment (see also Capone 2006).

Texture is defined as "the percentage of inclusions in the clay, their roundness, sphericity, and size distribution" (Peacock 1971:257). To this others have added measures of sorting (Orton et al. 1993:139, 239), as well as reporting the presence of voids and compactness of the clay (Capone 2006). For this analysis, observations were recorded for inclusion identity, frequency, shape, sorting, and size. For all attributes but identity and size, reference was made to a percentage inclusion estimation chart (Orton et al. 1993:238), an inclusion sorting chart (Orton et al. 1993:239), and a sphericity/roundness estimation chart (Orton et al. 1993:239).

Textural analysis usually relies on analysis of thin sections with a petrographic microscope. Large numbers of thin sections are both costly and destructive, however, so observations in this study were made on the sherds themselves, after a small clean break had been made, with a binocular microscope at 10× magnification (for a discussion see Orton et al. 1993:135–136). Prior to the larger analysis, a pilot study of a subsample of thin sectioned sherds ($n = 10$) was carried out using the estimation charts to demonstrate comparability in observations between petrographic and binocular microscopes. These sherds were selected from surface collections taken from the Lower Hidatsa (32ME10) and Sakakawea (32ME11) sites by Donald Lehmer and were provided by the State Historical Society of North Dakota. Sherds were selected to account for variability in the larger ceramic assemblage related to temper, vessel wall thickness, and production quality and to guarantee that two sherds did not come from the same vessel. For the broader binocular study, all body sherds larger than 0.5 inches in length (those that do not pass through a 0.5-inch screen) were analyzed from previously studied midden contexts at the different sites in an attempt to control for depositional life histories (not ritual or specialized depositional contexts; for more detail see Hollenback 2012).

A simple approach to measuring the size of inclusions was adopted here because a larger sample was desired (table 11.2; Orton et al. 1993: 141). To record grain size, the smallest and largest inclusions were measured at 10× using a micrometer for thin sections and a clear metric ruler under the binocular microscope for sherds. Intentionally added temper, which consisted of angular fragments of quartz, feldspar, and mica, typically ranged from 0.5 to 5.0 mm. The following size classes were employed: 0.5–1.0 mm, 0.5–2.0 mm, 0.5–3.0 mm, 0.5–4.0 mm, and 0.5–5.0 mm. This was sufficient to detect increases or decreases in temper size over time.

Sherds of utilitarian vessels from four of the five historic Awatixa village sites were analyzed for this study (fig. 11.2). While rim sherds are temporally diagnostic, the partially destructive nature of textural and temper analysis meant that permission was only obtained for analysis of body sherds (see table 11.2). The State Historical Society of North Dakota granted permission for 10 thin sections from Lower Hidatsa and Sakakawea sites to be analyzed with a petrographic

TABLE 11.2. Awatixa sherd samples

Site name	Dates	Epidemic period	Body sherds	Textural analysis	References
Lower Hidatsa (32ME10)	AD 1600–1782	Pre-1782	1,062	yes	Ahler and Benz 1980 Ahler and Weston 1981
Rock Village (32ME15)	AD 1782–1790s	Post-1782/Pre-1837	172	no	Hartle 1960 Lehmer et al. 1978
Sakakawea (32ME11)	AD 1790s–1837	Post-1782/Pre-1837	982	yes	Ahler and Benz 1980 Ahler et al. 1980
Taylor Bluff (32ME366)	AD 1837–1845	Post-1837	329	yes	Ahler et al. 1983 Ahler 1988
Like-A-Fishhook (32ML2)	AD 1845–1885	Post-1837	93	yes	Smith 1972

Note: see Hollenback 2012 for a detailed discussion of sites.

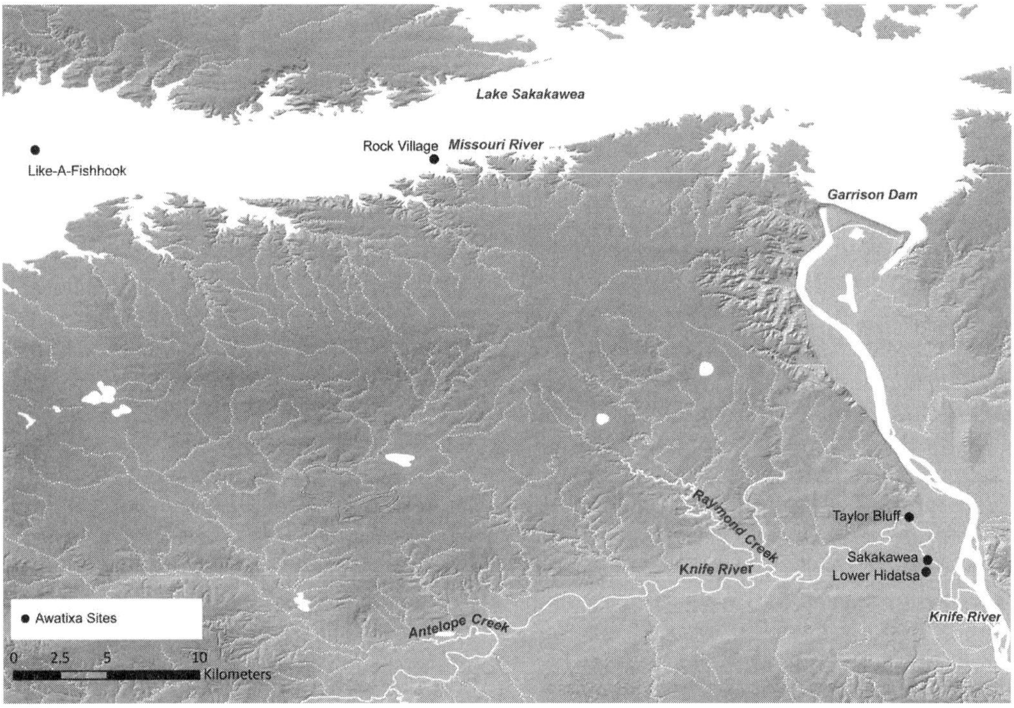

FIGURE 11.2. Historic Awatixa Hidatsa earthlodge village site locations, North Dakota (based on a map by Kathryn MacFarland).

microscope to build the comparative baseline (figs. 11.3 and 11.4). After comparability was established, all results reported here were based on observations made with a binocular microscope. Thin sections created for this analysis are housed at the State Historical Society of North Dakota in Bismarck. The larger assemblage of pottery sherds is housed at the Knife River Indian Villages National Historic Site in Stanton, North Dakota.

Results

Differences in sample sizes between the earlier and later sites (Lower Hidatsa/Sakakawea and Taylor Bluff/Like-A-Fishhook), with the exception of Rock Village, are a reflection of decreased

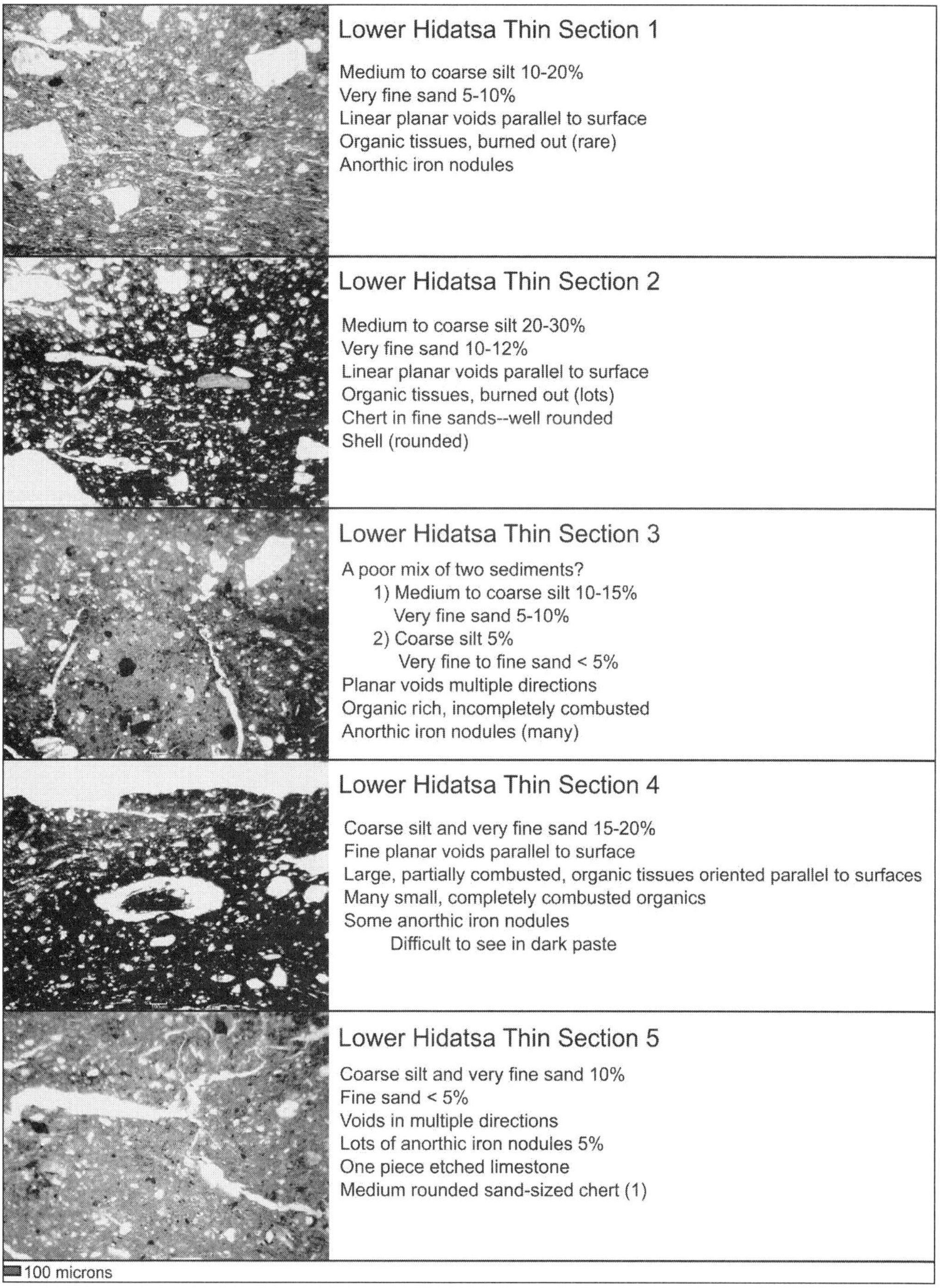

FIGURE 11.3. Thin sections of sherds from Lower Hidatsa (32ME10) in plane polarized light (40×).

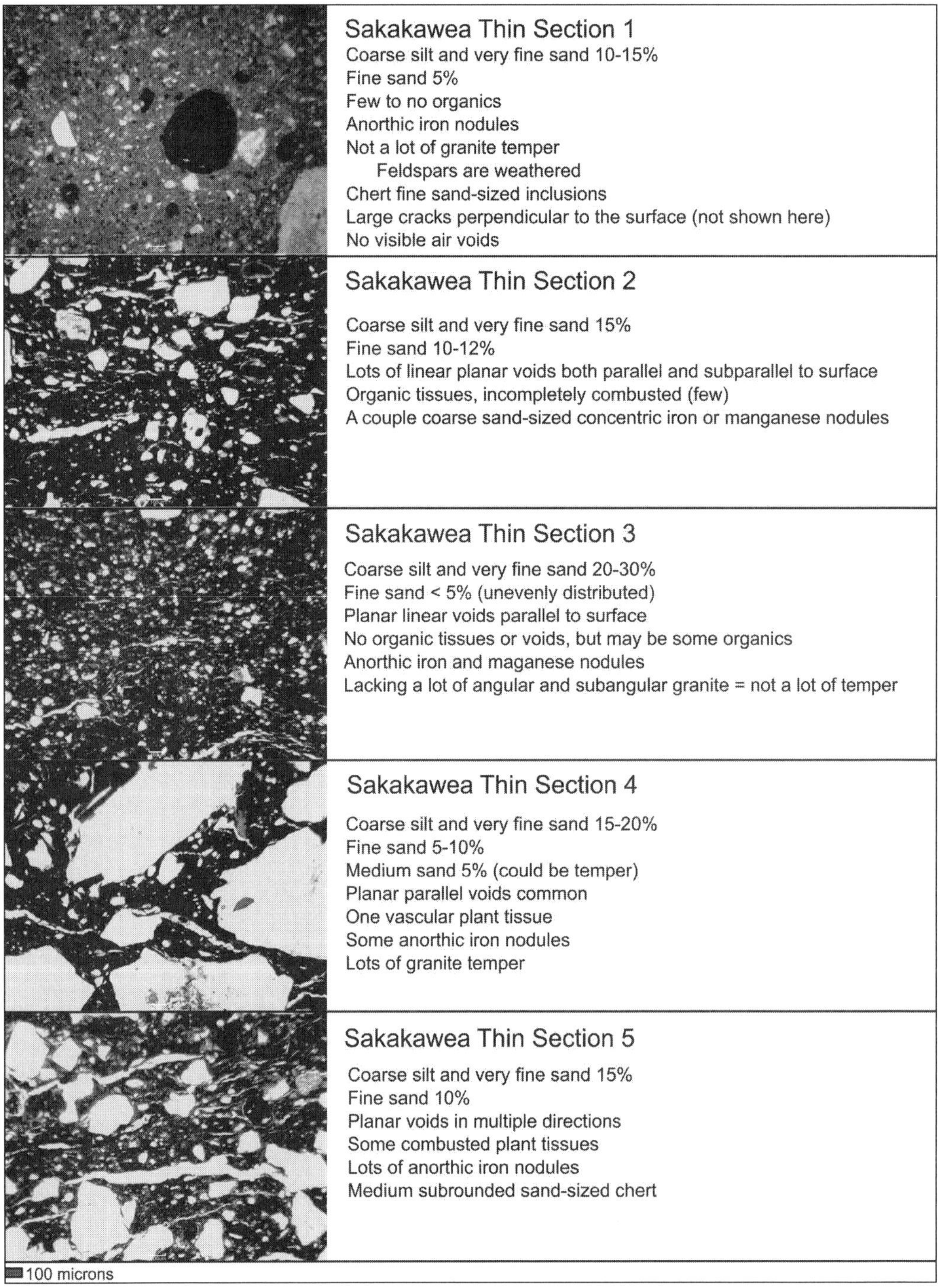

FIGURE 11.4. Thin sections of sherds from Sakakawea (32ME11) in plane polarized light (40×).

production in pottery after each epidemic (see table 11.2). Pottery production declined over time because of both the death of craft specialists in 1781–1782 and 1837–1838 and increased adoption of Euroamerican trade goods. It is important to note, however, that pottery production persisted through both epidemics with many continuities in practice and that pottery manufacture was observed on the reservation as late as the 1920s (Densmore 1923; Gilmore 1925). Despite the continuities in production, postepidemic pottery also differed in measurable ways.

Temper Acquisition: Inclusion Type and Shape

While it was hypothesized that potters might experiment with alternative tempers in post-epidemic contexts, all sherds examined here were tempered with pounded granite (quartz, feldspar, and mica). At 10× magnification granitic inclusions were angular to subangular with low sphericity, supporting the interpretation of intentionally crushed rock (see figs. 11.3 and 11.4). This dimension of the Awatixa ceramic technological tradition was stable: selection of temper before and after the epidemics did not change.

Temper Preparation: Inclusion Size, Density, and Sorting

It was also expected that grain size would increase over time, density would decrease, and sorting would become less even, in response to potters' decisions about time and labor allocation. These expectations were all supported.

Temper size increased over time (fig. 11.5). Pearson's Chi-Square confirms that this pattern is statistically significant ($p < 0.001$). The Awatixa data support an interpretation of decreased effort in temper preparation after the 1782 and 1837 epidemics. Lower Hidatsa, the pre-1782 epidemic site, exhibits a modal tendency toward fine temper (0.5–2.0 mm). After 1782 temper size shifts slightly upward in size mode, falling between 0.5–2.0 and 0.5–3.0 mm. Yet the most significant change occurred post-1845, when there were higher percentages of medium and coarse temper (0.5–5.0+ mm) relative to pre-1837 sites.

Percentage inclusions were recorded as 5, 10, 20, or 30 percent (Orton et al. 1993:238). Because of the low frequency of sherds with 30 percent temper, density classes were modified for statistical analysis: low, medium, and high (fig. 11.6). Examining trends in temper density by epidemic period reveals a decrease in vessels with high densities of temper and corresponding increases in medium and low density categories after 1782. After 1837 there were increases in both low and high densities, with the mode shifting toward low.

To examine these trends more closely, categories were created for each combination of inclusion size and density (low density, very fine temper; medium density, very fine temper; and high density, very fine temper). The assumption is that very fine and fine temper take more time and energy to produce than medium to coarse temper as more grinding is required. A corollary is that higher temper densities also require more effort. More temper required the collection of greater amounts of raw materials to process. The hypothesis is that medium and high densities of very fine and fine temper required the greatest amount of time and labor, whereas low densities of coarse temper require the least. Over time there should be a marked decrease in the former, with an increase in the latter.

Among the Awatixa very fine temper decreased overall over time, with decreases after 1782 and 1845 (fig. 11.7). For fine temper, a decrease in high density after 1782 was observed, but low and medium densities did not decrease until the move to Like-A-Fishhook in 1845. In contrast there was a general rise in medium-sized temper of all densities, with a dramatic increase in medium-size and medium density after 1782, correlating with the changes in fine temper. Coarse temper also slightly increased over time, with the biggest increase occurring after the move to Like-A-Fishhook. The increase in high-density coarse temper at Like-A-Fishhook was unexpected. Higher densities of large temper would be difficult to work and could cause structural problems, leading to increased breakage rates in the wares (Orton et al. 1993:115; Shepard

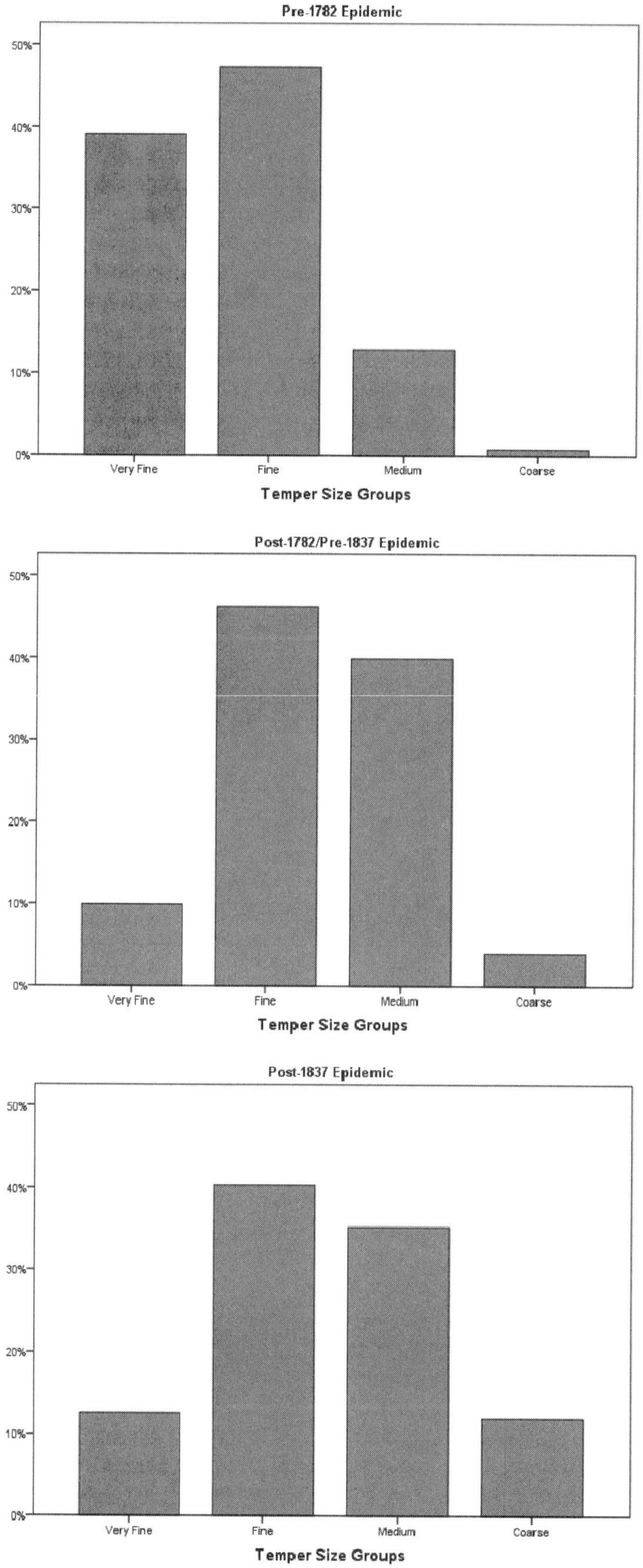

FIGURE 11.5. Temper size in body sherds from pre-1782, post-1782/pre-1837, and post-1837 epidemic Awatixa sites (very fine = 0.5–1.0 mm, fine = 0.5–2.0 mm, medium = 0.5–3.0 mm, and coarse = 0.5–4.0 mm or larger).

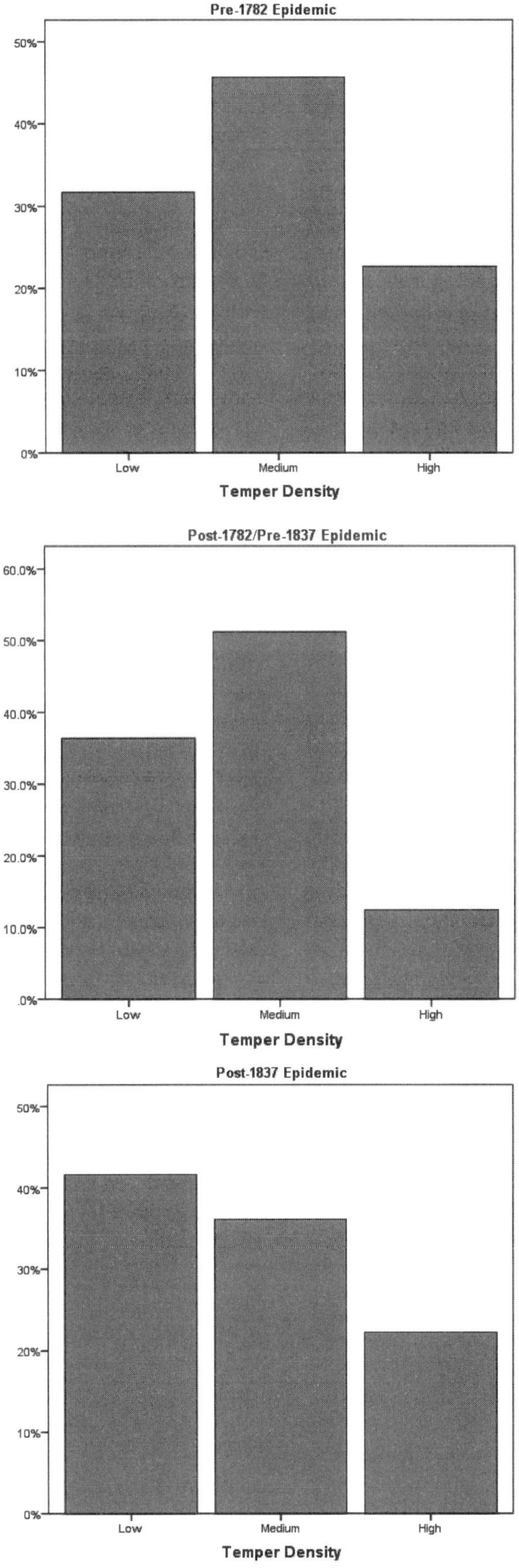

FIGURE 11.6. Temper density in body sherds from pre-1782, post-1782/pre-1837, and post-1837 epidemic Awatixa sites (low = 5 percent, medium = 10 percent, high = 20–30 percent).

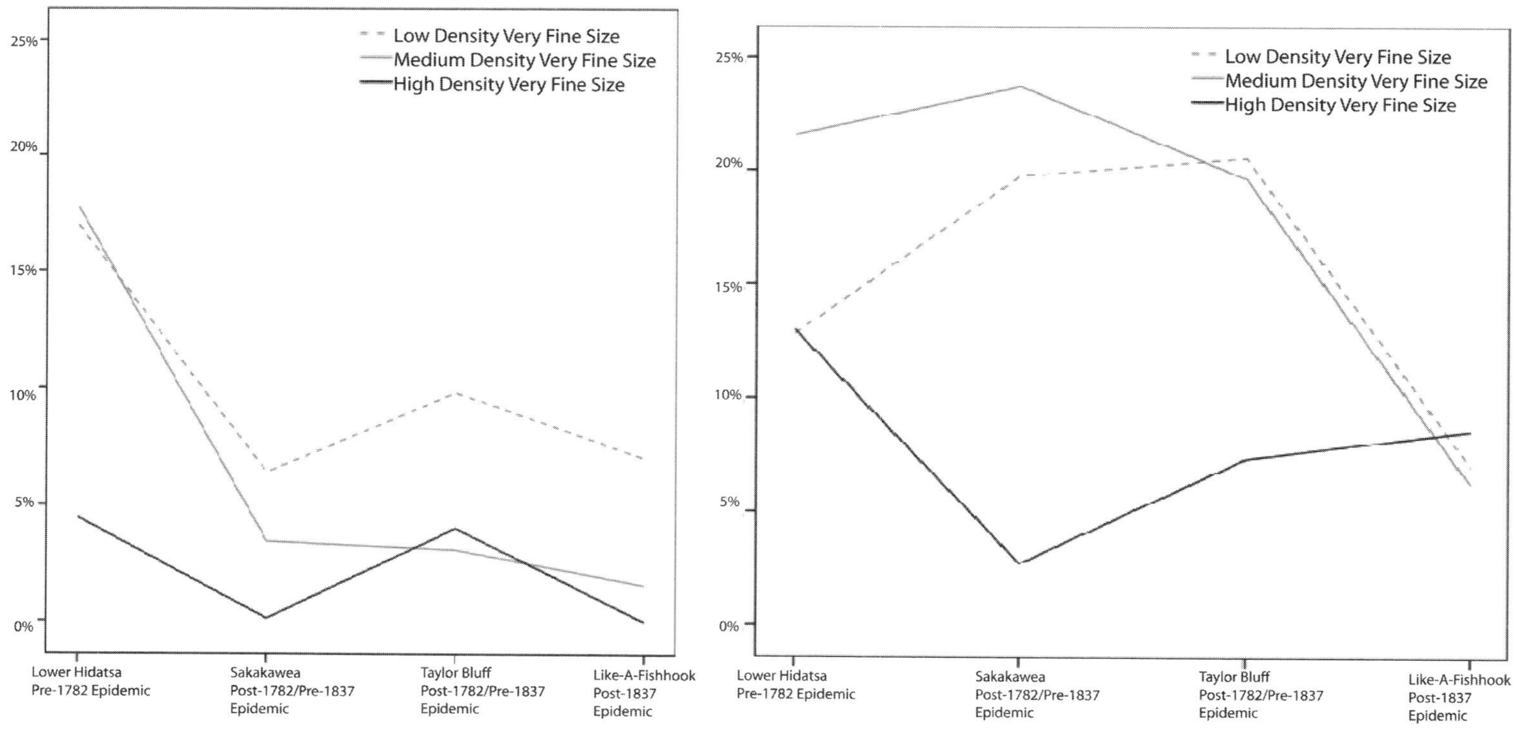

FIGURE 11.7. Temper size/density distributions in Awatixa body sherds.

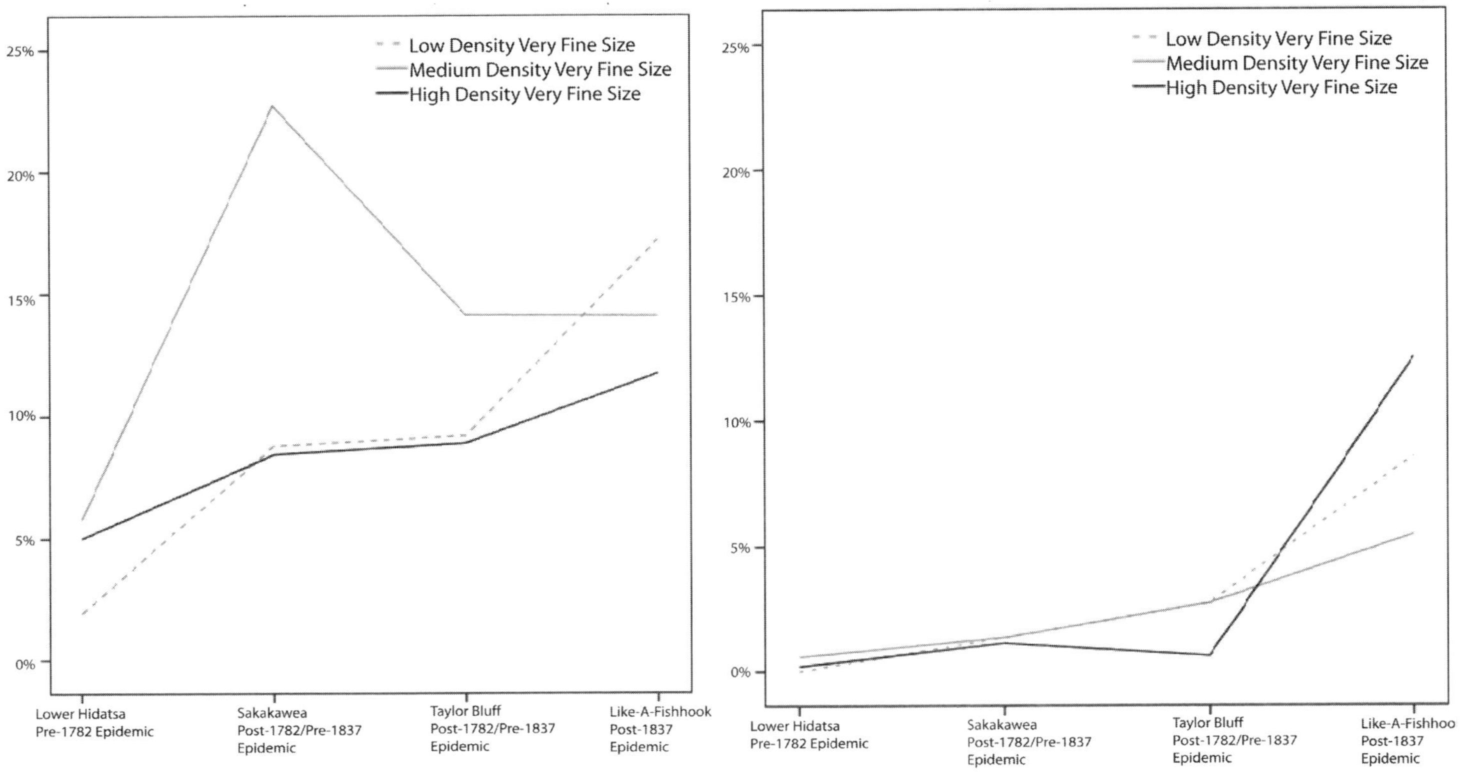

FIGURE 11.7. (cont'd.) Temper size/density distributions in Awatixa body sherds.

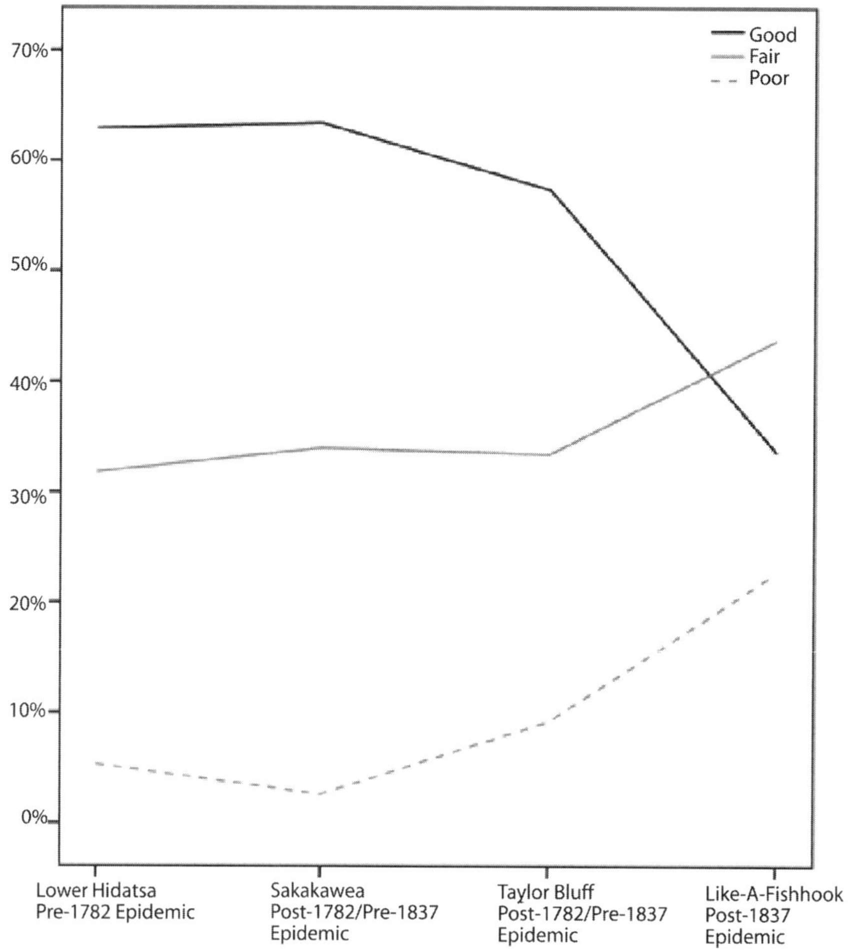

FIGURE 11.8. Temper sorting in body sherds from Awatixa sites.

1995:131). Yet greater amounts of temper decrease paste shrinkage as well as drying time (Orton et al. 1993:115).

Temper sorting reflects the evenness of the size of temper particles in a clay body (Orton et al. 1993:239), due to either the formation history of the temper deposits (sands) or temper preparation (length of time grinding or manually sorting). For ground materials, such as crushed sherd, shell, bone or rock, the assumption is that it takes a longer period of grinding to achieve evenly sorted inclusions. Alternatively, less time could be spent grinding materials if potters manually sorted temper prior to use. A number of methods could be used, including winnowing, sieving, or picking out larger fragments by hand. No historic or ethnographic evidence for these practices exists for the Hidatsa, but time and labor investments in both grinding and/or sorting are expected to decrease in postepidemic periods.

Little change in temper sorting took place between pre-1782 and post-1782 contexts (fig. 11.8). A change occurred circa 1837, however, with a slight decrease in "good" sorting with increases in "fair" and "poor." The most dramatic change occurred after 1837, with the move to Like-A-Fishhook Village. There was a decline in well-sorted temper, with the most pronounced increases in poorly sorted inclusions. Pearson Chi-Square indicates that this pattern is statistically significant ($p < 0.001$). These data are interpreted as supporting a conclusion that no pronounced reallocation of labor occurred after

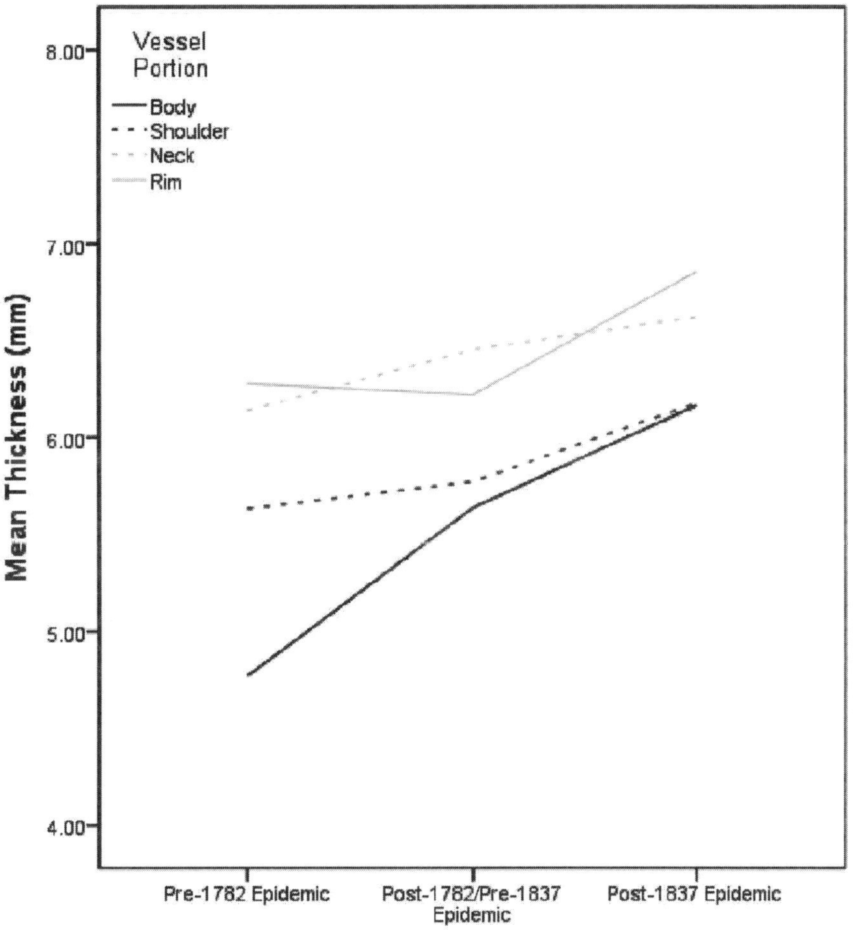

FIGURE 11.9. Mean thickness of Awatixa body, shoulder, neck, and rim sherds.

1782 in relation to temper grinding and sorting. Women spent the same amount of time on these activities as they had before the pandemic. But a significant change in behavior took place after 1837. From this time onward, women spent less time grinding and sorting than they had before the epidemic. Whether the changes observed at Like-A-Fishhook were related to the merging of technological traditions requires assessment of other pre-1845 Hidatsa, Mandan, and Arikara assemblages.

Discussion and Conclusion

Demographic disasters, such as large-scale epidemics, have five common impacts: depopulation, loss of cultural elements, effects on survivors, social and political changes, and population mobility and amalgamation (Smith 1987:58–60). For survivors, who may or may not be in a weakened state after an epidemic, the loss of other household and community members results in increased labor demands and fewer coping resources (for a detailed discussion see Hollenback 2012). This frequently leads to starvation among both sufferers and survivors (Smith 1987:59). Once an epidemic has passed, survivors must decide how to allocate their time and energy, as there are fewer people to undertake daily tasks and meet basic needs.

It was expected that the death of part-time craft specialists and the reallocation of Hidatsa labor after each epidemic would be visible in the ceramic technologies. Indeed, analysis of temper and other attributes such as vessel wall thickness (fig. 11.9) do indicate that women spent less time preparing and manufacturing traditional technologies in the periods after the epidemics, a process that Patricia Capone (2006:220) describes as an increase in "expediency." But this process was gradual, occurring over several decades. It is likely that some specialists survived each epidemic and thus that not all postepidemic potters were novices. The continued changes in Hidatsa pottery production probably resulted at first from the reallocation of women's time and labor and then from changes in expectations about the suitability and quality of "objects" driven by consumers and producers over time (Krause and Hollenback 2011).

The ubiquity and persistence of crushed, fire-cracked granite as temper were surprising. What explains the persistence of these materials? There are three explanations.

First, a number of potters likely survived each epidemic. The presence of well-made pots at all postepidemic sites could be explained as a result of curation of old pots or survival of experts.

Second, Hidatsa learning frameworks might not have been as "closed" and restrictive as early ethnographic accounts indicated. A survey of oral histories reveals that children and husbands helped potters collect raw materials (Bowers 1965:165; Cecilia Brown in Holland 1990:12; Malouf 2003:161). This detail is important because it may explain the continuity in certain ceramic attributes after the epidemics. Hidatsa pottery production was supposed to be a secret and sacred activity, yet the available information indicates that raw material collection and the location of procurement areas were not restricted knowledge. Family members assisted with this stage of the production sequence.

A final explanation is that fire-cracked granite may have been viewed as a sacred material integral to pottery production if it was indeed collected from sweatlodges, which are holy and culturally significant places. Capone (2006:229) notes several ethnographic cases in the American Southwest in which temper has symbolic meaning and cultural significance and imbues pots with sacredness. Indeed, pots themselves were seen as sacred objects among the Hidatsa (Hollenback 2012:321–328; Wilson 1910:289). The sacredness of both pots and temper and the use of pottery in daily household activities as well as ritual could explain the persistence.

One psychological impact of demographic disaster is ontological insecurity or a loss of the sense of order and continuity in one's life and identity (Giddens 1991:183–184; Laing 1965:39–61; Mellor 1993:12). Coping strategies for dealing with trauma attempt to restore a sense of normality and routine in life. As part of these strategies, what were once viewed as common practices and material objects take on new forms of significance (Enarson 2000; Kivland 2008). What may have been considered mundane or everyday activities prior to the onset of a disaster can become important parts of coping strategies after a hazardous event precisely because they are associated with social routines and can lead to reestablishment of ontological security. In traditional societies revitalization movements (Wallace 1956) and returns to tradition are common coping strategies for postdisaster ontological insecurity.

With the death of so many Hidatsa, especially in 1837–1838, came an overall loss of elements of culture, including the loss of "ancestral traditions," "traditional religious lore," and "genealogical lore" (Smith 1987:59–60). The diminution of this knowledge resulted in a process of "deculturation" or "simplification" (Saunders 2000:7–8). Depending on who in the community died, survivors may no longer have retained the knowledge of the appropriate ways to construct certain objects or perform certain rituals.

What becomes clear in the Awatixa case, however, is that potters tried to maintain many practices as forms of cultural coping. The textural analysis of temper has revealed both continuity and change in Hidatsa pottery production. The continuity may have been tied to the ritual significance of certain practices and objects; the changes most likely related to decisions about time and labor. Future studies of disaster-related technological change need to focus on the nuances of these processes if we are to understand how catastrophes affect the daily lives of survivors and their communities. This research would not have been possible without the combination of thin section analysis to establish a comparative baseline and the textural analysis of the sherds themselves to establish a large sample. Through this combined approach archaeologists can systematically explore behavioral changes related to pottery life histories that can be used to answer broader anthropological questions.

Acknowledgments

I thank Mary Ownby, Maria Masucci, and Isabelle Druc for inviting me to contribute to this volume. I owe Barbara Mills and Dick Krause a debt of gratitude for helping me think through concepts, methods, and arguments presented in this chapter. I also want to express my appreciation to T. J. Ferguson, Victoria Phaneuf, Christopher I. Roos, Michael B. Schiffer, and María Nieves Zedeño for commenting on earlier drafts. Funding for this research was provided by the Wenner-Gren Foundation and State Historical Society of North Dakota. Finally, I could not have completed this work without the assistance and support of individuals at the State Historical Society of North Dakota, the Knife River Indian Villages National Historic Site, and members of the Mandan, Hidatsa, and Arikara Nation.

References Cited

Ahler, Stanley A.
1988 *Archeological Mitigation at Taylor Bluff Village (32ME366), Knife River Indian Villages National Historic Site.* Department of Anthropology, University of North Dakota, Grand Forks.

Ahler, Stanley A., and Bruce F. Benz
1980 *Analysis of Controlled Surface Artifact Collections from the Sakakawea (32ME11) and Lower Hidatsa (32ME10) Village Sites.* Department of Anthropology, University of North Dakota, Grand Forks.

Ahler, Stanley A., Carl R. Falk, and Paul R. Picha
1983 *Archeological Mitigation for Small-Scale Construction Projects at the Taylor Bluff Site (32ME366), Knife River Indian Villages National Historic Site.* Department of Anthropology, University of North Dakota, Grand Forks.

Ahler, Stanley A., and Anthony A. Swenson
1993 KNRI and Upper Knife-Heart Region Pottery Analysis. In *The Phase I Archeological Research Program for the Knife River Indian Villages National Historic Site, Part III: Analysis of Physical Remains*, edited by Thomas D. Thiessen, pp. 1–172. National Park Service, Midwest Archeological Center, Lincoln, Nebraska.

Ahler, Stanley A., Thomas D. Thiessen, and Michael K. Trimble
1991 *People of the Willows: The Prehistory and Early History of the Hidatsa Indians.* University of North Dakota Press, Grand Forks.

Ahler, Stanley A., and Timothy Weston
1981 *Test Excavations at Lower Hidatsa Village (32ME10), Knife River Indian Villages National Historic Site.* Department of Anthropology, University of North Dakota, Grand Forks.

Ahler, Stanley A., Timothy Weston, and Keith D. McMiller
1980 *Cutbank Profiling and Test Excavations at Sakakawea Village (32ME11), Knife River Indian Villages National Historic Site.* Department of Anthropology, University of North Dakota, Grand Forks.

Bowers, Alfred W.
1965 *Hidatsa Social and Ceremonial Organization.* Smithsonian Institution Bureau of American Ethnology Bulletin 194. U.S. Government Printing Office, Washington, D.C.

Capone, Patricia
2006 Rio Grande Glaze Ware Technology and Production: Historic Expediency. In *The Social Life of Pots*, edited by Judith A. Habicht-Mauche, Suzanne L. Eckert, and Deborah L. Huntley, pp. 216–231. University of Arizona Press, Tucson.

Chardon, Francis A.
1997 *Chardon's Journal at Fort Clark, 1834–1839.* Edited by Annie H. Abel. University of Nebraska Press, Lincoln.

Chilton, Elizabeth S.
1996 Embodiments of Choice: Native American Ceramic Diversity in the New England Interior. Ph.D. dissertation, Department of Anthropology, University of Massachusetts, Amherst.

Crosby, Alfred
1976 Virgin Soil Epidemics as a Factor in the Aboriginal Depopulation in America. *William and Mary Quarterly* 33(2):289–299.
Curtis, Edward S.
1909 *The North American Indian, Volume 4*. University Press, Cambridge, MA.
Darvill, Tim, and Jane Timby
1982 Textural Analysis: A Review of Potentials and Limitations. In *Current Research in Ceramics: Thin-Section Studies*, edited by Ian Freestone, Catherine Johns, and Tim Potter, pp. 73–87. Occasional Paper No. 32. British Museum, London.
Densmore, Frances
1923 *Mandan and Hidatsa Music*. Smithsonian Institution Bureau of American Ethnology Bulletin 80. U.S. Government Printing Office, Washington, D.C.
Dobyns, Henry F.
1983 *Their Number Become Thinned: Native American Population Dynamics in Eastern North America*. University of Tennessee Press, Knoxville.
1993 Disease Transfer at Contact. *Annual Review of Anthropology* 22:273–291.
Enarson, Elaine
2000 "We Will Make Meaning Out of This": Women's Cultural Responses to the Red River Valley Flood. *International Journal of Mass Emergencies and Disasters* 18(1): 39–62.
Fenn, Elizabeth A.
2001 *Pox Americana: The Great American Smallpox Epidemic of 1775–82*. Hill and Wang, New York.
Giddens, Anthony
1991 *Modernity and Self-Identity: Self and Society in the Late Modern Age*. Stanford University Press, Stanford.
Gilmore, Melvin
1925 Arikara Uses of Clay and Other Earth Products. *Museum of the American Indian, Heye Foundation, Indian Notes* 2(4):283–289.
Hartle, Donald D.
1960 Rock Village: An Ethnohistorical Approach to Hidatsa Archaeology. Ph.D. dissertation, Columbia University, New York City.
Holland, Erik
1990 Oral Histories with Mandan, Hidatsa, and Arikara Members of the Fort Berthold Indian Reservation. Transcripts on file, Knife River Indian Villages National Historic Site, Stanton, North Dakota.
Hollenback, Kacy L.
2012 Disaster, Technology, and Community: Measuring Responses to Smallpox Epidemics in Historic Hidatsa Villages, North Dakota. Ph.D. dissertation, School of Anthropology, University of Arizona, Tucson.
Hollenback, Kacy L., and Michael Brian Schiffer
2010 Technology and Material Life. In *The Oxford Handbook of Material Culture Studies*, edited by Dan Hicks and Mary C. Beaudry, pp. 313–332. Oxford University Press, Oxford.
Jenkinson, Clay S.
2003 *A Vast and Open Plain: The Writings of the Lewis and Clark Expedition in North Dakota, 1804–1806*. State Historical Society of North Dakota, Bismarck.
Kivland, Chelsey Louise
2008 Hero, Eulogist, Trickster, and Critic: Ritual and Crisis in Post-Katrina Mardi Gras. In *Seeking Higher Ground*, edited by Manning Marable and Kristen Clarke, pp. 107–127. Palgrave Macmillan, New York.
Krause, Richard A.
1972 *The Leavenworth Site: Archaeology of an Historic Arikara Community*. University of Kansas Publications in Anthropology No. 3. University of Kansas, Lawrence.
2007 A Potter's Tale. In *Plains Village Archaeology*, edited by Stanley A. Ahler and Marvin Kay, pp. 32–40. University of Utah Press, Salt Lake City.
Krause, Richard A., and Kacy L. Hollenback
2011 A Brief Study of Arikara Ceramic Change. Paper presented at the 69th Annual Meeting of the Plains Anthropological Society. Tucson, Arizona.
Laing, Ronald D.
1965 *The Divided Self: An Existential Study in Sanity and Madness*. Penguin Books, Baltimore.
Lehmer, Donald J., W. Raymond Wood, and C. L. Dill
1978 *The Knife River Phase*. Department of Anthropology and Sociology, Dana College, Blair, Nebraska, and University of Missouri, Columbia.
Malouf, Carling I.
2003 The Native American, Bear-In-The-Water

(Hidatsa Warrior). In *People of the Buffalo: The Plains Indians of North America, Military Art, Warfare and Change: Essays in Honor of John C. Ewers*, vol. 1, edited by Colin F. Taylor and Hugh A. Dempsey, pp. 161–167. Tatanka Press, Wyk auf Foehr, Germany.

Matthews, Washington

1877 *Ethnography and Philology of the Hidatsa Indians.* Miscellaneous Publications of the U.S. Geological and Geographical Survey No. 7. U.S. Government Printing Office, Washington, D.C.

Mellor, Philip

1993 Death in High Modernity: The Contemporary Presence and Absence of Death. In *The Sociology of Death: Theory, Culture, Practice*, edited by David Clark, pp. 11–30. Blackwell Publishers, Oxford.

Middleton, Andrew P., Ian C. Freestone, and Morven N. Leese

1985 Textural Analysis of Ceramic Thin Sections: Evaluation of Grain Sampling Procedures. *Archaeometry* 27(1):64–74.

Mitchell, Mark David

2011 Continuity and Change in the Organization of Mandan Craft Production, 1400–1750. Ph.D. dissertation, Department of Anthropology, University of Colorado, Boulder.

Orton, Clive, Paul Tyers, and Alan Vince

1993 *Pottery in Archaeology.* Cambridge University Press, Cambridge.

Peacock, David P. S.

1971 Petrography of Certain Coarse Pottery. In *Excavations at Fishbourne 1961–1969*, edited by Barry Cunliffe, pp. 255–259. Reports of the Research Committee of the Society of Antiquaries of London No. 27, Vol. 2. W. S. Maney and Son, London.

Ramenofsky, Ann F.

1987 *Vectors of Death: The Archaeology of European Contact.* University of New Mexico Press, Albuquerque.

Rice, Prudence M.

1987 *Pottery Analysis: A Sourcebook.* University of Chicago Press, Chicago.

Robertson, Roland G.

2001 *Rotting Face: Smallpox and the American Indian.* Caxton Press, Caldwell, Idaho.

Saunders, Rebecca

2000 *Stability and Change in Guale Indian Pottery, AD 1300–1702.* University of Alabama Press, Tuscaloosa.

Schiffer, Michael Brian

1972 Archaeological Context and Systemic Context. *American Antiquity* 37:156–165.

Shepard, Anna O.

1995 *Ceramics for the Archaeologist.* 5th ed. Reprinted. Braun-Brumfield, Ann Arbor, Michigan.

Smith, G. Hubert

1972 *Like-a-Fishhook Village and Fort Berthold, Garrison Reservoir, North Dakota.* National Park Service, Washington, D.C.

Smith, Marvin T.

1987 *Archaeology of Aboriginal Culture Change in the Interior Southeast: Depopulation during the Early Historic Period.* University Presses of Florida, Gainesville.

Stewart, Frank H.

2001 Hidatsa. In *Handbook of North American Indians,* vol. 13 (part 1), edited by Raymond J. DeMallie, pp. 329–348. Smithsonian Institution, Washington, D.C.

Streeten, Anthony D. F.

1982 Textural Analysis: An Approach to the Characterization of Sand-Tempered Ceramics. In *Current Research in Ceramics: Thin-Section Studies*, edited by Ian Freestone, Catherine Johns and Tim Potter, pp. 123–134. Occasional Paper No. 32. British Museum, London.

Thiessen, Thomas D.

1993 Early Explorations and the Fur Trade at Knife River. In *The Phase I Archeological Research Program for the Knife River Indian Villages National Historic Site, Part II: Ethnohistorical Studies*, edited by Thomas D. Thiessen, pp. 29–46. National Park Service, Midwest Archeological Center, Lincoln, Nebraska.

Thornton, Russell

1987 *American Indian Holocaust and Survival: A Population History since 1492.* University of Oklahoma Press, Norman.

1997 Aboriginal North American Population and Rates of Decline, ca. A.D. 1500–1900. *Current Anthropology* 38(2):310–315.

Thornton, Russell, Tim Miller, and Jonathan Warren

1991 American Indian Population Recovery following Smallpox Epidemics. *American Anthropologist* 93(1):28–45.

Trimble, Michael K.
1985 Epidemiology on the Northern Plains: A Cultural Perspective. Ph.D. dissertation, Department of Anthropology, University of Missouri–Columbia.
1988 Chronology of Epidemics among Plains Village Horticulturalists: 1738–1838. *Southwestern Lore* 54:4–31.
1989 Infectious Disease and the Northern Plains Horticulturists: A Human Behavior Model. *Plains Anthropologist* 34(124, part 2: memoir 23):41–59.
1994 The 1837–1838 Smallpox Epidemic on the Upper Missouri. In *Skeletal Biology in the Great Plains: Migration, Warfare, Health, and Subsistence*, edited by Douglas W. Owsley and Richard L. Jantz, pp. 81–89. Smithsonian Institution Press, Washington, D.C.

Wallace, Anthony F. C.
1956 Revitalization Movements. *American Anthropologist* 58(2):264–281.

Wallaert-Pêtre, Hélène
2001 Learning How to Make the Right Pots: Apprenticeship Strategies and Material Culture, a Case Study in Handmade Pottery from Cameroon. *Journal of Anthropological Research* 57(4):471–493.

Wedel, Waldo R.
1957 Observations on Some Nineteenth Century Pottery Vessels from the Upper Missouri. In *Smithsonian Institution Bureau of American Ethnology Bulletin* 164, *Anthropological Papers* 51:87–114.

Weitzner, Bella (ed.)
1979 Notes on the Hidatsa Indians Based on Data Recovered by the Late Gilbert L. Wilson. *Anthropological Papers of the American Museum of Natural History* 56 (part 2): 181–322.

Wilson, Gilbert L.
1910 General Report. Manuscript on file, Gilbert L. and Frederick N. Wilson Papers, Minnesota Historical Society, St. Paul.
1912 Report. Manuscript on file, Gilbert L. and Frederick N. Wilson Papers, Minnesota Historical Society, St. Paul.

Wilson, Gilbert L., W. Raymond Wood, and Donald J. Lehmer
1977 Mandan and Hidatsa Pottery Making. *Plains Anthropologist* 22(76, part 1):97–105.

Witte, Stephen S., and Marsha V. Gallagher (eds.)
2012 *The North American Journals of Prince Maximilian of Wied: Volume III, September 1833–August 1834*. University of Oklahoma Press, Norman.

Wood, W. Raymond
1980 Plains Trade in Prehistoric and Protohistoric Intertribal Relations. In *Anthropology on the Great Plains*, edited by W. Raymond Wood and Margot Liberty, pp. 98–109. University of Nebraska Press, Lincoln.
1986 *The Origins of the Hidatsa Indians: A Review of the Ethnohistorical and Traditional Data*. Reprints in Anthropology 32. J & L Reprint Company, Lincoln, Nebraska.
2010 The Earliest Map of the Mandan Heartland: Notes of the Jarvis and MacKay 1791 Map. *Plains Anthropologist* 55(216):255–276.

Wood, W. Raymond, and Thomas D. Thiessen (eds.)
1985 *Early Fur Trade on the Northern Plains: Canadian Traders among the Mandan and Hidatsa Indians, 1738–1818: The Narratives of John MacDonnell, François-Antoine Larocque, David Thompson, and Charles McKenzie*. University of Oklahoma Press, Norman.

CHAPTER 12

Field-Based Experiments Replicating Ceramic Fabrics

Late Bronze Age Cookwares from Two Mediterranean Sites

Jerolyn E. Morrison *and* Mara T. Horowitz

Experiments designed to replicate ancient ceramic fabrics are valuable for testing hypotheses about the relationships between locally available potting materials (such as clay and temper) and the ancient production of ceramics (Moody et al. 2012:121, 122). The method employed in this study combines macroscopic and petrographic descriptions of ancient ceramics with the knowledge gained from the experience of producing pottery with raw materials. The result is a better understanding of production as a series of choices made by the ancient potters within the framework of available resources (Morrison 2010; Morrison and Archambeau 2004).

This chapter presents two case studies where samples of archaeological ceramics and raw materials from the immediate vicinity were examined macroscopically for similarities and raw materials judged to be consistent with the ancient fabrics were tested via replication experiments. A subsequent petrographic study characterizing and comparing the archaeological and experimental material tested the macroscopic characterizations done in the field and found that the experimental fabrics were indeed very similar to the archaeological originals.

These experiments were carried out at two locations: Mochlos, a Minoan coastal settlement in eastern Crete, and Tell Atchana (ancient Alalakh), located in the Amuq Valley on the Orontes River at the junction of Anatolia and the Levant (fig. 12.1). The versatility and strength of fabric replication experiments combined with petrographic analysis have contributed greatly to the ongoing study of ancient ceramic industries at these Bronze Age sites and allowed a clearer picture of how ancient potters could have engaged with the potting materials in their respective landscapes.

The Archaeological Sites

The sites of Mochlos and Tell Atchana have between them over a century of archaeological exploration (Seager 1909; Soles and Davaras 1992, 1994, 1996; Woolley 1955; Yener 2005, 2008, 2010a, 2010b). In 1908 Richard Seager embarked on the first explorations at Mochlos, revealing a series of wealthy Early Minoan (EM, before 3000–ca. 2000 BC) house tombs and portions of a long-lived Early to Late Minoan (before 3000–ca. 1500 BC) settlement on a small island that was once connected to the Cretan mainland by a narrow isthmus (Soles and Davaras 1992:413). An important discovery at Mochlos is the Late Minoan IB (LM IB, ca. 1600–1450 BC) Artisans' Quarter located on the coast directly across from the island and behind the modern village toward the Limenaria Cove (Soles 2003). Evidence for

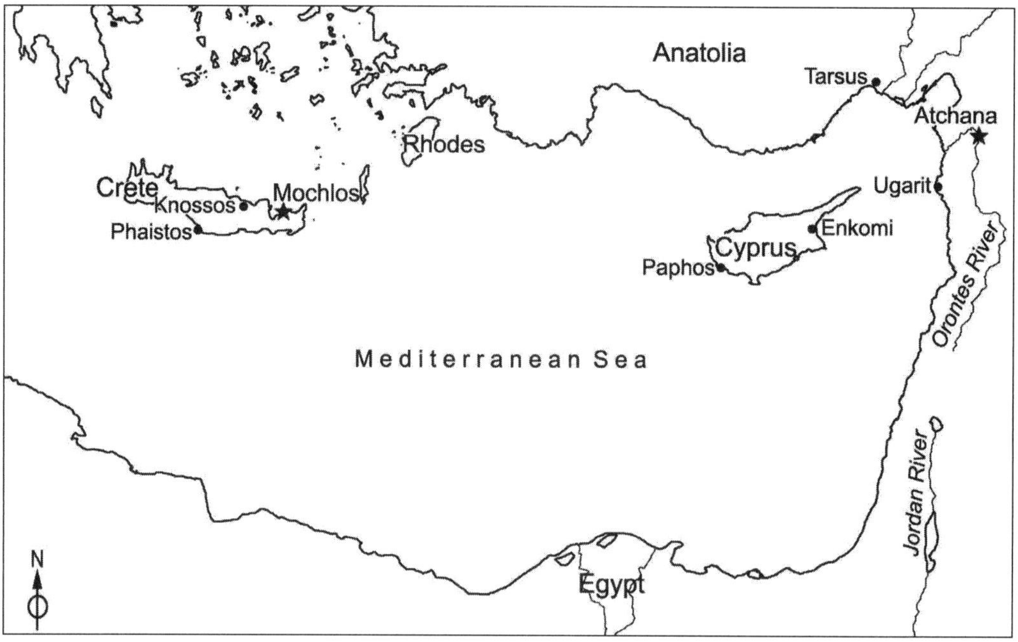

FIGURE 12.1. Mediterranean Sea and locations of Mochlos (Crete, Greece) and Tell Atchana (Hatay, Turkey) (prepared by Mara Horowitz).

large-scale pottery production was unearthed within this craft compound, including two kilns, several potters' wheels, and a proposed potters' pit where the ancient potters installed a wheel to produce their wares (Evely and Morrison 2010; Morrison and Park 2007:6, figure 1; Soles 2003:36–40, figures 20–22). Evidence from this sector along with supplementary material from the Neopalatial (Middle Minoan III–Late Minoan IB, ca. 1700–1450 BC) town located on the island was investigated during the replication experiments.

Across the eastern Mediterranean Sea, Sir Leonard Woolley began exploring Tell Atchana, the site of the Bronze Age city of Alalakh, in the 1930s (Woolley 1955). Alalakh was once the capital of a small principality of the Middle Bronze Age (MBA, ca. 2000–1600 BC) and Late Bronze Age (LBA, ca. 1600–1200 BC), sometimes referred to in contemporary texts as the Kingdom of Mukish (Yener 2010a:4). Throughout the city's 800-year lifespan, it presided over an important geographical crossroads linking Mesopotamia to Anatolia, the Levant, and the Mediterranean Sea. A long chronological sequence of massively built mud-brick palaces, temples, and fortifications has been unearthed. Among the new excavations is a LB I (ca. 1600–1400 BC) pottery workshop, identified by kilns and piles of ceramic wasters adjacent to a space located outside the palace precinct along the eastern edge of the tell (Area 3).

Cookware, Clays, and Regional Geology

Cookwares form a unique category within ceramic typologies due to the necessity that they have a resistance to thermal shock (Rice 1987: 228–232). Cross-culturally, cookwares are made in a variety of shapes, sizes, and fabrics. Potters produced cookwares either by using clay that had natural thermal dynamic properties or by creating a thermal dynamic clay body by mixing organic and/or inorganic tempering materials into one or more clays. Cookwares are of particular interest to the archaeologist, as they are intimately related to foodways. Cultural traditions, economic necessities, status considerations, and ecological parameters may all be examined from an interdisciplinary study of cookware that brings together specialists on ce-

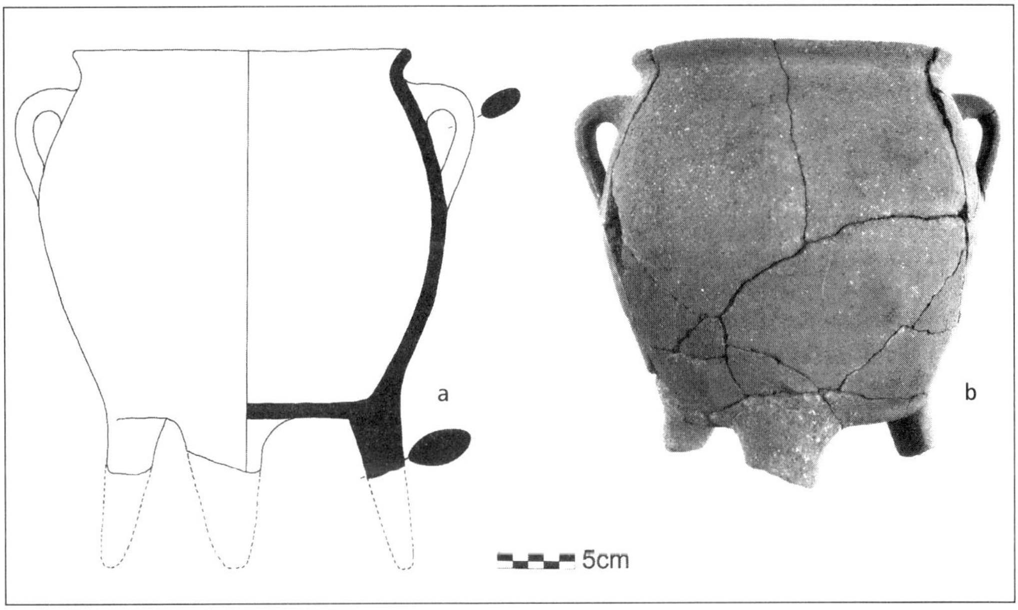

FIGURE 12.2. Mochlos LM I tripod cooking pots (after Barnard and Brogan 2003, figure 47): (*a*) line drawing; (*b*) photograph.

ramics (including modern potters), anthropology, archaeometry (such as residue analysis), and geoarchaeology (such as petrography).

At Mochlos ceramic specialists have established Bronze Age ceramic typologies that place the archaeological cookwares within the Neopalatial (Barnard and Brogan 2003) and the LM II-III periods (ca. 1450–1150 BC) (Smith 2010). Elongated and globular tripod cooking pots are a common Neopalatial and LM II-III cook pot type (fig. 12.2; Barnard and Brogan 2003:80–82, plate 25, figures 47–49; Smith 2010:114, 115, plate 33, figures 82, 83). A macroscopic and petrographic study of the Neopalatial cookware recovered from the town and the cookwares from the Artisans' Quarter (collaboration between Morrison and Eleni Nodarou, 2006–present) indicates that four recipes were used: (1) main phyllite, (2) semicoarse metamorphic, (3) semicoarse metamorphic with mica (fig. 12.3), and one vessel produced with a sedimentary fabric dominated by quartz and sandstone (table 12.1). The subrounded and subangular shapes of the 2–3 mm metamorphic and quartz inclusions within the coarse Neopalatial fabrics suggest that they were not crushed and added to the clay but occurred naturally. Macroscopic and petrographic analysis of the four cookware fabrics indicates that their primary components are consistent with the East Cretan Phyllite-Quartzite series (Barnard 2003; Day et al. 2003; Nodarou 2003, 2010; Smith 2010; collaboration between Morrison and Nodarou 2006–present).

The Mochlos plain lies in a tectonic valley flanked on the eastern and western side by the Ornos Mountain range (Nodarou 2010:4; Soles 2003:155, figure 11.1). The plain consists of crystalline limestone, with the southern side dominated by the Phyllite-Quartzite series of Permian-Triassic age (fig. 12.4). This series is characterized by dark gray, greenish, or purple phyllites with intercalations of sandstone, quartzite, limestone, and dolomite (Nodarou 2010:4, 5). The plain is also composed of Pleistocene fluvial deposits containing gravels, pebbles, and red sands along with Miocene marine formations (marls and sandstones) (Nodarou 2010:4). As demonstrated by work with sourcing materials at Mochlos for mud-brick production (Nodarou et al 2008), the closest exposed material of this nature is located approximately 1 km east of the Artisans' Quarter in the Limenaria Cove.

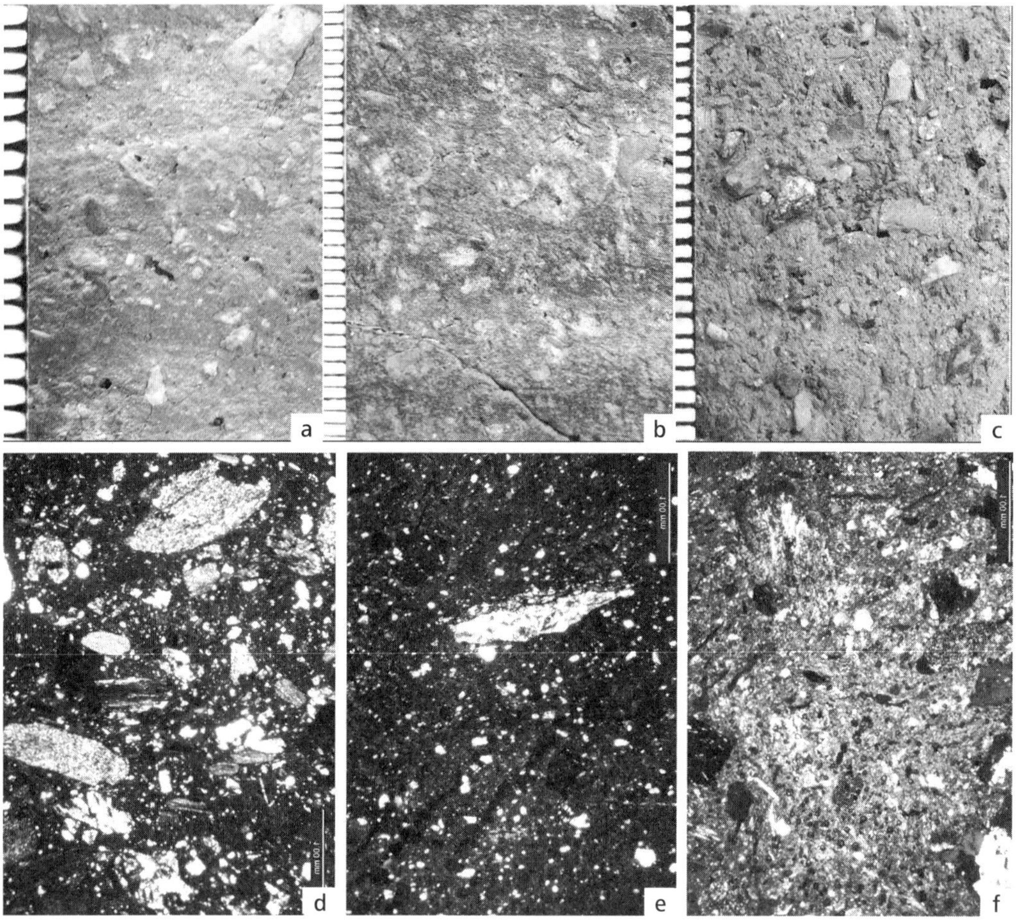

FIGURE 12.3. Mochlos LM I cookwares: (*a*) main phyllite, (*b*) semicoarse metamorphic, (*c*) semicoarse metamorphic with mica. Scale in mm. Photographs by Jerolyn Morrison. Photomicrographs (XPL): (*d*) main phyllite, (*e*) semicoarse metamorphic, (*f*) semicoarse metamorphic with mica. Field of view 4 mm.

Cookwares within the Tell Atchana ceramic assemblage form a distinct category in the site's typology. The most common form is a round-bottomed cooking jar or cook pot with a slight neck and folded or flared rim (fig. 12.5). It was produced in multiple sizes with rim diameters ranging from ca. 20 to 35 cm. While the cook pot shape changed very little through time, there is a marked difference between the earlier and later fabrics. Macroscopic and petrographic analyses have revealed that potters in the Middle Bronze levels produced cook pots using a coarse ferrous fabric, tempered with crushed clear calcite and fired to a mottled dark brown-black color. The exterior surfaces bear the distinctive marks of smoothing with water: rag marks, drips, and the separation of finer particulates of clay on the surface of the pot. Named Mineral Ware for its temper, this cookware fabric has not yet been the subject of experiments, because the search for the source of both the clay body and the calcite is ongoing.

In the transition to and early part of LB I (ca. 1625–1575 BC), a new cookware was introduced that had a medium-fine, calcareous, micritic clay body (fig. 12.6d), similar to the typical plainware ceramics of the site, but heavily tempered with crushed shell (see table 12.1). The color of the vessels ranges from peach-pink to tan-yellow (fig. 12.6a and b), with variable sand-

TABLE 12.1. Macroscopic Ceramic Fabric Analysis descriptions of the Mochlos and Atchana archaeological cookwares

	Mochlos			Tell Atchana	
Fabric name	main phyllite	semicoarse metamorphic	semicoarse metamorphic with mica	crushed-calcite mineral	crushed-shell
Texture	coarse	coarse	coarse	coarse	very coarse
Inclusion sorting	moderate	poor	poor	moderate	moderate
Inclusion size	2–3 mm	< 2 mm	< 2 mm	< 2 mm	2–4 mm
Inclusion shape	subrounded, subangular	subrounded, subangular	subrounded, subangular	angular	angular
Inclusion type	red-purple phyllite (+); milk-white quartz (-)	red-purple phyllite (+); milk-white quartz (+)	red-purple phyllite (+); milk-white quartz (-); silver mica (-)	clear calcite mineral (+)	gray-bluish shell (+)
Munsell (2001)	red, 2.5yr 5/6	red, 2.5yr 5/6	red, 2.5yr 5/6	pale red, 2.5yr 7/2	very pale brown and light yellowish brown, 10yr 7-6/4; reddish yellow, 5yr 7/6
Authors' color name	dark red	dark red	dark red	red-gray	pale tan-peach

Note: (+) = quantity of inclusions present within the fabric matrix (Kemp 1985:17): (1) predominant (> 70%), (2) dominant (50–70%), (3) frequent (30–50%), (4) common (15–30%); (-) = quantity of inclusions present within the fabric matrix (Kemp 1985:17): (1) few (5–15%), (2) very few (2–5%), (3) rare (0.5–2%), (4) very rare (< 0.5%).

iness, and water-smoothing was still applied (fig. 12.6c). The two fabrics (Mineral Ware and Shellware) existed side-by-side for a brief period in early LB I, after which Shellware became preferred and the Mineral Ware disappeared.

The reason for the change from a fabric with crushed calcite temper to a fabric with shell temper is not yet understood. Historical evidence reveals that the levels of Amuq Lake can change rapidly and dramatically (Wilkinson 2000; Yener 2005). It is not yet known if the advent of shell-tempered cookware coincided with a change in the environment that might have resulted in an increase in freshwater mussel availability (proximity due to rising lake and marsh levels) and possibly the consumption of mussels at Alalakh and the smaller sites in the Amuq Valley where Shellware is also found. Textual evidence from the seventeenth century BC palace (Level VII), (Woolley 1955) details how Alalakh and these subsidiary sites formed a political, cultural, and economic system (Magness-Gardiner 1994) but unfortunately makes no mention of mussels or changes in the lake. The experiments detailed in this chapter included preliminary experiments to explore possible evidence for cooking of the shells. Strontium analysis of local human skeletal remains is planned to search for changes from a terrestrial to a marine diet.

Tell Atchana's geological surroundings are complex (fig. 12.7). The Orontes, Afrin, and Kara Su Rivers drain a large catchment of variable geology into the Amuq basin, historically dominated by marshlands and a large shallow lake (Lake Antioch). The various river catchments include a variety of sedimentary series from the Mesozoic and Cenozoic eras, an ophiolite sequence, intrusive and extrusive basaltic flows, and even a Cambrian limestone formation (Ministry of Petroleum and Mineral Resources 1986; General Directorate of Mineral Research and Exploration 2002). Tectonic uplift along the Dead Sea and Anatolian fault lines in the Cenozoic era created the Amuq basin, a flat plain with a shallow lake that expands and contracts due to climatic shifts and changes location in the plain

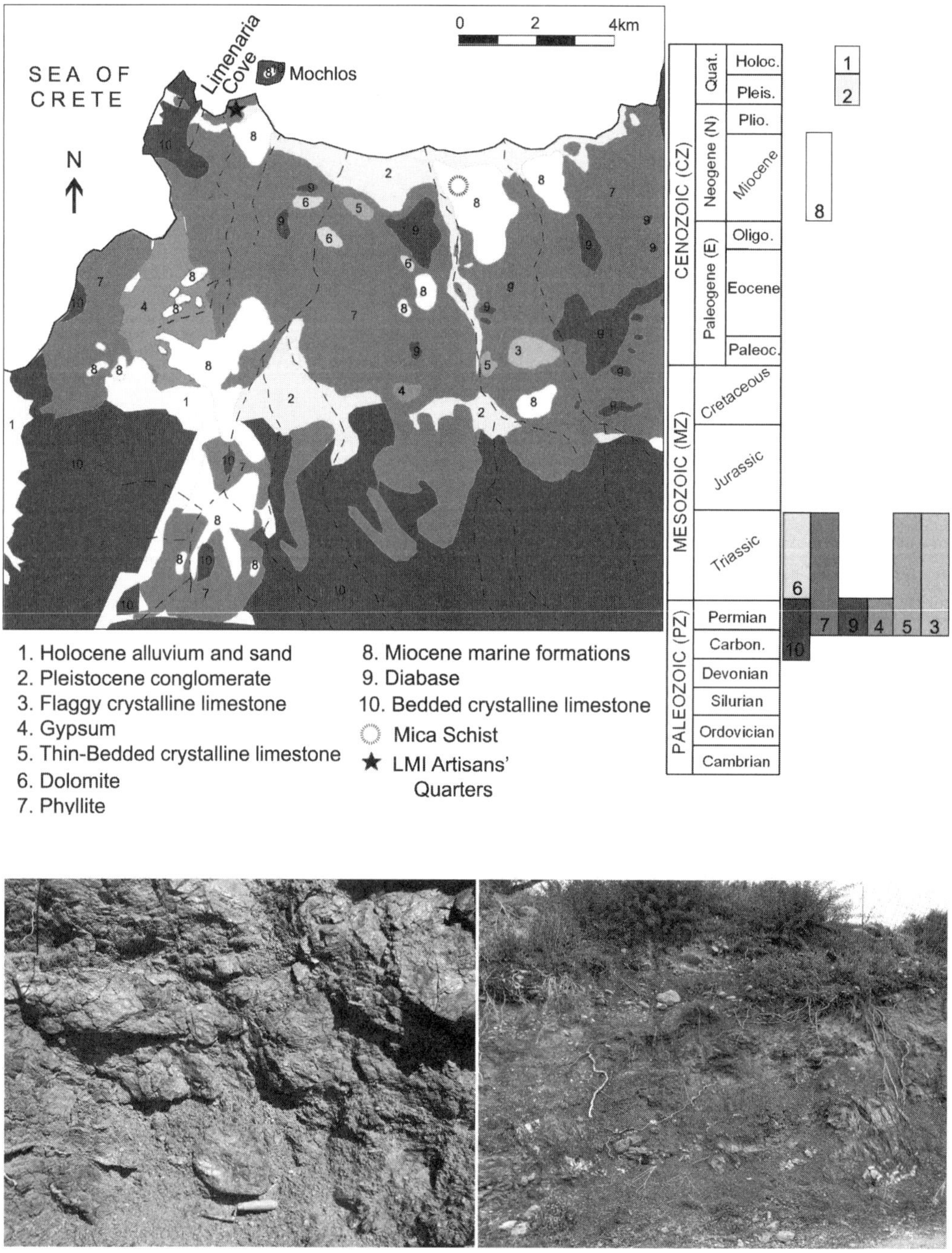

FIGURE 12.4. Geological map of the Mochlos area, eastern Crete (illustration by Mara Horowitz after Papastamatiou et al. 1959). *Below*: photographs by Jerolyn Morrison of weathered silver mica schist.

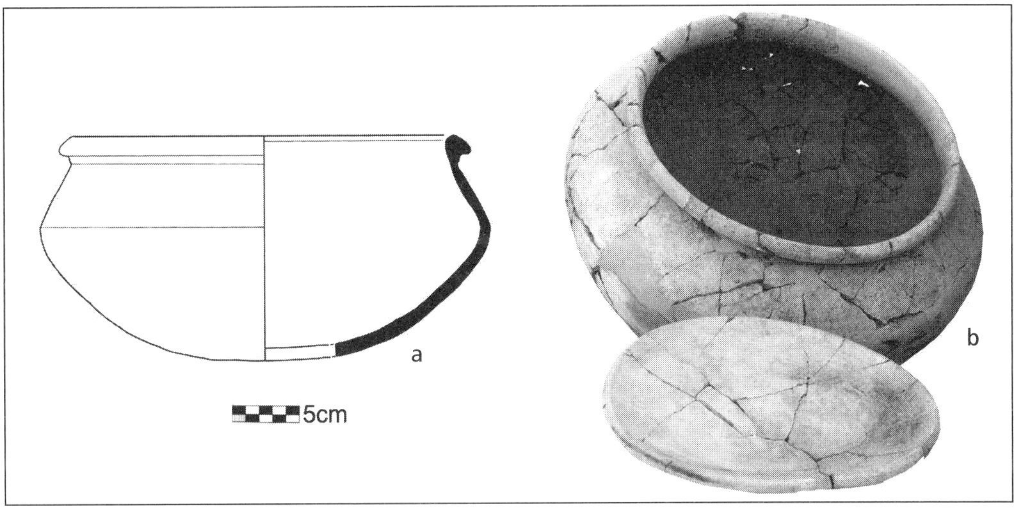

FIGURE 12.5. Tell Atchana LB I Shellware cooking jars: (*a*) illustration by Sarah Hawley, (*b*) photograph by Murat Akar.

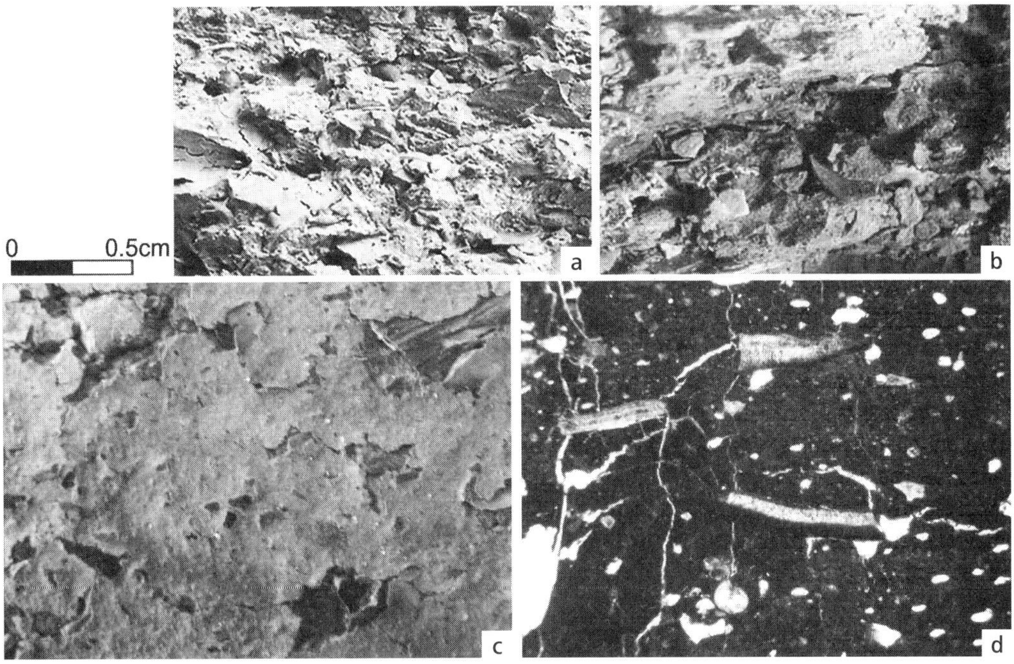

FIGURE 12.6. Tell Atchana LB I Shellware: (*a*) pink-peach fabric; (*b*) tan-yellow fabric section; (*c*) tan-yellow fabric surface showing water smoothing. Photographs by Mara Horowitz. Photomicrographs (PPL): (*d*) Shellware. Field of view 2.6 mm. Photograph by Simon Groom.

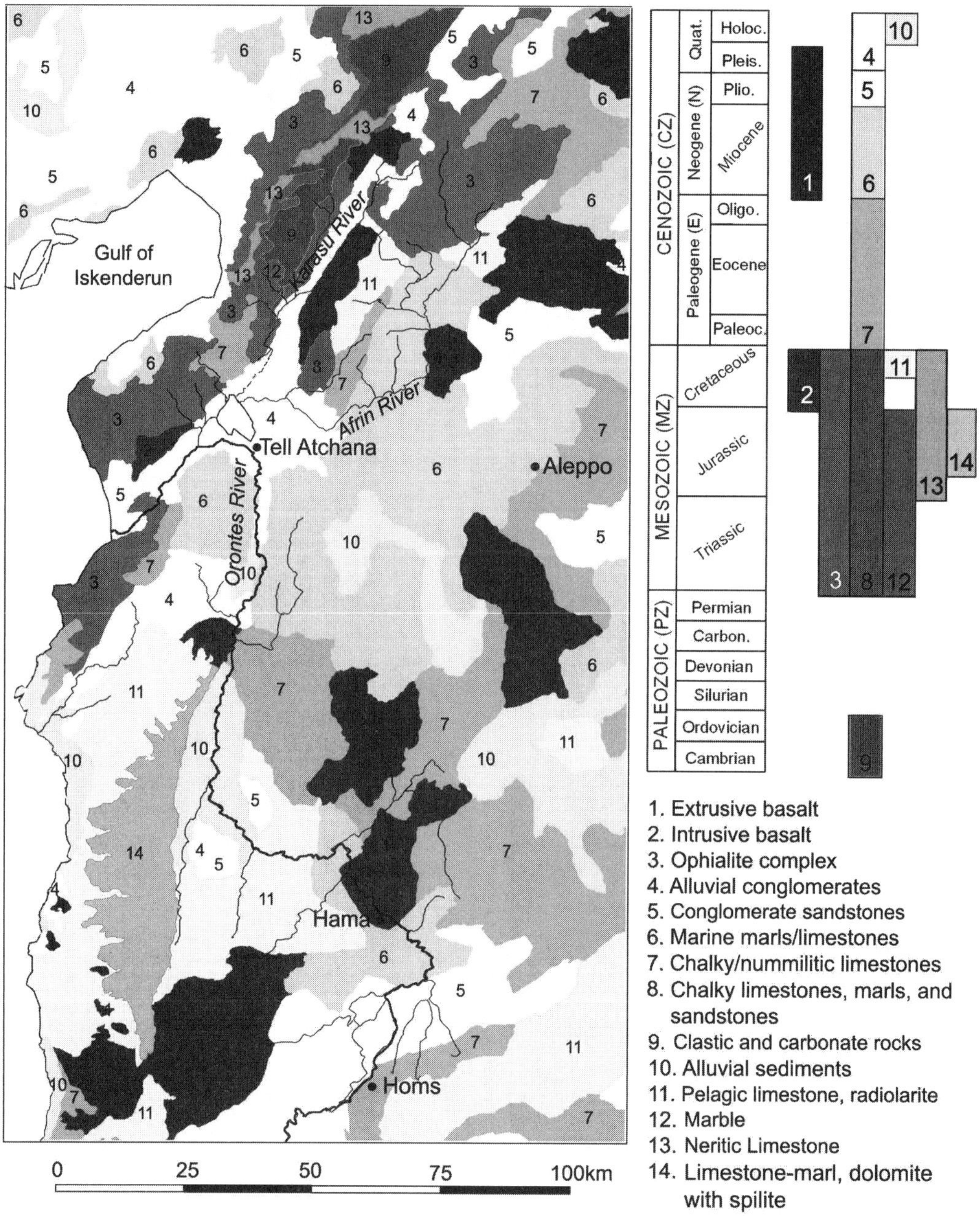

FIGURE 12.7: Geological map of the Amuq Valley catchment and adjacent areas. Illustration by Mara Horowitz (based on Ministry of Petroleum and Mineral Resources, 1986 and General Directorate of Mineral Research and Exploration 2002).

from time to time because of sedimentation and the uplift and subsidence of landforms (Wilkinson 2000; Wilkinson et al. 2001:211–213). The clay resources on the Amuq plain in the vicinity of Atchana are currently obscured by up to 6 m of sediment accumulated since ca. 1200 BC; in the Bronze Age access to local substrate clay deposits would have been easier (Yener et al. n.d.). Today the 3–4-m-deep cuts of local canals and the Orontes riverbed are the only access to local clays.

Methodology

The basic steps of pottery production include first locating and processing the potting materials then forming, drying, and firing the vessel (Leach 1976). Variable and repetitive actions within these steps are dependent upon the environmental conditions and cultural frameworks in which the potter worked, the technology and techniques used to form a pot, the intended function of the vessel, the physical makeup of the potting materials, skill levels, and potter's preferences (Morrison 2010). Ancient potting practices are better understood when archaeologists relate pots to human behavior within this paradigm. The interpretations of such relationships should be expressed as pathways of possibilities and probabilities within this behavioral model, however, rather than as definitive laws of human action (Morrison 2010). These pathways help identify and better understand cultural systems by mapping out how material use and the production process changed over time (Grave et al. 2008:1974–1975).

Experiments that replicate ancient fabrics explore the types of potting materials available in the landscape and how they might have been manipulated (Moody et al. 2012; Morrison and Archambeau 2004). To begin to understand an ancient fabric, we must first have a fabric description that includes texture composition of the clay matrix, type and shape of inclusions, and the nature of voids to determine whether organic tempering materials could have been used. Potentially analogous raw materials are then identified in the landscape through survey. The aim of the experiments is to determine how these components may have been altered, if at all, from their natural state and then combined in some proportion to create the archaeological fabric (Moody et al. 2012; Morrison and Archambeau 2004).

While petrographic characterization using qualitative analysis of the archaeological cooking ware fabrics from Mochlos and Tell Atchana by the appointed petrographers of the independent projects was ongoing simultaneously, the experimental stage of the project (selection and processing of materials) proceeded, using macroscopic characterizations of the pottery fabrics and the local geology. Approximately 50 petrographic slides of the ancient cooking wares from Mochlos and Tell Atchana and 10 petrographic slides of the replicated fabrics from both sites were examined and are housed independently. The slides produced from Mochlos excavations and the geological clay samples from Crete are housed at INSTAP Study Center for eastern Crete. The Tell Atchana materials are stored at the Alalakh Excavations dig house depot in Tayfur Sökmen Köy, Reyhanlı, Hatay, Turkey. It was an intentional decision to choose the materials for the replication experiment before the petrographic analysis was complete. We wanted to attempt material sourcing by examining the respective landscapes the way a potter would, which is looking for materials that are suitable for making pottery.

The LM I and LB I cookwares at Mochlos and Tell Atchana are made from coarse and very coarse fabrics, making their primary characteristics (texture, color, and reaction to hydrochloric acid) and tempering components (shell and mineral fragments) easily identifiable to the naked eye, although a 10×–30× hand lens was also used in some cases for further investigation (see table 12.1, figs. 12.3 and 12.6). Thereafter the experimental materials were also subjected to petrographic analysis and compared to the archaeological originals. Petrographic analysis of all samples was qualitative, focusing on type of inclusions, their amount and sizes, and paste texture.

Experiment

The 2008 field-based experiments were designed to follow the basic steps of pottery production. While the premise of the experiments at Mochlos and Tell Atchana was the same, each team formulated specific research questions for their respective body of material. The Mochlos experiment sought to locate and test the potting materials exposed in the Limenaria Cove to gain a better understanding of the local Neopalatial production of cookwares. One goal of the Tell Atchana project was to understand why the cookware tradition changed from Mineral Ware to Shellware at the beginning of the LBA; the focus of 2008 study was the Shellware, however, since a local source of the calcite material and ferrous clay has yet to be located.

Locating and Processing Potting Materials

While exposed clays with low-grade metamorphic materials are located in various ravines across the Mochlos plain, the closest to the LM IB Artisans' Quarters are in the Limenaria Cove (Barnard 2003:5; Nodarou et al. 2008) (fig. 12.8c). New roads that expose the clays have been cut across the ravines in this area due to recent development. One clay, called Purple Phyllite Hill (PPH), when unfired is purple with red-purple phyllite, milky-white quartz, and additional inclusions. The other clay, Development Red (DR), when unfired is red-orange with silver, green, brown and purple phyllite, milky-white quartz, and additional inclusions.

Inclusions identical to those in the ancient fabrics are present in the clays exposed in the Limenaria Cove, but these inclusions are more plentiful and typically larger. If the ancient potters used these clays, they would have had to remove the larger inclusions (ca. 4–15 mm) by sieving the clay through holes ca. 2–3 mm in size. While little archaeological evidence supports the hypothesis that ancient potters used a wet-sieving method as opposed to dry crushing to process their clay, this method was employed in the replication experiments to prevent the removal of the finer clay particles by the strong summer winds, as might have happened if dry crushing and sieving was used (fig. 12.8a and b).

The modern irrigation canals that flank the site at Tell Atchana reveal 3–6 m of variable coarse sedimentary deposit above a fine calcareous clay deposit. The color of this clay when unfired is greenish tan. Aplastic sand-sized (1⁄16–2 mm) inclusions naturally occurring within the clay are macroscopically similar to those within the matrix of the ancient Shellware: round and subangular particles of red, white, gray, black, yellowish-clear, and brown, which petrography revealed to be serpentine, olivine, gabbro, radiolarian chert, magnetite, shell fragments, and fossils (Groom et al. n.d.). A wet-sieving method was employed to remove large stones and organic material from the clay prior to use. Crushing and rehydrating was not necessary, because the clay was already saturated from the canal water (fig. 12.9a and b).

Based on a preliminary inspection of the crushed shell fragments in the archaeological material by Dr. Canan Çakırlar (staff zooarchaeologist at Tell Atchana), freshwater mussels (*Unionides* spp.) were identified as the most likely candidate. Modern shells belonging to this genus were collected from the local canals to be used as temper to replicate the Shellware. The angular shape of the ca. 2–5 mm shell fragments in the LB I Shellware fabric (see table 12.1 and fig. 12.6) indicates that the potters crushed this tempering agent before it was added to the clay (Rye 1981:37).

One hypothesis for why this fabric appears in the LB levels is that local availability and/or consumption of mussels increased, so a side experiment was conducted that explored how heat treatment (such as baking and boiling in water) affects the shells' appearance and fracture patterns when crushed. Testing this hypothesis was complicated by the fact that eating mussels is not the only reason to heat shell prior to crushing it for temper. Shells are formed of calcium carbonate (often a mix of calcite and aragonite) and organic polymers that act as a binder (Feathers 2006:91–92). At ca. 300–400°C, any aragonite with its block-shaped crystalline lattice transforms to calcite with a plate-like lattice, leading to a slight expansion (Feathers 2006:92). It is thus desirable to heat the shells prior to their in-

FIGURE 12.8. Collecting and processing Mochlos clay: (*a*) removing unwanted materials (such as rocks, twigs, and leaves) from the clay using water sieving method (2–3 mm screen), (*b*) storing cleaned clay in a pithos for aging until pinch pots are produced, (*c*) collecting clay from a road cut.

clusion in a clay body so that they do not expand during firing of a vessel.

The change in crystalline structure (aragonite to calcite) effected by heating should be observable in the fracture patterns of the crushed shell even after the shell is (re)fired in a ceramic vessel. Different forms of heating (burning/roasting versus boiling in water) may also affect the fracture patterns, as boiling dissolves the organic polymers that bind the calcite of the shell. The experiment proceeded to test whether the shells were heated before crushing, even though the reason for the heating would remain uncertain (cooking the mussels for consumption versus preheating the temper to prevent expansion). The shells were cleaned with water and divided into three samples: (1) shells receiving no heat treatment before crushing, (2) shells boiled in water (ca. 100°C) for forty-five minutes before crushing (fig. 12.9c), and (3) shells baked in an electric oven on high (ca. 250°C) for forty-five minutes. After the treatment the shells were left to cool overnight.

The following afternoon a modern basalt pestle and mortar similar to those found in the Late Bronze levels of Atchana was used to crush

FIGURE 12.9. Collecting and processing Atchana clay and temper: (*a*) kneading the raw clay, (*b*) wet-sieving, (*c*) boiling shells, (*d*) crushed shells after boiling. Photographs by Murat Akar.

the shells. The baked shells were the easiest to crush; they quickly reduced to a sparkly pinky-brown dust, so that some care had to be taken to leave fragments of a suitable size for temper. Both the boiled and non-heat-treated shells were crushed using relatively little force and tended to be more of a grayish-blue-to-white color. The elongated, angular shapes of all the crushed shells were macroscopically similar to one another and similar to the ancient shell temper (fig. 12.9d). The experiment proceeded using all three prepared tempers.

Making, Drying, and Firing Pinch-Pots

In the initial stages of experimentation it is best to limit variables by focusing on how the clay responds to basic forming techniques (Moody et al. 2012; Morrison and Archambeau 2004). Pinch-pots are practical because they do not take advanced skills and tools to make and have no weakened areas created by construction methods (seams from coil-joins, which can compromise the vessel's structure during drying and firing). They are formed by hand using the pulling method. Pinch-pots for each experimental fabric were produced and numbered via incision on the bottom.

The Mediterranean summer climate is hot and dry, with abundant wind and daytime temperatures usually exceeding 28–30°C in eastern Crete and 38°C in the Amuq Valley. To ensure that the experimental pots did not crack from drying too fast, they were placed in the shade once they were manufactured. The next evening the vessels were ready to be fired. This is a testament to the suitability of the local clays for potting, as some clays will crack during drying due to either their chemical and physical composition and/or how rapidly the vessel is dried (Rhodes 1973; Rye 1981).

At Mochlos the experimental pots were fired in an oxidized-neutral atmosphere created by an electric kiln to determine if the colors of the collected clays remained distinct from one another after firing. The firing time was five hours and temperatures reached 700–850°C (Rhodes 1973; Rye 1981). Our justification for using this modern firing tool was to determine whether the fired clays macroscopically resembled the color of the LM cookware. Cookware has not been found in assemblages that are associated with LM kilns (Shaw et al. 2001; Soles 2003). The question of how ancient Mochlos cookwares were fired remains unanswered, but experimentation and ethnographic evidence indicates that vessels produced using Cretan clays could have been fired using pit- or ground-firing techniques (Moody et al. 2012; Morrison and Archambeau 2004).

Experimental archaeology has shown that shell-tempered pottery cannot be fired above ca. 600–800°C because the calcite that makes up the shell will break down into calcium oxide (lime) and carbon dioxide; the resulting lime will rehydrate during use and expand dramatically, leading to severe spalling and breakage of the vessel (Feathers 2006:92). To reduce the damaging effects of lime spalling potters can add salt or saltwater or submerge their fired vessels in water for a few days using a technique called "docking." The pithoi-potters in Crete practice the docking technique today (Morrison Ph.D. work 2007–present). At Tell Atchana the yellow-tan, peach-pink, and brown-black mottling of the LB I Shellware cooking jars suggests that a pit- or ground-firing method was used (though some of the brown-black staining is probably due to postfiring use of cookware). To replicate these conditions, the experimental pots were fired in a traditional pita oven or "tandır," consisting of a large beehive-shaped clay firing chamber built into a mudbrick structure. An air hole at the bottom, at ground level, created a draft. The top of the installation was open. This arrangement should closely mimic traditional pit or ground firing. The firing procedure consisted of the following steps:

1. Preparation of a bed of coals using dung (fig. 12.10a);
2. Positioning of the pots on top of coals, lip-side down;
3. Placement of dried cotton plants and pieces of wood over the pots and hot coals (fig. 12.10b);
4. Complete fuel consumption when the pots were fired (fig. 12.10c);
5. Pots left in the oven to cool overnight before removing (fig. 12.10d).

The firing took about eighty minutes. Unfortunately, no thermocouple was available to measure the temperature range of the fire. Based on the dull "clinky" sound of the fired pots and the fact that the fired clay did not rehydrate after soaking in water for an hour, we conclude that the firing temperatures reached at least 650°C (Rhodes 1973; Rye 1981).

Results

The primary question for the Mochlos experiment was whether the Limenaria Cove clays could be used to produce vessels that resemble the LM IB cookwares from the Artisans' Quarter and Neopalatial town. The two clays (PPH and DR) each produced pinch-pots that were easily formed and fired, but the PPH has a much "greasier" feel and is more plastic than the DR, making it easier to form a pot (table 12.2).

The dark red color of the fired PPH clay paste (fig. 12.11c) is characteristic of all three metamorphic LM IB cookpot fabrics. But the dominant inclusions (red-purple phyllite and milky-white quartz; fig. 12.11d and e) are only matched macroscopically and petrographically by those within the main phyllite (fig. 12.3a and d) and the semicoarse metamorphic fabrics (fig. 12.3b and e). Either another clay source with inclusions of silver mica schist remains to be found that produced the semicoarse metamorphic with mica fabric or additional potting materials (such as mica schist) were collected, processed, and mixed with the clay. There is a small exposure of weathered silver mica schist in the western end of the plain (see fig. 12.4). The dark red-orange of the fired DR clay matrix is somewhat similar in color to the LM IB cookwares, but the dominant inclusions (silver, green, brown, and purple phyllite) are not characteristic of them. Rather,

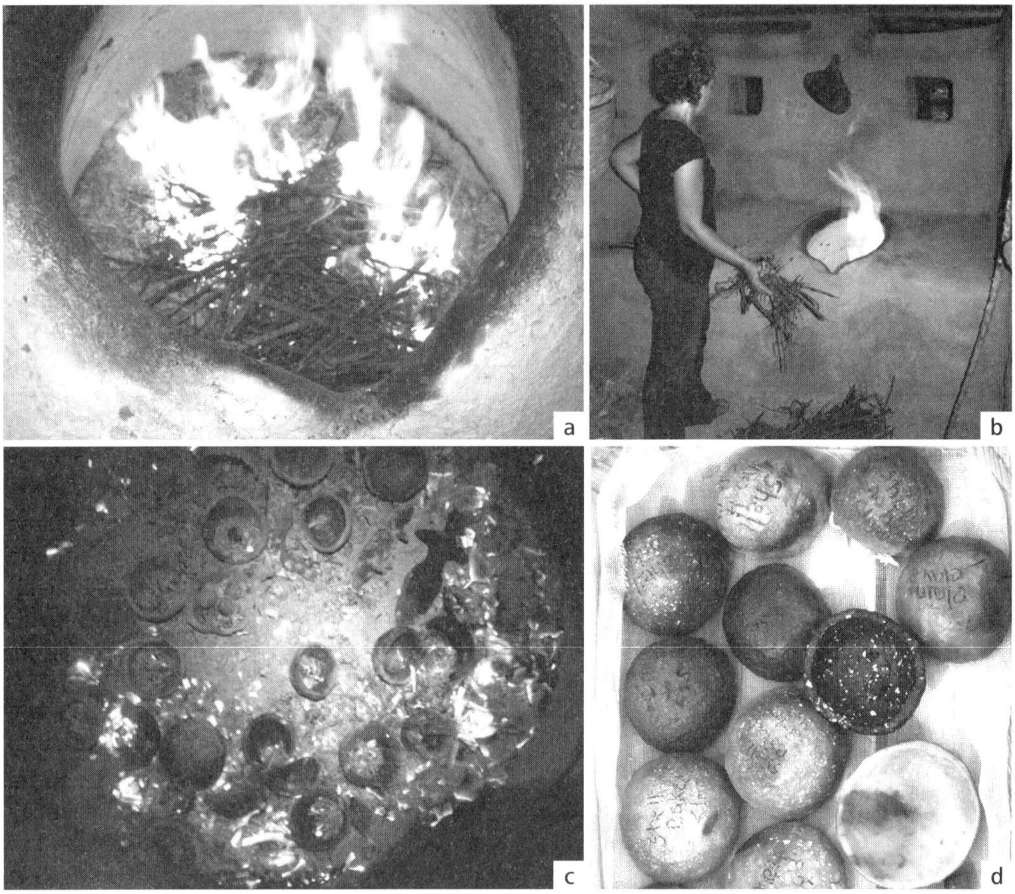

FIGURE 12.10. Firing the Tell Atchana pinch-pots: (*a*) preparing a bed of coals using dung and sticks, (*b*) placing dried cotton plants and pieces of wood over the pots and hot coals, (*c*) fired pots after the fuel has burned away, (*d*) pots removed from the oven after cooling over night. Photographs by Murat Akar.

they closely resemble the LM II-III wares found in the Mochlos settlement (Nodarou 2010:5, 7, 8, plate 1A).

In regard to the cultural or technological implications of this apparent discontinuity in choice of raw materials to produce cookwares, it is important to consider that while the tripod cooking pot shape is typical of both LM IB and LM II-III Mochlos, the LM II–III vessels are larger in size on average (Barnard and Brogan 2003; Smith 2010; Morrison Ph.D. work 2007–present). This increase in cook pot size may relate to either the quantity of food being prepared or the number of people being served at one time. Functional experiments now being carried out will test the suitability of DR and PPH for making larger cook pots by re-creating and using similar vessels.

The primary question for the experiments at Tell Atchana was whether the clay and mussel shells collected from the local canal be used to produce a fabric that is analogous to the ancient LB I Shellware. The experimental results were found to be very similar to the sandier, more yellow-tan end of the Shellware spectrum. When fired, the color of the clay is macroscopically pale tan-yellow (see table 12.2, fig. 12.12a, b, and c) and has a slightly gritty texture. The finer, more peach-pink variety of Shellware might have been the result of more intensive clay processing, clay mixing, and/or a different firing technique (updraft kilns are known from the site: see Yener

TABLE 12.2. Macroscopic Ceramic Fabric Analysis descriptions of the experimental Mochlos and Atchana cookwares

	Mochlos		Tell Atchana		
Fabric name	Purple Phyllite Hill	Development Red	Crushed Unaltered Shell	Crushed Boiled Shell	Crushed Baked Shell
Texture	coarse	coarse	very coarse	very coarse	medium
Inclusion sorting	poor	poor	poor	poor	good
Inclusion size	2–3 mm	2–3 mm	2–3 mm	2–3 mm	< 1 mm
Inclusion shape	subrounded, subangular	subrounded, subangular	angular	angular	subangular
Inclusion type	red-purple phyllite (+); milk-white quartz	green, silver, brown, purple phyllite (+); milk-white quartz (+)	grayish-blue, white freshwater shell	grayish-blue, white freshwater shell	pinky-brown freshwater shell
Munsell (2001) of unfired clay	10r 4/3, 5/3	2.5yr 4/8, 5/8	2.5y 7/2, 7/3	2.5y 7/2, 7/3	2.5y 7/2, 7/3
Authors' color name of fired clay	purple	red-orange	gray-green-brown	gray-green-brown	gray-green-brown
Munsell (2001) of fired clay	red, 2.5yr 5/6	reddish yellow, 5yr 6/6	very pale brown and light yellowish brown, 10yr 7-6/4; reddish yellow, 5yr 7/6	very pale brown and light yellowish brown, 10yr 7-6/4; reddish yellow, 5yr 7/6	very pale brown and light yellowish brown, 10yr 7-6/4; reddish yellow, 5yr 7/6
Authors' color name of fired clay	dark red	dark red-orange	pale tan; peach	pale tan; peach	pale tan; peach

Note: (+) = quantity of inclusions present within the fabric matrix (Kemp 1985:17): (1) predominant (> 70%), (2) dominant (50–70%), (3) frequent (30–50%), (4) common (15–30%); (-) = quantity of inclusions present within the fabric matrix (Kemp 1985:17): (1) few (5–15%), (2) very few (2–5%), (3) rare (0.5–2%), (4) very rare (< 0.5%).

2010b). Microscopically, the brown, micritic experimental fabric was very similar in texture and color to the archaeological original and contained a similar range of mineral inclusions (fig. 12.12d, e, f, g, h, and i). The only differences seemed to be in the overall amount of sand-grade mineral grits, as noted macroscopically. As previously stated, the petrographic analysis was ongoing while the field-based fabric experiment was being conducted, but the macroscopic methods utilized here (Moody et al. 2012) provided some information on the amount of material to add.

A secondary question for the Atchana experiments was whether it could be determined (given that intense heat affects the physical properties of the mussel shells) if the mussels were preheated, perhaps indicating cooking and consumption, prior to the shells being crushed for temper. Of the three experimental tempers, the boiled shell was the most clearly different under the microscope (fig. 12.12f–g).

Microscopically, a circular gray mottling appeared in one shell fragment, but the fragmentation pattern commonly seen in many fragments was irregular (for example, the fracture is both longitudinal and perpendicular to the plane), not strongly linear (for example, longitudinal to the plain) like the untreated and baked shell. This might be related to the dissolving of structural organic polymers by the boiling water. This is curious, however, because we assumed that these polymers would also be dissolved during the firing process of the pottery. The untreated shell (fig. 12.12d–e) was the most homogeneous in color and was a stronger, almost bluish gray

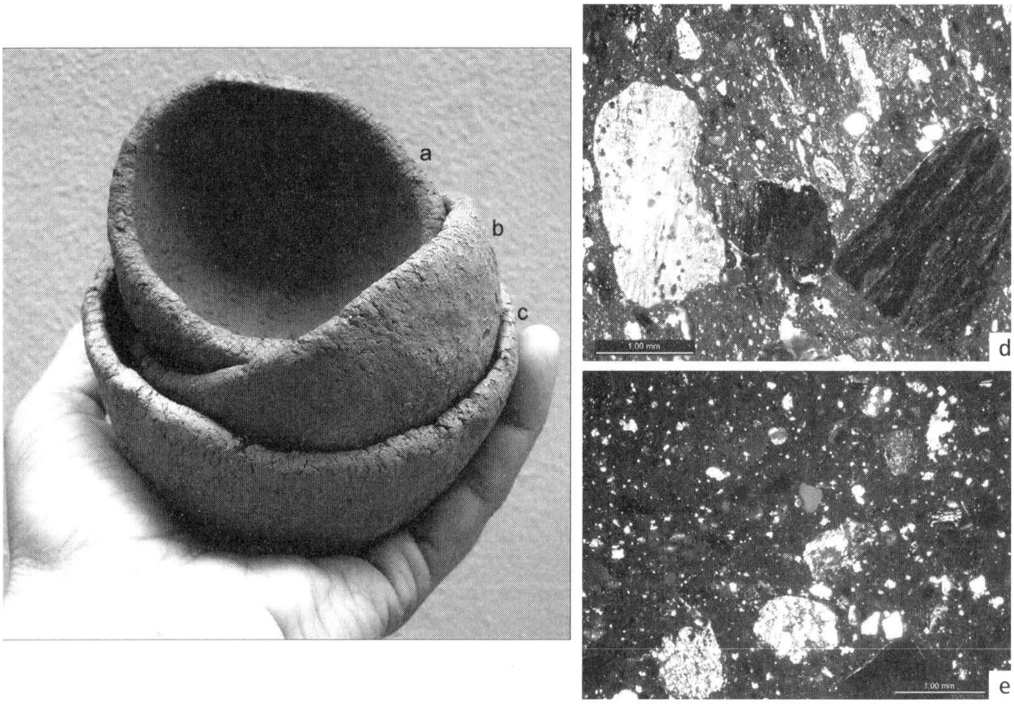

FIGURE 12.11. Mochlos experimental cookwares: (*a*) Development Red, (*b*) Development Red and Purple Phyllite Hill mix, (*c*) Purple Phyllite Hill. Photographs by Jerolyn Morrison. Photomicrographs (XPL): (*d*) Development Red, (*e*) Purple Phyllite Hill. Field of view 4 mm.

under the microscope with a fracture pattern that was longitudinal to the plain. The baked shell (fig. 12.12h–i) was strongly fragmented in a linear pattern (for example, longitudinal to the plane) and was a lighter, more brownish gray. It also exhibited a lattice-like patterning on some of the fragments.

While the three experimental tempers clearly differed from each other, it is more difficult to judge which of them is most similar to the archaeological material. As seen in fig. 12.6d, linear banding is present in Shellware temper as well as lattice-like patterning, which could suggest affinity with the baked shell experimental temper. But further exploration of archaeological Shellware is now needed to provide a reliable sample for comparison, and the possibility of chronological differences should also be considered. Thus the results of the secondary experiment on Atchana shell temper are equivocal at the present time, with clear differences in the experimental material but only the preliminary suggestion of a match to archaeological material.

These experiments demonstrated that it is possible to create a pot from local materials that strongly resembles the LB I Shellware. The Atchana team is currently exploring the landscape for sources of gritty ferrous clays and clear calcite crystals to replicate experimentally the cook pot fabrics of the MBA Mineral Ware type. The long use of these rounded cooking pots, in first one and then the other fabric, may have implications for the continuity of cultural practices in the Amuq area amid the backdrop of tumultuous political events in the second and early first millennium BC.

Conclusions

By integrating field-based fabric experiments with petrographic analysis, a clearer picture of how ancient potters at Mochlos and Tell Atchana could have engaged with the potting materials

FIGURE 12.12. Tell Atchana experimental cookwares: (*a*) crushed-baked shell, (*b*) crushed-boiled shell, (*c*) crushed-unaltered shell. Photographs by Jerolyn Morrison. Photomicrographs (PPL): (*d* and *e*) untreated shell, (*f* and *g*) boiled shell, (*h* and *i*) baked shell. Field of view 3 mm. Photographs by Allan Gilbert.

in their respective landscapes has developed. These experiments demonstrated that fabrics resembling the archaeological ceramics could be produced with locally available raw materials. Although the choices made during the replication process may or may not have been exactly those of the ancient potters, the physical possibilities and limitations of these resources are better understood.

Defining material choices is one way to explore how culture can affect ancient technologies. For example, it is curious that at Mochlos LM IB potters living on the coast had access to abundant marine materials yet did not produce shell-tempered fabrics. Instead they used particular deposits of clay with metamorphic inclusions to create versatile fabrics that were used to produce a range of vessels (Barnard and Brogan 2003). The ancient Tell Atchana potters also lived and worked adjacent to an abundant source of shellfish (in their case a large freshwater lake), and they chose to begin using crushed freshwater mussel shells as a tempering agent at the end of the MB period.

Experiments to characterize and replicate archaeological ceramic fabrics do not have to be overly complicated, cost an enormous amount of money, or take a long time to produce useful information. In fact these sorts of experiments have many advantages. They can be used as preliminary tests to determine what more advanced work needs to be carried out and can also help archaeologists gain information about the local potting materials in their research area for a better definition of the local assemblage, production technology, and ancient potters' choices. The combination of petrographic, macroscopic, and experimental methods creates a strong framework relating the geological landscape and functional properties of raw materials to the human element that dictated the choices made in the course of ceramic production.

Acknowledgments

We would like to thank Costis Davaras, Jeffrey Soles, K. Aslıhan Yener, Jennifer Moody, Sherry Fox, Simon Groom, Murat Akar, Canan Çakırlar, and Lillian Green (who also read the original paper at the SAA in 2009) for their support of experimental archaeology. Funding and other considerations were provided by INSTAP, the Fund for Amuq Valley Archaeological Expeditions, Koç University, Mustafa Kemal University, the Turkish Ministry for Culture and Tourism, and the Fulbright Foundation. Special thanks to Dr. Allan Gilbert for his advice and use of the Fordham University laboratory.

References Cited

Barnard, Kellee
2003 A Macroscopic Analysis of the Neopalatial Fabrics. In *Mochlos IB: Period III, Neopalatial Settlement on the Coast: The Artisans' Quarter and the Farmhouse at Chalinomouri: The Neopalatial Pottery*, edited by Kellee Barnard and Thomas Brogan, pp. 3–11. INSTAP Academic Press, Philadelphia.

Barnard, Kellee, and Thomas Brogan (eds.)
2003 *Mochlos IB: Period III, Neopalatial Settlement on the Coast: The Artisans' Quarter and the Farmhouse at Chalinomouri: The Neopalatial Pottery*. INSTAP Academic Press, Philadelphia.

Day, Peter, Louise Joyner, and Maria Relaki
2003 A Petrographic Analysis of the Neopalatial Pottery. In *Mochlos IB: Period III, Neopalatial Settlement on the Coast: The Artisans' Quarter and the Farmhouse at Chalinomouri: The Neopalatial Pottery*, edited by Kellee Barnard and Thomas Brogan, pp. 13–32. INSTAP Academic Press, Philadelphia.

Evely, R. Doniert G., and Jerolyn Morrison
2010 The Minoan Potter's Wheel: A Study in Experimental Archaeology. *Proceedings of the 6th International Congress on the Archaeology of the Ancient Near East, Volume 1*, edited by Paolo Matthiae, Frances Pinnock, Lorenzo Nigro, and Nicolo Marchetti, pp. 283–288. Harrassowitz, Wiesbaden.

Feathers, James
2006 Explaining Shell-Tempered Pottery in Prehistoric Eastern North America. *Journal of Archaeological Method and Theory* 13(2):89–133.

General Directorate of Mineral Research and Exploration
2002 *Geological Map of Turkey: Hatay*. General Directorate of Mineral Research and Exploration, Ankara, Turkey.

Grave, Peter, Lisa Kealhofer, Ben Marsh, and Marie-Henriette Gates
2008 Using Neutron Activation Analysis to Identify Scales of Interaction at Kinet Höyük, Turkey. *Journal of Archaeological Science* 35:1974–1992.

Groom, Simon, Paul Bown, Mara T. Horowitz, and Bryan Cockrell
n.d. Fossils and Extrusives in Simple Ware: The Microstructure of the Local Ceramic Tradition at Late Bronze Alalakh (Turkey). Forthcoming.

Kemp, Rob A.
1985 *Soil Micromorphology and the Quaternary*. Quaternay Research Association Technical Guide 2. Quaternay Research Association, Cambridge.

Leach, Bernard
1976 *A Potter's Book*. Faber and Faber, London.

Magness-Gardiner, Bonnie
1994 Urban-Rural Relations in Bronze Age Syria: Evidence from Alalaȟ Level VII Palace Archives. In *Archaeological Views from the Countryside: Village Communities in Early Complex Societies*, edited by Glenn M. Schwartz and Steven E. Falconer: pp. 37–47. Smithsonian Institution Press, Washington, D.C.

Ministry of Petroleum and Mineral Resources
1986 *Geological Map of Syria, 1:1,000,000*. 2nd ed. Ministry of Petroleum and Mineral Resources, Damascus, Syrian Arab Republic.

Moody, Jennifer, Jerolyn Morrison, and Harriet Robinson
2012 Earth and Fire: Cretan Potting Traditions and Replicating Minoan Cooking Fabrics. In *Philistor: Studies in Honor of Costis Davaras*, edited by Eleni Mantzourani and Philip Betancourt, pp. 119–132. INSTAP Academic Press, Philadelphia.

Morrison, Jerolyn
2010 A Potter's Note about the Use of Raw Materials to Produce Minoan Cooking Pots. *Kentro Newsletter* 13:6–11.

Morrison, Jerolyn, and Marie Archambeau
2004 Replicating Minoan Ceramic Fabrics: Experiments in Production and Usage. *La Tinaja Ceramic Newsletter* 15:2–15.

Morrison, Jerolyn, and Douglas Park
2007 Throwing Small Vessels in the LM IB Mochlos Potter's Pit. *Kentro Newsletter* 10:6–10.

Munsell, Albert H.
2001 *X-Rite Munsell Soil Color Book*. Revised ed. New Windsor, New York.

Nodarou, Eleni
2003 Appendix B: A Petrographic Analysis of a Clay Sample from the Field Adjacent to the Artisans' Quarter. In *Mochlos IB: Period III, Neopalatial Settlement on the Coast: The Artisans' Quarter and the Farmhouse at Chalinomouri: The Neopalatial Pottery*, edited by Kellee Barnard and Thomas Brogan, pp. 171–174. INSTAP Academic Press, Philadelphia.
2010 Petrographic Analysis of the Late Minoan III Ceramics. In *Mochlos IIB: Period IV, The Mycenaean Settlement and Cemetery: The Pottery*, edited by R. Angus K. Smith, pp. 3–14. INSTAP Academic Press, Philadelphia.

Nodarou, Eleni, Charles Frederick, and Anno Hein
2008 Another (Mud)brick in the Wall: Scientific Analysis of Bronze Age Earthen Construction Materials from East Crete. *Journal of Archaeological Science* 35:2997–3015.

Papastamatiou, John, Dimitrios Vetoulis, Athan Tataris, John Bornovas, Georgios Christodoulou, and George Katsikatsos
1959 *Geological Map of Greece: 1:50,000, Ierapetra Sheet*. Institute for Geology and Subsurface Research, Athens.

Rhodes, Daniel
1973 *Clay and Glazes for the Potter*. Chilton Book Company, Philadelphia.

Rice, Prudence M.
1987 *Pottery Analysis: A Sourcebook*. University of Chicago Press, Chicago.

Rye, Owen
1981 *Pottery Technology: Principles and Deconstruction*. Australian National University Press, Washington, D.C.

Seager, Richard
1909 Excavations on the Island of Mochlos, Crete, in 1908. *American Journal of Archaeology* 13:273–303.

Shaw, Joseph, Aleydis Van de Moortel, Peter Day, and Vassilis Kilikoglou
2001 A LM IA Kiln in South-Central Crete. *Hesperia Supplement* 30. American School of Classical Studies at Athens, Princeton, New Jersey.

Smith, R. Angus K. (ed.)
2010 *Mochlos IIB: Period IV, The Mycenaean Settlement and Cemetery: The Pottery*. INSTAP Academic Press, Philadelphia.

Soles, Jeffrey (ed.)
2003 *Mochlos IA: Period III, Neopalatial Settlement on the Coast: The Artisans' Quarter and the Farmhouse at Chalinomouri, The Sites*. INSTAP Academic Press, Philadelphia.

Soles, Jeffrey, and Costis Davaras
1992 Excavations at Mochlos, 1989. *Hesperia* 61:413–445.
1994 Excavations at Mochlos, 1990–1991. *Hesperia* 63:391–436.
1996 Excavations at Mochlos, 1992–1993. *Hesperia* 65:175–230.

Wilkinson, Tony J.
2000 Geoarchaeology of the Amuq Plain. *American Journal of Archaeology* 105:1–51.

Wilkinson, Tony J., Elizabeth S. Friedman, E. Ercan Alp, and Anton P. J. Stampfl
2001 The Geoarchaeology of a Lake Basin: Spatial and Chronological Patterning of Sedimentation in the Amuq Plain, Turkey. In *Recherches Archéométriques, Cahiers d'Archéologie du CELAT*, edited by Michel Fortin, pp. 211–226. Laval University, Quebec.

Woolley, C. Leonard
1955 *Alalakh: An Account of the Excavations at Tell Atchana in the Hatay, 1937–1949*. Oxford University Press, Oxford.

Yener, K. Aslıhan
2008 Aççana Höyüğü 2007 Yılı Çalışmaları. *Kazı Sonuçları Toplantısı 30*. General Directorate of Antiquities and Museums, Ankara.
2010a Introduction. In *Tell Atchana, Ancient Alalakh, a Bronze Age Capital in the Amuq Valley, Turkey. Volume I: The 2003–2004 Excavation Seasons*, edited by K. Aslıhan Yener, pp. 1–6. Koç University Press, Istanbul.

Yener, K. Aslıhan (ed.)
2005 *The Amuq Valley Regional Project, Vol. 1: Surveys in the Plain of Antioch and Orontes Delta, Turkey, 1995–2002*. Oriental Institute Publications No. 131. Oriental Institute of the University of Chicago, Chicago.
2010b *Tell Atchana, Ancient Alalakh, a Bronze Age Capital in the Amuq Valley, Turkey. Volume I: The 2003–2004 Excavation Seasons*. Koç University Press, Istanbul.

Yener, K. Aslıhan, Murat Akar, and Mara T. Horowitz
n.d. *Tell Atchana, Ancient Alalakh in the Late Bronze II Period. Volume II: Excavations 2006–2010*. Forthcoming.

CHAPTER 13

Point/Counter Point II

The Accuracy and Feasibility of Digital Image Techniques in the Analysis of Pottery Tempers Using Sherd Edges

PATRICK C. LIVINGOOD *and* ANN S. CORDELL

Digital imaging analysis has been proposed as an efficient alternative to traditional petrography for some applications (Daniels and Lipo 2008; Gonzalez and Woods 2007; Hansen 2000; Livingood 2007, 2010; Russ 2006; Velde and Druc 1998; Whitbread 1991; see Reedy 2006 for a review of applications). In 2009 advocates of each method decided to join forces to put this hypothesis to the test in the measurement of temper size and abundance in pottery thin sections (Livingood and Cordell 2009). That study concluded that under the right circumstances digital image analysis was a viable means of measuring temper and could be much faster and cheaper. Furthermore, these advantages could be maximized by using digital image analysis to measure inclusions present in a sherd edge rather than in a thin section. This chapter evaluates that claim and compares the use of digital image analysis of exposed sherd edges with traditional microscopic petrography to measure shell temper in a sample of 25 Mississippian Period sherds from the Moon Site in Arkansas (fig. 13.1). We find that under the right circumstances accurate measurement of temper in sherd edges is possible but that this approach is probably useful for only a small number of temper types with high visual contrast to the matrix.

In the first paper (Livingood and Cordell 2009) digital imaging analysis and traditional petrographic point-counting were conducted on a small sample of thin sections of grog and shell–tempered Mississippian Period pottery from the Pevey site (22Lw510), Lawrence County, Mississippi (Livingood 2010; Livingood and Cordell 2009). The test sample consisted of 4 thin sections selected from a larger sample of 29 analyzed using digital image analysis (Livingood 2007). The relative merits of the two techniques were evaluated in terms of accuracy, precision, cost, and time. It was agreed that both methods make valuable contributions to the understanding of pottery in archaeology. Petrographic methods have a much broader application, are recommended as a precursor to any proposed digital analysis, and have a much better capacity to permit fine distinctions. Digital image analysis of thin sections was a viable means of conducting temper analysis and might be preferable under certain circumstances. The advantage increases if the process can be more easily automated. Automation is facilitated when the samples are relatively homogeneous and if the particles of interest have a high visual contrast or particular regularities in size, shape, color, or texture. If automation or partial automation is feasible then a digital approach is preferable for the analysis of large sample sizes. It is also preferable when the goal is to return complex metrics about the inclusions (Livingood and Cordell 2009).

The research presented in this chapter exam-

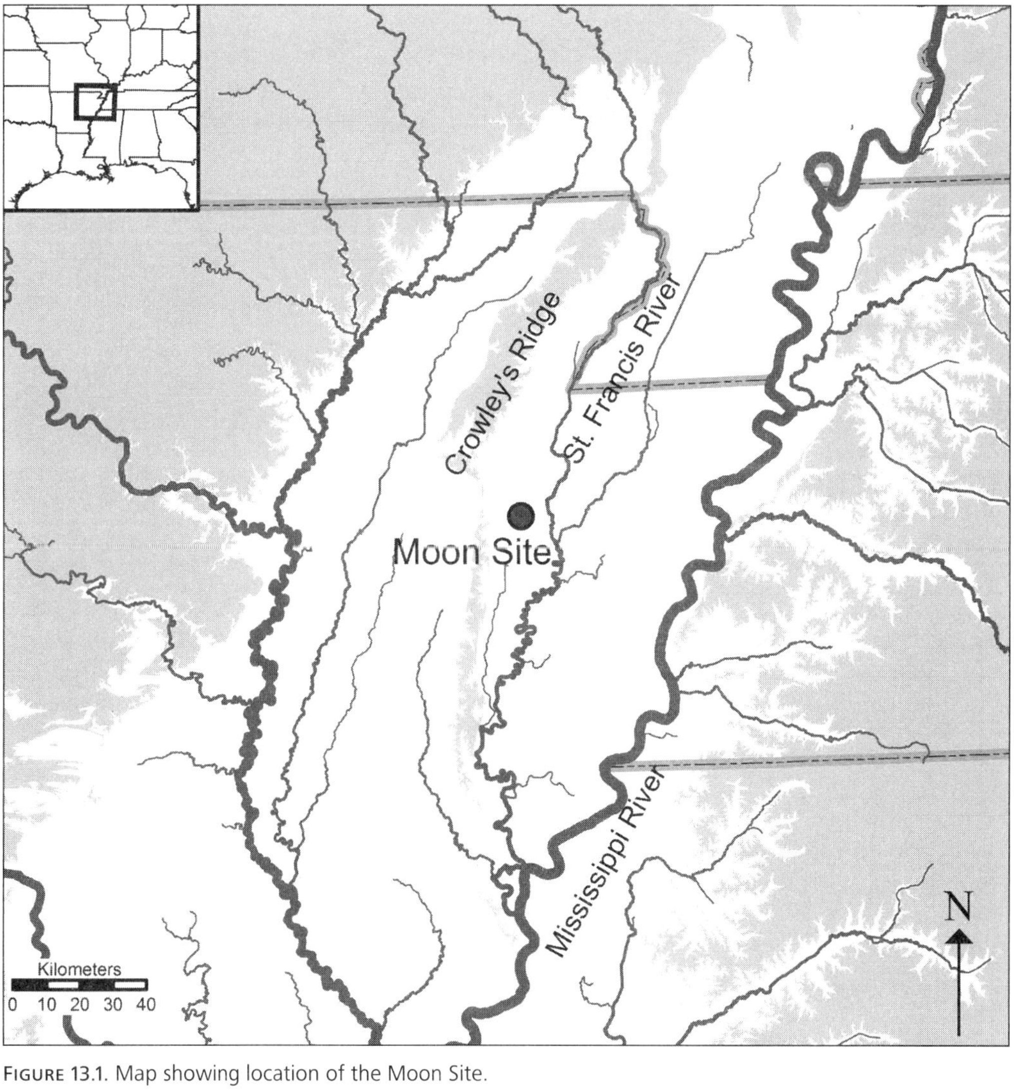

FIGURE 13.1. Map showing location of the Moon Site.

ined the observation that it should be possible to skip the expensive and time-consuming process of thin section preparation entirely if the intent is solely to measure temper and the temper is highly visible. In order to assess this, we analyzed sherd edges with highly visible freshwater mussel shell temper, which is present in late prehistoric ceramics in the southeastern United States. Digital analysis of shell temper in sherd edges has been done before (Daniels and Lipo 2008; Pereira 2012), but to our knowledge this is the first study to test its accuracy by comparing the results with traditional petrography.

Petrographic Point-Count Analysis

The pottery samples were sent to Spectrum Petrographics for thin sectioning. Thin-section preparation included vacuum embedding (with EPOTEK 301), standard slide format (27 by 46 mm) and thickness (30 μ), and acrylic mounting with a permanent glass coverslip. Petrographic point-count analysis was conducted in order to quantify the relative abundance of shell temper and other inclusions in the thin sections. As with the 2009 study, a Leitz Laborlux 11 Pol petrographic microscope with a mechanical stage was used to conduct the Glagolev-Chayes

TABLE 13.1. Raw petrographic point counts

Sample number	Total count	Voids	Matrix	Total aplastics	Shell	Clay lumps, iron nodules	Silt	Very fine quartz	Fine quartz	Medium quartz	Quartzite	Feldspars, other	Organic, siliceous
1	242	8	189	53	9	14	14	1	4	3	1	6	1
2	227	15	160	67	34	3	11	6	3	2	1	7	—
3	259	22	204	55	28	11	6	3	2	2	—	3	—
4	274	14	218	56	44	1	9	1	—	—	—	1	—
5	284	25	199	85	41	8	7	4	12	3	3	7	—
6	367	30	255	112	69	12	10	7	3	1	1	8	1
7	218	14	164	54	25	14	6	3	1	—	1	4	—
8	207	8	162	45	24	7	6	3	1	—	1	3	—
9	256	10	208	48	15	2	17	3	2	3	1	5	—
10	311	18	222	89	37	23	13	6	3	—	1	5	1
11	209	7	155	54	29	9	7	—	5	1	—	3	—
12	211	6	147	64	18	11	5	9	10	1	3	7	—
13	285	13	205	80	42	1	16	4	6	6	—	5	—
14	295	13	208	87	29	4	29	6	7	6	1	5	—
15	252	10	198	54	19	17	8	5	—	1	—	2	2
16	222	8	156	66	29	19	9	1	3	—	—	5	—
17	210	15	151	59	35	3	7	1	2	2	2	7	—
18	206	6	166	40	25	4	7	1	—	1	—	1	1
19	244	10	186	58	26	2	9	9	2	2	3	4	1
20	250	6	184	66	28	1	11	5	4	4	7	5	1
21	223	8	163	60	24	22	10	—	—	2	—	2	—
22	299	12	240	59	24	4	14	2	1	3	2	7	2
23	267	16	192	75	68	5	—	—	—	—	—	2	—
24	298	12	209	89	65	17	5	—	—	—	—	2	—
25	301	15	213	88	44	26	10	4	1	1	—	2	—

point-counting procedure (Galehouse 1971:389), following recommendations by Stoltman (1989, 1991, 2001). A counting interval of 1 mm by 1 mm with 10× magnification was used. Each point or stop of the stage was assigned to one of the following categories: clay matrix, nontemper voids, shell-temper voids, silt-sized particles, clay lumps, ferric nodules or concretions, and very fine through very coarse quartz and other aplastics of varying compositions. Size of temper and aplastics was estimated with reference to the Wentworth Scale (Rice 1987:38). A total count of 100–300 points, excluding voids, is required to ensure a reliable sample (Stoltman 2001:306). A minimum of 200 points is especially recommended (Stoltman, personal communication, 2001), which was used for this study. For cases in which fewer than 200 points were counted ($n = 2$), the thin section was rotated 180° on the mechanical stage and counted a second time.

The data generated are actual counts of tempers and aplastics within specified size ranges (table 13.1). Percentages can be calculated that represent reliable estimates of particle abundance and facilitate comparison between samples (table 13.2; Stoltman 1989:151–152). Temper and matrix compositions were also determined. The petrographic analysis permitted identification of accessory constituents and peculiarities of the matrix paste. Point-count data were used to calculate a variation of Stoltman's (2001:314) sand size index for siliceous sands, and this index was also adapted to provide a shell-size index.[1] All petrographic analyses were carried out by Cordell in the Florida Museum of Natural History Ceramic Technology Laboratory (FLMNH-CTL). SAS 9.2 for Windows (SAS Institute, Inc. 2008) was used for computer analysis and statistical comparisons of data.

Principal Constituents of the Paste

The principal constituent in the samples is shell temper, evident exclusively as platy voids left from the dissolution of shell. Some shell voids are filled with debris that might hinder detection with digital scanning. Shell temper data are summarized in table 13.3. When other data are considered (shell abundance, sand abundance, and so forth) shell size was always extremely variable.

Other common constituents include silt and sand-sized particles, ferric nodules and concretions, and clay lumps. Silt in these samples is primarily a particle size category, but composition is dominated by quartz and other silicate minerals. Sand is composed primarily of quartz but also includes quartzite, feldspars, and a variety of rock fragments: chert, sandstone, possible rhyolite, and unidentified plutonic igneous or metamorphic lithics with felsic composition. These constituents are generally subrounded and so low in frequency that they must be natural to the source clay or alluvially deposited detrital grains incidental to sand tempers. Counts of these constituents were included in sand totals for calculating sand percentage and sand size indices.

Most of the sand grains are subrounded and rarely exceeded medium in size (see table 13.3). Yet particle size variability indicates poor sorting among sand constituents, and frequency is under 10 percent in most cases. These data support the likelihood that the sand constituents were naturally present in the clay source(s). Percentage of shell temper in the sample is inversely related to percentage of sand.

Clay lumps are generally subrounded and may have resulted from incomplete mixing during paste preparation. Ferric nodules and concretions are generally considered natural constituents, but some could also be incidental to detrital tempers. Both these types of inclusions generally lacked angular features characteristic of grog or sherd temper. Relative abundance of ferric nodules and clay lumps was generally low in most cases, further suggesting that they are natural. In contrast to sand constituents, the ferric nodules and clay lumps occur in predominantly larger particle sizes (see table 13.3). Ferric nodules and clay lumps were combined for purposes of statistical analyses.

Accessory Constituents

Other constituents that are generally occasional or rare include polycrystalline quartz or quartzite; plagioclase, microcline feldspars and perthite; mica, mostly muscovite; and mafic minerals, mostly hornblende and epidote (see table 13.3). Counts of these constituents were

Table 13.2. Petrographic point-count percentages

Sample number	% Voids	% Matrix	% Total aplastics	% Shell temper	% Clay lumps, ferric nodules	% Silt	% Very fine sand	% Fine sand	% Medium sand	% Total sand	% Quartz	% Quartzite, feldspars, other
1	3	78	22	4	6	6	1	3	2	6	3	3
2	6	70	30	15	1	5	4	1	3	8	5	4
3	8	79	21	11	4	2	2	1	1	4	3	1
4	5	80	20	16	< 1	3	1	—	—	1	< 1	< 1
5	8	70	30	14	3	2	2	7	1	10	7	4
6	8	70	30	19	3	3	4	2	< 1	5	3	2
7	6	75	25	12	6	3	3	< 1	1	4	2	2
8	4	78	22	12	3	3	2	2	—	4	2	2
9	4	81	19	6	1	7	2	2	2	6	3	2
10	6	71	29	12	7	4	4	1	—	5	3	2
11	3	74	26	14	4	3	< 1	3	1	4	3	1
12	3	70	30	8	5	2	6	8	< 1	14	10	5
13	4	72	28	15	< 1	6	2	2	3	7	6	2
14	4	70	30	10	1	10	3	3	2a	8	6	2
15	4	79	21	8	7	3	2	< 1	< 1	3	2	1
16	4	70	30	13	9	4	1	2	1	4	2	2
17	7	72	28	17	1	3	2	2	2	7	2	4
18	3	81	19	12	2	3	1	—	< 1	1	1	< 1
19	4	76	24	11	1	4	4	2	2	8	5	3
20	2	74	26	11	< 1	4	3	4	3	10	5	5
21	4	73	27	11	10	4	1	—	1	2	1	1
22	4	80	20	8	1	5	2	1	3a	5	2	3
23	6	72	28	26	2	—	—	1	—	1	—	1
24	4	70	30	22	6	2	< 1	< 1	—	1	—	1
25	5	71	29	15	9	3	2	1	< 1	3	2	1
Mean %	4.1	74.2	22.8	12.9	3.7	3.2	2.2	2.0	1.2	5.3	4.7	2.2

Note: for samples 14 and 22 two coarse sand grains were included with medium sand in calculations of percentages; for sample 17 one coarse sand grain was included with medium sand in calculations of percentages.

TABLE 13.3. Summary of petrographic composition, abundance, and particle size data

Principal constituents	Abundance (%) range	Abundance (%) mean	Particle size index range	Particle size index mean	Wentworth sizes range	Wentworth sizes modal
Shell temper voids	4–26	13	1.77–3.10	2.31	very fine–pebble	fine 23% medium 23% coarse 36%
Silt[a]	< 1–10	3.0	NA	NA	0.0039–0.0625 mm	NA
Sand	1–14	4.5	0.50–1.50	0.98	very fine–medium	very fine 43% fine 36% medium 21%
Quartz	0–10	4.7	NA	NA	very fine–medium	very fine 42% fine 36% medium 21% coarse < 1%
Other sand[b]	< 1-5	2.2	NA	NA	very fine–coarse	very fine 44% fine 36% medium 17% coarse 3%
Ferric nodules and clay lumps	< 1–10	4.6	NA	NA	very fine–granule	coarse 34% very coarse-granule 19% fine 20% medium 17% very fine 10%
Mica	P to 3–5	est. 1	NA	NA	very fine–fine	very fine
Sponge spicules	P to 1–3	NA	NA	NA	silt	silt
Phytoliths	P to 1–3	NA	NA	NA	silt	silt
Diatoms	A-P	NA	NA	NA	silt	silt

Note: A = absent; P = present.
[a] Primarily quartz, lesser feldspars, ferric nodules, siliceous microfossils.
[b] Includes polycrystalline quartz, feldspars (plagioclase and microcline), rock fragments, mafic minerals.

included in sand totals for calculating sand percentage and sand size indices. They may be naturally occurring constituents of the source clay(s) or incidental to sand tempers. Bone temper was observed in one sample, as an incidental constituent.

Other constituents in most samples include siliceous microfossils, specifically sponge spicules and opal phytoliths and more rarely diatoms (see table 13.3). Sponge spicules are needle-shaped rods composed of silica that formed the skeletal support for some freshwater sponge genera (Borremans and Shaak 1986). Opal phytoliths are botanical microfossils composed of silica (Rapp and Mulholland 1992). Diatoms are unicellular algae with ornate cell walls made of silica (Round et al. 2007). Remnants of circular valves of a freshwater centric diatom, such as *Cyclotella* (Round et al. 2007:144), and faint remnants of freshwater pennate diatoms were observed. Diatoms generally were observed in samples that also contained occasional sponge spicules or phytoliths. These constituents are considered natural to the source clay(s) and can only be detected with petrographic methods.

It is important to note that not all aspects of the petrographic analysis have counterparts in the digital analysis. The mineralogy and lithologies of sand constituents might not be resolved with digital scanning in this study but should at least be distinguished from shell temper. Clay lumps and ferric nodules might be determined in terms of particle size. Bone temper would probably be resolved as shell temper in digital analysis. Mica in these samples, which rarely exceeded very fine in size, might be too small

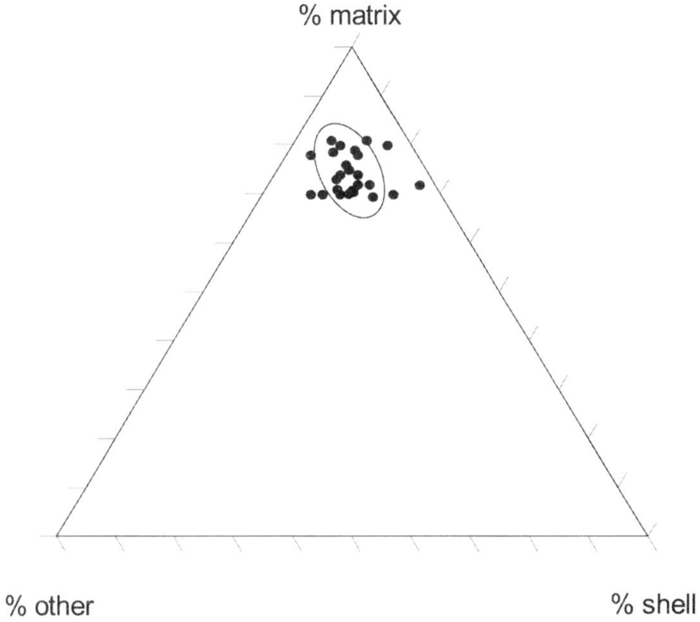

FIGURE 13.2. Ternary plot of bulk composition of thin sections from petrographic analysis. The circle indicates the closely related samples.

to be identified with digital scanning. Silt and siliceous microfossils would most certainly not be detected by digital scanning methods.

Resource and Temper Categories

Ternary diagrams (after Graham and Midgley 2000) of point-count data are useful for the recognition of trends and were constructed to examine bulk compositional and textural variability in the sample (Stoltman 1989). Bulk composition within the sample is shown in fig. 13.2. This shows close clustering for most cases, with matrix being by far the predominant category in the sample.

Given this close clustering and considering all the data for naturally present constituents, one potential source clay is postulated, designated Resource A. This resource generally contains rare to occasional mica, siliceous microfossils, and silt, sand, and ferric nodules/clay lumps in variable but low quantities. It accounts for all but one of the pottery samples. To examine further trends within this gross resource grouping, nontemper (nonshell) aplastic composition of constituents was also plotted (fig. 13.3). This diagram compares percentages of silt, siliceous sands, and ferric nodules/clay lumps, with matrix counts excluded from the calculations. This is a comparison of granulometry as well as of composition. The diagram shows that Resource A can be broken down into two main categories (A1 and A2) and outliers. Resource A1 is characterized by about 5 percent silt/sand and 10 percent ferric/clay inclusions, while A2 contains more silt/sand and less ferric/clay constituents (table 13.4). The differences in mean percentages of nonshell constituents are statistically significant (table 13.5). Outliers represent extremes on both ends. Together these main groups and outliers may represent compositional variability within a given clay deposit. Another resource, B, is represented by a single micaceous paste sherd, indicating a clay source with noticeable amounts of mica. The silt/sand and ferric/clay constituents for B are intermediate between means established for resource groups A1 and A2. It is not expected that digital scanning methods will provide such information about potential clay sources represented by the sherd samples.

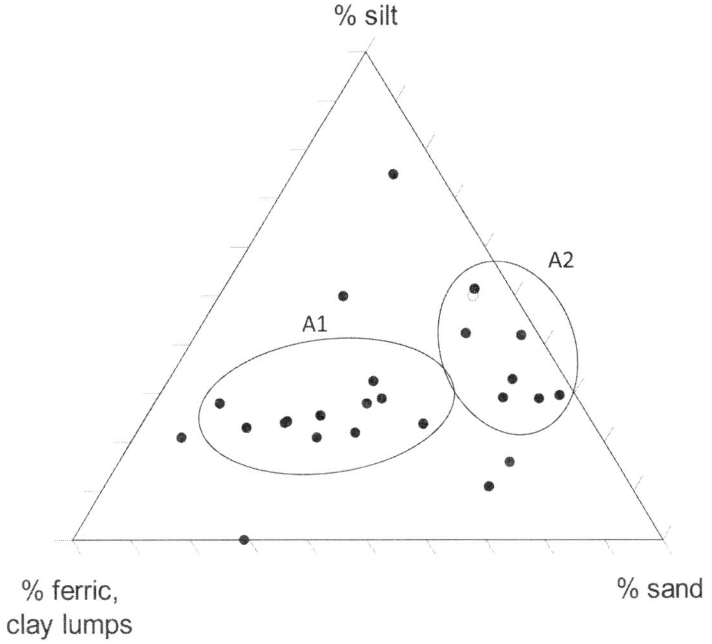

FIGURE 13.3. Ternary plot of nonshell aplastic composition from the petrographic analysis. The plot indicates two groupings labeled here as A1 and A2.

TABLE 13.4. Petrographic point-count percentages for resource groups for bulk (calculated excluding shell temper) and aplastic composition (calculated excluding shell temper and matrix)

Bulk

Resource group	Sample size	% Matrix	% Silt and sand	% Clay lumps, ferric nodules
A1	11	84.2	4.5	11.2
A2	7	84.9	8.3	6.7
B	1	78.0	9.0	12.0
Outliers	6	88.5 (± 7.9)	5.5 (± 6.7)	6.0 ± 2.6)

Aplastic

Resource group	Sample size	% Silt	% Sand	% Clay lumps, ferric nodules
A1	11	25.6	30.1	44.4
A2	7	36.7	55.9	7.3
B	1	50.0	43.0	7.0
Outliers	6	28.8 ± 28.1)	34.3% (± 25.1)	36.8% (± 27.4)

In order to identify ware groupings within the sample, aplastic compositional variability was plotted, excluding the matrix component. The three components plotted consist of percentages of shell temper, sand (plus silt), and ferric nodules plus clay lumps (fig. 13.4). In this diagram it is apparent that the 25 sherds sort into roughly four "ware" categories, characterized by decreasing amounts of shell temper. Ware Group A is characterized by an abundance of shell temper and very low quantities of other aplastics; it consists of outliers in terms of resource

TABLE 13.5. Statistical comparisons of resource groups A1 and A2 (for data presented in table 13.4) and ware groups (for data presented in table 13.6)

Comparison of resource groups		*t*-test statistic	Degrees of freedom	Attained significance level
A1 vs. A2 for bulk composition	% silt and sand	$t = -4.37$	df = 16	$p = 0.0005$
	% clay lumps, ferric	$t = 3.64$	df = 16	$p = 0.0022$
A1 vs. A2 for aplastic composition	% silt	$t = -3.76$	df = 16	$p = 0.0017$
	% sand	$t = -5.35$	df = 16	$p < 0.0001$
	% clay lumps, ferric	$t = 8.61$	df = 16	$p < 0.0001$

Comparison of ware groups		*t*-test statistic		Degrees of freedom	Attained significance level	
		bulk	aplastic		bulk	aplastic
A vs. B	% shell	$t = 4.10$	$t = 5.50$	df = 12	$p = 0.0015$	$p = 0.0001$
	% silt and sand	$t = -4.52$	$t = -5.45$	df = 12	$p = 0.0007$	$p = 0.0001$
	% clay lumps, ferric	$t = -1.71$	$t = -2.38$	df = 12	$p = 0.1134$	$p = 0.0348$
A vs. C	% shell	$t = 3.28$	$t = 6.83$	df = 8	$p = 0.0112$	$p = 0.0001$
	% silt and sand	$t = -8.85$	$t = -7.44$	df = 8	$p < 0.0001$	$p < 0.0001$
A vs. D	% shell	$t = 4.99$	$t = 8.79$	df = 5	$p = 0.0041$	$p = 0.0003$
	% silt and sand	$t = -6.76$	$t = -9.67$	df = 5	$p = 0.0011$	$p = 0.0002$
B vs. C	% silt and sand	$t = -7.69$	$t = -8.05$	df = 16	$p < 0.0001$	$p < 0.0001$
	% clay lumps, ferric	$t = 4.36$	$t = 5.28$	df = 16	$p = 0.0005$	$p < 0.0001$
B vs. D	% shell	$t = 3.54$	$t = 4.47$	df = 13	$p = 0.0036$	$p = 0.0006$
	% silt and sand	$t = -7.74$	$t = -13.86$	df = 13	$p < 0.0001$	$p < 0.0001$
	% clay lumps, ferric	$t = 1.60$	$t = 1.80$	df = 13	$p = 0.1339$	$p = 0.0953$
C vs. D	% shell	$t = 3.25$	$t = 5.41$	df = 9	$p = 0.0100$	$p = 0.0004$
	% silt and sand	$t = -2.17$	$t = -3.49$	df = 9	$p = 0.0580$	$p = 0.0068$
	% clay lumps, ferric	$t = 2.08$	$t = -2.07$	df = 9	$p = 0.0670$	$p = 0.0683$

groupings (table 13.6). Ware group D is characterized by an abundance of silt/sand, and lesser quantities of shell temper and ferric/clay lumps. It is variable with respect to resource groupings. Ware groups B and C are intermediate between A and D in terms of the prominence of shell temper. Ware group B is characterized by similar proportions of silt/sand and ferric nodules/clay lumps, whereas ware group C has more silt/sand and lesser ferric/clay lumps. Ware groups B and C correspond closely to resource groups A1 and A2, respectively. These may represent paste recipes in which differing quantities of shell temper were added to clay, depending on the amounts of naturally present silt, sand, and other constituents. An alternative interpretation involves adding differing quantities of shell and sand, depending on the amounts of naturally present silt and other constituents. The legitimacy of these ware groups is corroborated by statistically significant differences in percentages of the aplastic components for most comparisons (tables 13.5 and 13.6).

Digital Imaging Analysis

The first step in the digital imaging analysis process was to prepare the sherds for photography. Sherd edges were flattened and polished in using a pair of Micro-Mark Professional Model Maker's Disk Sanders that are capable of very

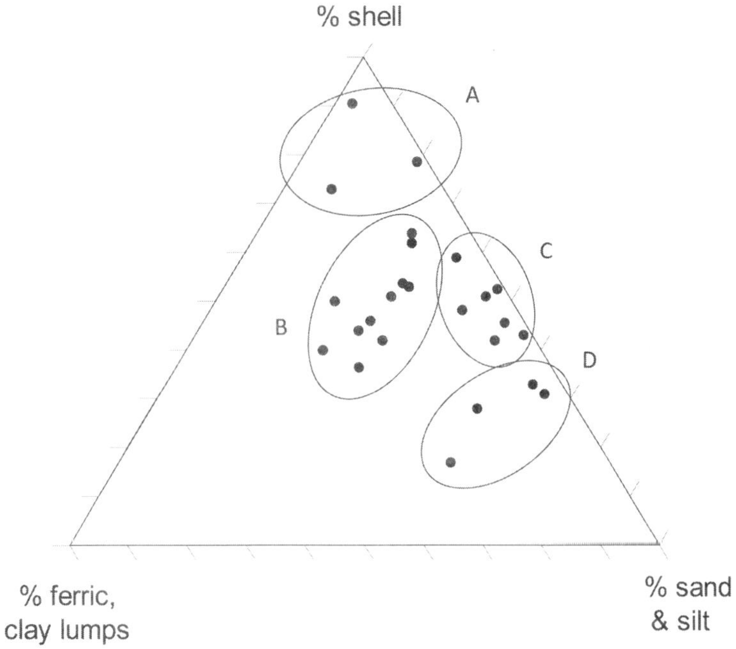

FIGURE 13.4. Ternary plot of the aplastic composition from the petrographic analysis showing the ware groups A, B, C, and D.

TABLE 13.6. Petrographic point-count percentages for ware groups for bulk composition and aplastic composition (excluding matrix)

Ware group	Sample size	% Matrix	% Shell temper		% Silt and sand		% Clay lumps, ferric nodules	
			bulk	aplastics	bulk	aplastics	bulk	aplastics
A	3	74.0	21.3	81.0	2.7	10.3	4.0	9.3
B	11	74.6	12.6	49.3	6.7	27.6	5.8	23.2
C	7	73.4	13.0	48.7	12.0	46.6	1.4	4.6
D	4	74.8	7.0	27.2	14.8	59.5	3.2	13.2

low-speed rotation (approximately 300 rpm). First, the edge is flattened using standard 360 grit sandpaper and then the edge is polished using 20 micron diamond microfinishing film. Finally, the edge is cleaned of dust using a blast of compressed air. Because an existing flat edge was selected for imaging, the amount of material removed was very minimal for most specimens. Certainly, this approach is significantly less destructive than removing a piece for thin-sectioning.

Next the sherd edge was placed on a specially designed stage at a fixed distance from a Canon EOS-1Ds Mark III digital camera equipped with a Canon 100 mm Macro lens. The camera was tethered to a PC and Breeze Software's DSLR Remote Control Camera software was used to control camera operation. Three separate images were taken of each sherd edge. The first (fig. 13.5, image 3) was taken with light from a Canon MR-14EX ring flash attached to the lens of the camera, so that the light is from the direction of the lens. This light has the effect of completely removing any shadows visible on the sherd edge. The second and third images (fig. 13.5, images 1 and 2) were taken with light from a pair of

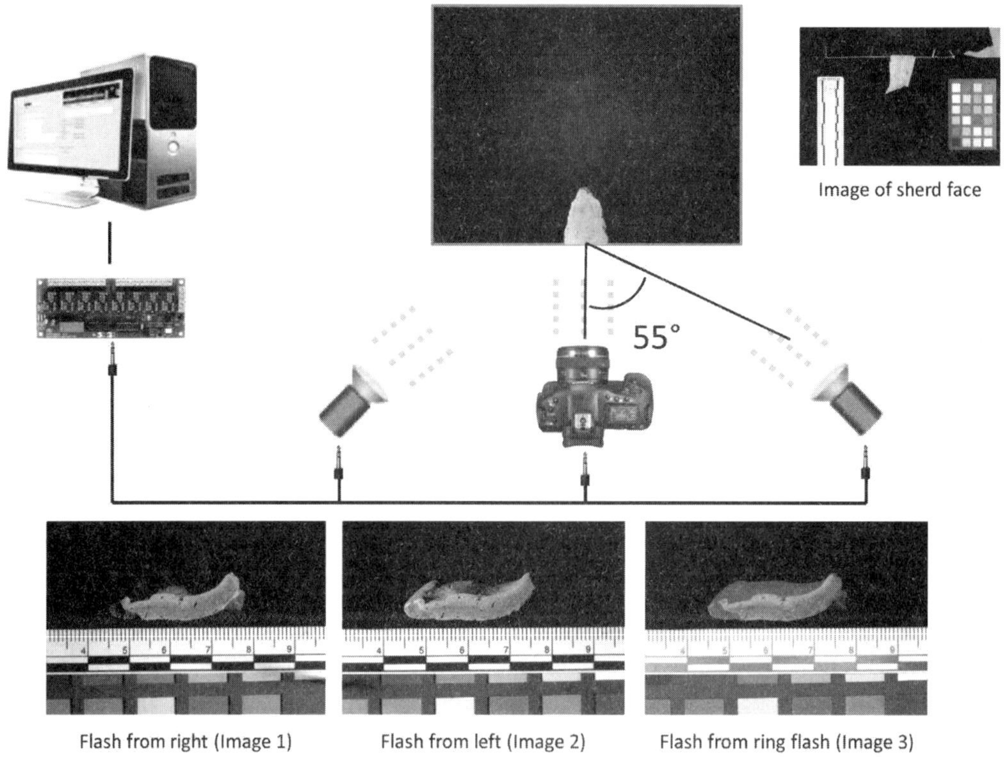

FIGURE 13.5. Diagram of the arrangements of camera and flash units and example images produced of the sherd edge and face with the apparatus.

Alienbees B800 studio flash units placed 55 degrees left and right of the sherd-to-camera axis but in the same horizontal plane as the sherd and camera. These shots provide light that rakes the sherd edge from the left and right respectively and casts shadows over holes and voids. A NCD R810PRO relay board provided a computer interface to control the flash units from the PC. In addition, a photograph of the sherd face (the vessel exterior or decorated surface) was taken using a Canon EOS 40D camera (fig. 13.5) although it was not essential to the measurement of the temper. The entire data collection apparatus was shrouded by black fabric to reduce the amount of incidental light (see fig. 13.5 for a diagram of the apparatus).

The entire process of gathering the four images was automated so that a single user command would trigger the collection and storage of the images. The image collection was launched from the same Microsoft Access form that was used to enter basic provenience data about each sample. The automated data collection process took less than 15 seconds. The data entry and sherd processing steps averaged an additional four minutes per sherd. In fact the larger study that this chapter is a part of collected data on over 20,000 sherds from the Moon Site. The 25 sherds used in this study were randomly selected from the larger collection for petrographic analysis after the image analysis was complete.

The analysis of these images to measure the shell temper was accomplished using Reindeer Graphic's Fovea Pro software and Adobe Photoshop CS4. The basic steps of analysis involved (1) manipulating the image so that the features of interest are emphasized with the highest possible visual contrast, (2) thresholding these images to produce Boolean images in which all pixels that are part of the feature of interest are coded in black and all other pixels in white, and finally (3) measuring the features in the image, which is a highly automated output from the data (Russ 2006). The image enhancement was achieved by selecting color channels from the original images that best depict the images in question and combining them to produce high visual contrast. It may also be necessary to remove noise from background features, to correct for issues introduced by the image acquisition process, and to sharpen the image as appropriate to bring out a strong visual contrast. The thresholding step takes all pixels that meet certain characteristics (frequently the lightest or darkest pixels in the processed image) in order to produce a Boolean image. Finally, all features (block of contiguous black pixels) are counted and measured using a few dozen metrics including size, width, breadth, shape, orientation, diameter, aspect ratio, and others.

Analyses of sherd edges are inherently much more limited than the analyses that can be performed on a thin section. For example, grain compositions that are distinguished on the basis of birefringence or other optical properties in petrography are difficult if not impossible to detect digitally. Microscopic scale porosity of the samples would also be difficult to distinguish in a sherd edge. The digital image analysis of the sherd edge works by identifying features with high visual contrast and voids that are revealed by the different use of light. Therefore the characteristics mapped and measured in this pilot study are area of the sherd edge, areas of existing shell, shell voids, and nonshell voids (although extant shell was almost completely absent in these 25 sample images). The sherd edge is easy to map against the black background of the photographic stage. The shell, when present in small amounts, is easy to detect because it has high image intensity values. There is a danger that other inclusions with high luminosity and white or nearly white color may have been categorized as shell by this method. For example, bone, which was observed in one of the samples using petrography, is likely to be miscategorized as shell. It is therefore important to prescreen the sample using other methods, such as petrography, for possible confounding inclusions. In this case, bone is so rare that its presence and miscategorization leads to fairly negligible error.

Most of the shell in the sherds was leached out by postdepositional processes, leaving simple laminar voids. These voids are mapped by comparing the three images collected. First, the images taken with light coming from the sides (see fig. 13.5, images 1 and 2) are combined so that at each pixel, the images are compared, and the darker pixel of the two is retained. This produces a composite image (fig. 13.6, labeled image 4) in which the shadows from the left and right light sources are effectively combined. This image (image 4) is then compared to the image taken using the ring flash (image 3). Any location that is significantly darker in the combined image (image 4) than the ring flash image (image 3) represents a void in the edge of the sherd and is mapped as such (image 5). Shell voids in image 5 are distinguished from nonshell voids on the basis of aspect ratio of 2 or greater (elongated shapes). This is a somewhat crude approach that probably miscategorizes some small number of voids but seems to be accurate in the whole. Even for a trained petrographer, the determination of the nature of a void is open to interpretation, and we would expect two qualified petrographers to differ somewhat in their assignments.

For this particular study, the only comparison being made between the digital image analysis and the petrographic analysis is the abundance of shell temper. Other studies have shown that the orientation of shell as measured in the sherd edge is of value in studying vessel construction (Daniels and Lipo 2008), whereas Carla Pereira (2012) has found value in examining shape characteristics of shell. Several more measurements produced from the digital image analysis will be utilized for future research, including the size distribution of the shell, the orientation of the shell, and potentially the color of the paste.

Comparison of Results

In our earlier study (Livingood and Cordell 2009) we weighed the advantages and disadvantages of traditional microscopic and digital image analysis approach to measuring temper inclusions in pottery thin sections. For clarity we follow the same template in this chapter for comparing the digital analysis of sherd edges and the

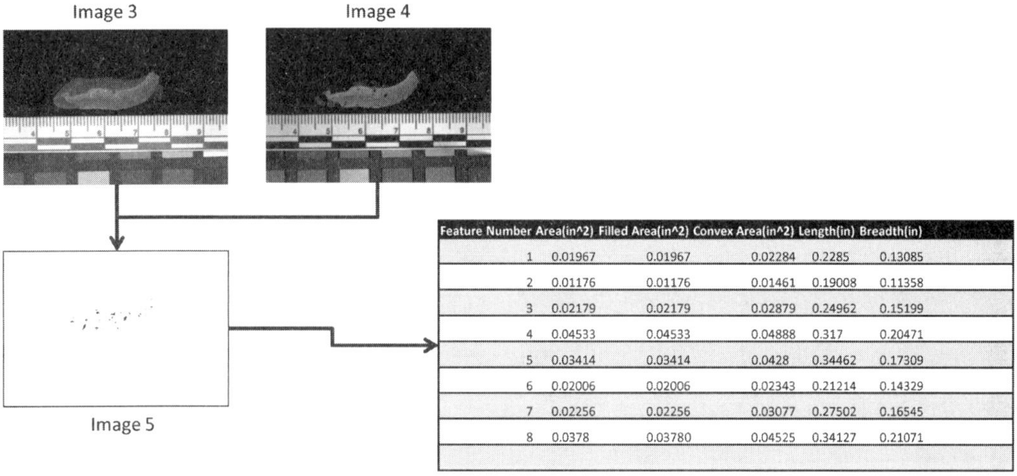

FIGURE 13.6. Simplified depiction of the digital analysis steps. Images 3 and 4 are compared, and the difference is the basis for the binary image map of the shell voids. These features are then automatically measured, producing a spreadsheet with metrics about all of the features.

microscopic analysis of thin sections. There are numerous reasons to expect that these methods should not yield identical measurements of shell abundance. First and foremost, while each method considered the same sherd, it was impossible for these methods to examine precisely the same plane through the sherd. Sherd edges were photographed for the digital analysis; then pieces were cut for thin sectioning, such that a precise correspondence between sherd edge and thin section is not possible. Second, most of the measurements are based on shell voids, which are open to some level of subjective interpretation (see the discussion below).

The basis for comparison of petrographic and digital percentages also differs. Petrographic percentages represent reliable estimates of particle abundances. In this study percentage of shell temper was calculated using the total number of point counts of shell divided by the total number of point counts (excluding nonshell voids). In contrast, the digital shell percentage is calculated as shell divided by the area mapped as the polished sherd edge.

Table 13.7 shows the comparison of shell abundance measurements of the two methods. Figure 13.7 shows this comparison in graphical form.[2] In general the methods agree fairly well, accounting for the possible sources of error. Just to put these values in perspective, the petrographic analysis of these samples averaged measuring 255 point counts per sample. Thus if a single measurement of shell was added or subtracted from any sample, on average this would change the shell abundance by 0.39 percent. Hence a difference of 0.3 percent shell measured in Sample 1 is the magnitude of a single point count difference and is thus fairly small.

For the most part, there appear to be few systematic differences in the measurement, though as a group the digital method is ever so slightly likely to undercount shell. The largest differences by magnitude are found in Samples 6, 23, and 25, all of which have differences between 3 and 4 percent. By magnitude of the total shell present, the biggest differences are in Samples 6, 10, 16, 18, and 25, wherein the difference between the two values as a percentage of the petrographic measurement exceeds 15 percent.

In developing the digital image analysis algorithm, we anticipated several possible problems with systematic error that are evaluated here. One of our initial concerns for this project was that the smallest inclusions may be systematically undercounted because of resolution and focus issues. This was an issue for the smallest birefringent particles in the initial study that compared digital and petrographic point-counting

TABLE 13.7. Percent difference between digital image analysis and petrographic data for shell abundance (percentage of the difference value as a percentage of the petrographic measurement)

Sample number	Petrographic shell percent	Digital shell percent	Difference	Percentage difference
1	3.72	3.42	0.3	8.1
2	14.98	13.96	1.02	6.8
3	10.81	10.67	0.14	1.3
4	16.06	16.51	−0.45	−2.8
5	14.44	16.32	−1.88	−13.0
6	18.80	22.03	−3.23	−17.2
7	11.47	11.28	0.19	1.7
8	11.59	10.28	1.31	11.3
9	5.86	5.15	0.71	12.1
10	11.90	13.97	−2.07	−17.4
11	13.88	13.80	0.08	0.6
12	8.53	7.38	1.15	13.5
13	14.74	14.24	0.5	3.4
14	9.83	10.49	−0.66	−6.7
15	7.54	7.74	−0.2	−2.7
16	13.06	11.03	2.03	15.5
17	16.67	18.23	−1.56	−9.4
18	12.14	10.07	2.07	17.1
19	10.66	11.38	−0.72	−6.8
20	11.20	12.27	−1.07	−9.6
21	10.76	9.59	1.17	10.9
22	8.03	7.07	0.96	12.0
23	25.47	28.97	−3.5	−13.7
24	21.81	22.65	−0.84	−3.9
25	14.62	18.07	−3.45	−23.6

* Negative percent indicates that the digital result was less than the point-count result.

approaches to the study of thin sections (Livingood and Cordell 2009). In this chapter, however, there does not appear to be a systematic undercount of shell based on particle size. Calculating the correlation between discrepancy (difference between the shell percentages) and the petrographically measured number of shell in the very fine or fine size category results in a very low positive correlation ($r = 0.10$).

Another concern was possible systematic error related to the largest shell inclusions. In this case the correlation between discrepancy (difference between the shell percentages) and the petrographic counts of very coarse or larger shell is still small ($r = 0.21$) but indicates that larger particles could be causing more error. This could be explained by a few factors. First, we would expect more overall variance in the sources of error for sherds with large shell because of quantization issues. A void has to be categorized either as a shell void or as a void with a different origin, and a single subjective decision (see below) can have a significant outcome on the results if the void is large. A sherd with lots of large voids creates more opportunities for higher divergence with more of these judgments. Second, large very closely spaced shell particles are more likely to be accidentally measured as one mega-particle using the digital image analysis method. This has been observed in a

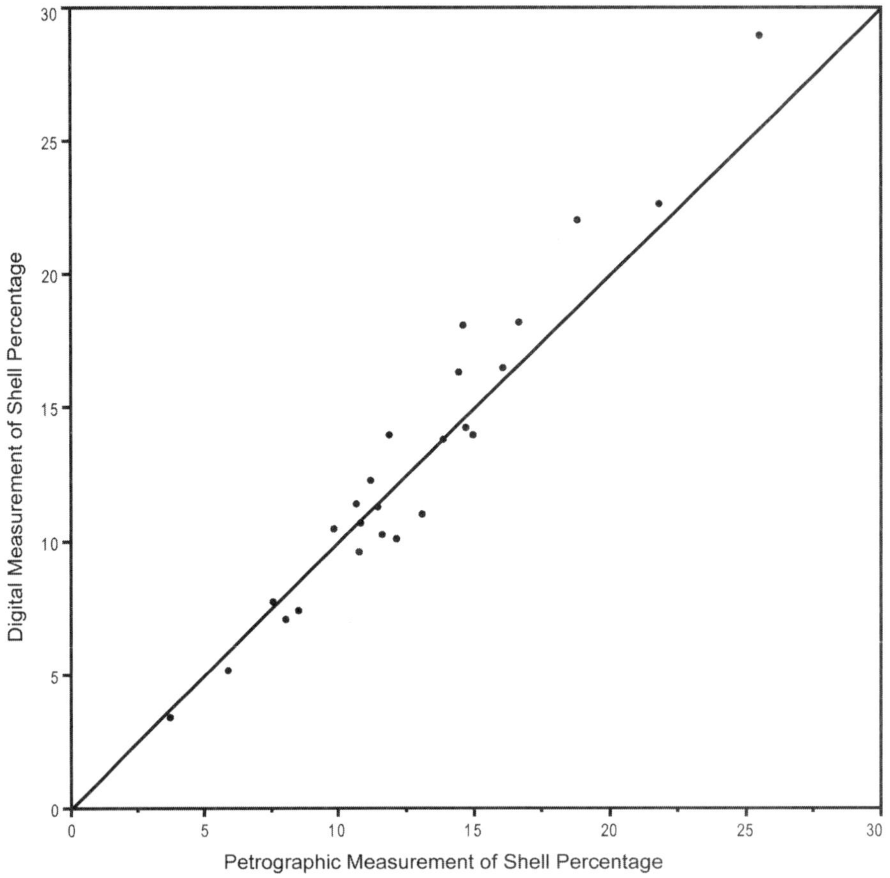

FIGURE 13.7. Bivariate plot of the measurement of shell percentage as obtained from petrographic analysis of a thin section and digital image analysis of a sherd face. The line indicates $r = 1$; points closer to the line show more exact agreement between the measurements.

few instances where the gap between the shell voids is very small and is effectively ignored by the algorithm, thus artificially inflating the measurement of shell percentage slightly.

A final concern was that voids in the matrix might be systematically miscategorized as shell or vice versa. The correlation between the percentage of voids in the petrographic samples and the discrepancy between the two methods is a moderate $r = 0.50$. This suggests that 25 percent of the differences between the digital and petrographic measurements can be accounted for by the void frequency in the sample as measured by the petrographic method. This is the largest potential source of identifiable error. While this may indicate a systematic problem, we are inclined to view this as a product of the somewhat subjective nature of categorizing voids as being the result of leached shell or of some other process. In support, the petrographic results showed a smaller correlation between void and shell frequencies ($r = 0.41$). While this may indicate that sherds with more shell also were more porous, it more likely points to the subjective nature of categorizing voids. The differences between the digital and petrographic analyses can therefore be seen as error that is inherent in the process.

Costs

The total estimated minimum cost per sample for the petrographic method is about $120, which includes about $20 for thin section

TABLE 13.8. Comparison of costs

Expense categories	Digital imaging analysis of temper in sherd edge	Petrographic point counts
Thin-section preparation	NA	about $20 per sample
Analysis charges	NA	$75–$150 per sample
Supplies	5–8 cents per sample for consumables, including sandpaper, polishing film, and hard drive space	NA
Equipment and software	~$22,000 for one-time expense of camera, computers, studio flash, ring flash, and software	a few hundred to several thousand dollars for purchase of microscope and mechanical stage, if not accessible

TABLE 13.9. Comparison of analyst time

Digital imaging analysis of temper in sherd edge	Petrographic point counts
4 minutes of relatively untrained labor (performed by students) to prepare each sherd, perform data entry, and capture images	60 minutes of analyst time per sample (depends on area of thin section, counting interval, and minimum number of counted points required)
10–15 minutes of computer processing per sample	
several dozen hours to prepare initial laboratory setup and create macros	

preparation and analysis charges of $75 to $150 per sample. The digital image analysis procedure here involves high fixed equipment costs of over $22,000, but low per-sample costs such as sandpaper, lapping paper, and hard drive space (5–8 cents per sample) plus labor costs of four minutes per sample (table 13.8).

Time Considerations

The time required to perform the digital image analysis also involves large fixed initial investments in time to refine the procedure and algorithms but low per-sample costs (table 13.9). It is extremely economical for large sample sizes, as in the Moon site assemblage, but would be prohibitive with small or idiosyncratic samples. Dozens of hours were spent in this instance assembling the equipment and creating the macros to do the analysis, but the per-sample analysis time includes just four minutes of data capture plus five–ten minutes of computer processing time. The digital analysis of sherd edges approach makes it possible to consider sample sizes in the hundreds or thousands.

The point-counting procedure required an average of sixty minutes of analysis time per sample, which is of course related to the size or area of the thin section, the counting interval, and minimum number of points or number of times that the thin section must be counted. An additional fifteen–thirty minutes per sample was expended to record other observations. Thus the total average analysis time was just over sixty minutes per sample. Neither set of totals includes time spent on data entry, tabulation, or write-up.

Conclusion

This discussion reveals several advantages to petrographic point-counting and petrographic compositional analysis (table 13.10). The established point-counting methods have acceptable accuracy, and petrographic compositional analysis draws upon over a century of geologic literature and decades of application to archaeological data. Therefore it is the preferred technique for analyzing smaller sample sizes, highly heterogeneous samples, or samples about which

TABLE 13.10. Comparison of advantages and disadvantages

Petrographic analysis of thin section	
Advantages	• proven standard of accuracy in microscopic compositional analyses • only technique suitable and appropriate for analyzing smaller, highly heterogeneous samples, or samples about which there is no prior knowledge • best technique for discriminating compositions of birefringent grains and for resolving silt and very fine particle sizes • especially adept at correct identification of similar-looking constituents • recommended as a precursor to a proposed digital image analysis study
Disadvantages	• may be more time-consuming • may be more expensive
Digital imaging analysis of thin section	
Advantages	• can provide many more measurements on particle size, shape, and location • much easier to share the steps and the results of the analysis with others because of its digital nature • may be faster under some circumstances
Disadvantages	• the resolution must be matched to the smallest and least visually distinct feature of interest (in this study the resolution selected was unable to resolve silt and very fine particle sizes of birefringent grains and compositions of certain tempers and other inclusions)
Digital imaging analysis of sherd edge	
Advantages	• the fastest and lowest-cost method
Disadvantages	• the technique is not appropriate to detect all types of inclusions or tempers

there is no prior knowledge. It is the best technique for discriminating compositions of birefringent grains and for resolving silt and very fine particle sizes. In this study it was the only method that can resolve clay resource differences among the samples.

For cases in which the temper has a high visual contrast from the matrix in terms of composition or shape, such as with shell and shell voids, there is an option to measure the temper on a prepared sherd edge. This is several orders of magnitude less expensive and faster than traditional petrographic point-counting. We can conceivably approach gathering data from hundreds of sherds using this method. Based on these results, it can measure the temper with acceptable accuracy. It has gathered data only on temper, however, and has not gathered data on any of the other inclusions, voids, or clay properties, which can sometimes be a very important part of the analysis. Although not discussed here, this digital analysis of sherd edge method has provided a rich set of information regarding shell metrics, including length, breadth, orientation, and others for every inclusion with no added analysis cost.

Unfortunately, for the southeastern United States there are only a few temper inclusions that lend themselves to being measured in the sherd edge alone. Carbonate inclusions such as limestone could be enhanced and mapped with the addition of a stain such as alizarine or hematoxylin. Fiber temper could be mapped as voids, and sponge spicules would be measurable with enough magnification or high enough resolution. The use of ultraviolet or infrared light may make it possible to analyze inclusions such as ash or bone (Verhoeven 2008). Rock fragments and coarse/very coarse inclusions will only be measurable when they have high visual contrast. It is unlikely that this method will ever be appropriate for the measurement of grog tempers, because grog is so often hard to distinguish from the matrix even under a microscope and is so variable in color and appearance when it is visible, which makes automated mapping challenging (Cuomo di Caprio and Vaughn 1993). Thin-section analysis may always be required for sand and any nonopaque and birefringent minerals. As stated in a previous paper (Livingood and Cordell 2009), for samples where thin-section analysis is required, we may still choose between traditional microscopic and digital analysis options. The advantages of each are considered in that paper.

In closing, we view these not as competing methods but rather as complementary tools in archaeology. Petrographic methods remain the tool with the broadest application and would be recommended as a precursor to any proposed digital analysis, having a much better capacity to permit fine distinctions. As discussed, digital image analysis of sherd edges may be preferable when the particles of interest are easily recognizable using automated methods and when the temper has high visual contrast. This study has demonstrated that the archaeologist now has the option to use digital image analysis to measure the temper directly without the need of a thin section for certain select tempers.

Acknowledgments

We are grateful for the opportunity to contribute to this volume. A shorter version of this paper was presented in the symposium "Petrography's Continued Role in Ceramic Studies: New Advances and Debates" at the 77th Annual meeting of the Society for American Archaeology. We wish to thank the editors of this volume for their thoughtful comments, which significantly improved the quality of our chapter, and the National Center for Preservation Technology and Training for funding this research. The Moon Site ceramic collection was on loan from the Arkansas Archaeological Survey and is housed there. The data collected during the digital image analysis are available from the senior author. The thin sections are curated at the FLMNH-CTL.

Notes

1. The calculations for computing sand and shell size indices differ from Stoltman's (2001:314) formula in that very fine grains are given a value of 0.5 instead of 1.
2. A complete list of measurements from the digital image analysis is on file with the authors.

References Cited

Benn, David W.
1998 A Fortified Mississippian Period Village in Poinsett County, Arkansas. In *Changing Perspectives on the Archaeology of the Central Mississippi Valley*, edited by Michael. J. O'Brien and Robert Dunnel, pp. 225–257. University of Alabama Press, Tuscaloosa.

Benn, David W. (ed.)
1992 *Excavations at the Moon Site (3PO488), a Middle Mississippian Village in Northeastern Arkansas.* Center for Archaeological Research, Report 780. Southwest Missouri State University, Springfield, Missouri.

Borremans, Nina T., and Graig D. Shaak
1986 A Preliminary Report on Investigation of Sponge Spicules in Florida "Chalky" Paste Pottery. *Ceramic Notes* 3:125–131.

Cuomo di Caprio, Ninina, and S. Vaughn
1993 An Experimental Study in Distinguishing Grog (Chamotte) from Argillaceous Inclusions in Ceramic Thin Sections. *Archeomaterials* 7(1):21–40.

Daniels, James, and Carl Lipo
2008 Digital Image Processing of Shell Temper Variability in Late Prehistoric Ceramics. Poster presented at the 73rd Annual Meeting of the Society for American Archaeology, Vancouver, British Columbia.

Galehouse, J.
1971 Point Counting. In *Procedures in Sedimentary Petrology*, edited by Robert E. Carver, pp. 385–408. Wiley Interscience, New York.

Gonzalez, Rafael C., and Richard E. Woods
2007 *Digital Image Processing*. 3rd ed. Prentice Hall, Harlow, New Jersey.

Graham, David J., and Nicholas G. Midgley
2000 Graphical Representation of Particle Shape Using Triangular Diagrams: An Excel Spreadsheet Method. *Technical Communication in Earth Surface Processes and Landforms* 25:1473–1478.

Hansen, Eric Floyd
2000 Ancient Maya Burnt-Lime Technology: Cultural Implications of Technological Styles. Ph.D. dissertation, Department of Anthropology, University of California, Los Angeles.

Livingood, Patrick
2007 Plaquemine Recipes: Using Computer-Assisted Petrographic Analysis to Investigate Plaquemine Ceramic Recipes. In *Plaquemine Archaeology*, edited by Mark Rees and Patrick Livingood, pp. 108–126. University of Alabama Press, Tuscaloosa.
2010 *Mississippian Polity and Politics on the Gulf Coastal Plain: A View from the Pearl River, Mississippi*. University of Alabama Press, Tuscaloosa.

Livingood, Patrick C., and Ann S. Cordell
2009 Point/Counter Point: The Accuracy and Feasibility of Digital Image Techniques in the Analysis of Ceramic Thin Sections. *Journal of Archaeological Science* 36:867–872.

Pereira, Carla
2012 Stylistic Variability within Assemblages of Late Prehistoric Pottery in Lower Mississippi River Valley. Poster presented at the 77th Annual Meeting of the Society for American Archaeology, Memphis, Tennessee.

Rapp, George, Jr., and Susan C. Mulholland
1992 *Phytolith Systematics: Emerging Issues*. Advances in Archaeological and Museum Science Vol. 1. Plenum Press, New York.

Reedy, Chandra L.
2006 Review of Digital Image Analysis of Petrographic Thin Sections in Conservation Research. *Journal of the American Institute for Conservation* 45(2):127–146.

Rice, Prudence M.
1987 *Pottery Analysis: A Sourcebook*. University of Chicago Press, Chicago.

Round, Frank E., Raymond M. Crawford, and David G. Mann
2007 *The Diatoms: Biology and Morphology of the Genera*. Cambridge University Press, Cambridge.

Russ, John C.
2006 *The Image Processing Handbook*. 4th ed. CRC Press, Boca Raton, Florida.

SAS Institute, Inc.
2008 SAS 9.2 for Windows. SAS Institute, Cary, North Carolina.

Stoltman, James B.
1989 A Quantitative Approach to the Petrographic Analysis of Ceramic Thin Sections. *American Antiquity* 54(1):147–160.
1991 Ceramic Petrography as a Technique for Documenting Cultural Interaction: An Example from the Upper Mississippi Valley. *American Antiquity* 56(1):103–121.
2001 The Role of Petrography in the Study of Archaeological Ceramics. In *Earth Sciences and Archaeology*, edited by Paul Goldberg, Vance T. Holliday, and C. Reid Ferring, pp. 297–326. Kluwer Academic/Plenum Publishers, New York.

Velde, Bruce, and Isabelle C. Druc
1999 *Archaeological Ceramic Materials: Origin and Utilization*. Springer-Verlag, Berlin.

Verhoeven, Geert
2008 Imaging the Invisible Using Modified Digital Still Cameras for Straightforward and Low-Cost Archaeological Near-Infrared Photography. *Journal of Archaeological Science* 35(12):3087–3100.

Whitbread, I. K.
1991 Image and Data processing in Ceramic Petrology. In *Recent Developments in Ceramic Petrology*, edited by Andrew P. Middleton and Ian C. Freestone, pp. 369–386. Occasional Paper No. 81. British Museum, London.

CHAPTER 14

Ceramic Petrography

Integration, Adaptation, and Innovation

Ian K. Whitbread

Ceramic petrography plays a pivotal role in technological and provenance studies of archaeological ceramics. It bridges fieldwork and laboratory contexts, from hand specimen examination with a hand lens or stereomicroscope to thin-section analysis using a polarizing microscope. As part of a range of techniques used to analyze ceramic materials it can be argued that petrography has at times been underutilized in favor of instrumental analyses (Ownby and Heidke, this volume). The technique has never been fully replaced, however. While recent developments in instrumental mineralogical analysis hold considerable potential they are an enhancement rather than a substitute (Knappett et al. 2011).

An earlier collection of papers illustrates how ceramic petrography has contributed to recent investigations of technology and exchange networks (Quinn 2009). The case studies presented here build on this foundation but significantly advance the debate by exploring the vital role of ceramic petrography as an integral component within a flexible package of analytical methods that can be applied to current archaeological problems. This chapter examines some of these issues, drawing out the strengths and weaknesses of ceramic petrography and exploring some of the common themes raised in this volume for advancing our understanding of past societies and their social organization through ceramic technology and exchange.

Integrating Petrography and Chemistry

Integrating ceramic petrography with chemical analysis has been advocated since Shepard's pioneering work in the 1930s (Thompson 1991), because the two approaches are complementary in their strengths and weaknesses. The principal contributions of each can be summarized in a brief overview. Petrography is most effective at analyzing coarse-grained ceramic fabrics, typically those of utilitarian pottery and cooking ware, as it focuses on the compositions of rock and mineral inclusions and fabric micromorphology. Petrographers draw on a range of supporting materials, including reference ceramics, geological resources (fieldwork, maps and literature), and ethnographic and experimental studies, in addressing questions of provenance and technology (Quinn 2013:153–203). In contrast, the ceramic material is normally converted to powder during sample preparation for bulk chemical analysis such as ICP, WDXRF (wavelength dispersive X-ray fluorescence), or NAA (Pollard et al. 2007:58, 107, 128), thereby removing morphological features. The value of the method, however, lies in its measuring a wide range of elemental concentrations with high precision, and most significantly those of the provenance-sensitive trace elements. It is therefore very effective at characterizing fine-grained ceramics but less so for materials containing numerous trace element bearing inclusions (Day et al. 1999; Mommsen 2004). Bulk chemical analysis is mainly used for provenance determination by establishing the presence or absence of elemental matches with reference ceramics of known origin. It is also valuable in technological studies for comparing raw clays with ceramics and seeking evidence of clay levigation (Kilikoglou et al. 1988; Storey 1988).

Until the 1990s materials analyses of archaeological ceramics were strongly focused, with some notable exceptions (Beck 1987), on issues of exchange and trade using provenance determination. This situation is illustrated in Jones's (1986) major review of Greek and Cypriot ceramic analyses. The importance of ceramic technology is certainly recognized, as is the role of petrography for analyzing coarse pottery, but most analyses in the review addressed issues of provenance and generally favored finer pottery for which well-established styles of decoration and shape could provide a chronological and distribution framework that was scientifically testable. As noted above, bulk chemical analysis is best suited to this type of material and research question. Its application was further aided by the rapid development of computers and statistical programs that could be used to process and interpret the considerable quantities of numerical data being generated.

A paradigm shift developed in ceramic materials studies during the 1990s, when implications arising from the socially embedded nature of technology became more fully realized (for example, Loney 2000; Pfaffenberger 1988; Sillar and Tite 2000; Stark 1998; Whitbread 2001); that is, technology carries significant potential for investigating people, not just pots. Exchange and trade remained important issues, but ceramic technology could contribute directly to wider studies of social organization in the past, such as defining identities and social boundaries. This potential was clearly demonstrated in ethnographic studies, as in the correlation of language groups, learning networks, and methods of pottery production in southern Cameroon (Gosselain 1998). More importantly for petrography, this reorientation extended to the subject matter. Utilitarian ceramics were rarely the focus of provenance determination, but as artifacts of daily use they became directly relevant to social analysis. Through these developments, more than in any previous period, petrography plays a key role given its strengths in both provenance and technological studies.

Increasing interest in technology has also impacted the application of chemical analysis to ceramics. Spatially resolved techniques such as the electron microprobe (Pollard et al. 2007: 109–113) trade the sensitivity of bulk chemical methods for the ability to characterize specific locations within ceramic microstructures, opening new avenues to technological and provenance studies. For example, the technique was used to measure major element concentrations of glass shards in Maya ceramics from El Pilar (Ford et al., this volume). Masucci et al. (this volume) report on its use to identify compositional distinctions between pumice fragments in Guangala finewares from coastal Ecuador, possibly indicating a change in sources. Spataro and Villing (2009) similarly combined petrography and SEM-EDX (scanning electron microscopy with energy dispersive X-ray spectroscopy) to characterize and determine the provenance of mortaria (Roman grinding/mixing bowls) in the eastern Mediterranean. More recent developments such as bench-top micro-XRF (μ-XRF) for spatially resolved analyses and portable XRF (pXRF) (Stovel et al., this volume) have brought chemical analysis into the small laboratory and into the field, where thin section and macroscopic methods were previously favored for reasons of cost and maintenance. Archaeological research with these new instruments requires careful evaluation and development (Speakman and Shackley 2013), but they are eroding long-established boundaries between laboratory and field analyses and generating new opportunities for integrating petrography and chemical analysis.

Provenance determination by both petrography and chemical analysis relies heavily on the quality of ceramic reference materials of known origin. These incorporate the technological traits of clay body preparation and firing that geological samples lack. The most secure reference ceramics are those that are unequivocally associated with ceramic production at a specific location. Without this evidence the origin of reference ceramics is generally based on assumptions, such as similarities in their styles of decoration or shape to ceramics of known origin, the predominance of similar ceramics within specific assemblages, or a supposition that objects of a particular type would not have been exchanged. In the case of petrographic analysis, the mineralogical composition of inclusions can also be used to test potential sources of ceramics against regional geology. Predicting likely ceramic sources based on elemental concentrations of clays is extremely difficult and largely restricted to reviewing previous analyses. In part this capability of petrography prompted the research of Burke et al. (this volume) because the principal NAA study of Early Bronze Age ceramics in southern mainland Greece was unable to source chemical groups that did not match the reference material or to differentiate between technological and provenance-related traits in the ceramics. Ownby's examination of Tanque Verde Red-on-Brown pottery exchange in the Tucson Basin (this volume) also used petrography to verify NAA results by generating reference materials for provenance determination. In this case she combined ceramic petrography with previous extensive petrofacies modeling in the region to characterize raw materials in specific locations. These studies show how petrography may assist chemical provenance determination for ceramics containing at least some distinctive inclusions, compensating in part for deficiencies in the reliability of reference materials. Indeed, Ford et al. (this volume) advocate using petrography to assess ceramic materials at the outset of integrated studies, thereby enabling chemical analyses to be more effectively targeted.

From Assemblage to Laboratory and Back: Macroscopic and Microscopic Analyses

Several chapters in this volume draw attention to the essential role of macroscopic analysis for studying ceramic fabrics in hand specimen. Heidke (this volume) in particular makes a strong case for using this approach, but he notes that it has become less common with developments in methods of instrumental analysis. Hand specimen studies were central to Shepard's research (Heidke, this volume; Shepard 1956; Thompson 1991) and later echoed by Peacock (1977), whose methodology is incorporated within the freely available guidelines for analyzing and publishing prehistoric ceramics in the United Kingdom (Prehistoric Ceramics Research Group 2010). It is the most effective method for processing the large quantities of ceramics accumulated by major field projects, achieving adequate discrimination with relatively rapid assessments but relying on minimal equipment.

Macroscopic analysis is commonly used in selecting ceramic samples for laboratory examination. The more powerful chemical and petrographic methods that require these facilities are indispensable when addressing many archaeological questions, but they can usually process only relatively small numbers of samples. This means that sampling rarely corresponds to the proportions of different wares within an assemblage. More often it is aimed at representing their range, ideally with enough samples to determine the degree of fabric variation within each ware. If groups resulting from laboratory analyses were not evident in the preliminary macroscopic sorting, then a concerted effort is necessary to extrapolate these new results back to the assemblage. This procedure is necessary if the representation of laboratory-defined groups is to be ascertained at the assemblage level as a basis for wider archaeological inference. Given the size of many ceramic assemblages, methods of macroscopic analysis still offer the best solution to this problem.

Hand specimen examination can present several difficulties. Published macroscopic descriptions do not always make it clear whether inclusion identifications were based purely on hand specimen observation or derived in part from thin-section analysis. The latter can give a false impression of the resolving power and reliability of hand specimen study if it is not declared. All analytical methods require a period of training and accumulated experience, but formal training in hand specimen examination of rocks and minerals is rare (Dietrich and Skinner 1979). Grain boxes with representative rock and mineral fragments (Heidke, this volume) or high-quality macrophotographs with accompanying explanations are key investments for maintaining accuracy and precision in macroscopic studies.

It is productive to have macroscopic groups verified by thin-section analysis (Heidke, this volume). This is an aim of the study of Early Bronze Age ceramics in southern mainland Greece by Burke et al. (this volume). The andesite ware that they discuss is a good example of what can be achieved. It probably originated from the island of Aegina, off the coast of Athens. Aeginitan ceramics, notably cookingware, were widely exported to the Greek mainland and Aegean islands during the Bronze Age and further afield in the Classical period. These ceramics have recently been the subject of a substantial and detailed laboratory analysis (thin section, chemical, and SEM) that also incorporated a conscious effort to include macroscopic descriptions to aid field archaeologists in identifying the occurrence and distribution of Aeginetan ceramics elsewhere in the Mediterranean region (Gauss and Kiriatzi 2011:47–67).

One way to encourage use of macroscopic analysis is to explore ways of enhancing its application. Livingood and Cordell (this volume) had mixed success in applying digital image analysis to sherd edges, but their research both scrutinizes the complex nature of macroscopic analysis and, more importantly, identifies conditions under which digital image analysis could be developed into a valuable asset. Their work highlights the limited capabilities of digital image analysis, compared to the human eye and brain, in observing and interpreting visual information. As they conclude, the way forward probably lies in developing applications that exploit its strengths, such as using high-contrast features to characterize fabrics, rather than attempting to replicate traditional macroscopic observations. Heidke (this volume) takes a different approach, using macroscopic analysis with thin-section verification and statistical data processing to show how the large quantities of ceramics held in the Hohokam assemblage at Honey Bee Village could be rigorously investigated in terms of functional analysis. Innovative approaches such as these encourage reevaluation of well-established techniques such as macroscopic analysis and in so doing enhance the potential for integrating petrography with developing methods of field-based instrumental analysis.

Petrographic Description and Characterization

Description (the information recorded during an analysis) is the foundation for all that follows. For chemical analysis it is expressed in terms of elemental detection limits, accuracy, and precision as defined by the particular instrument in use and subject to the experience of analysts operating it. In the case of petrography, be it hand specimen or thin section examination, description comes directly from the analysts' interpretation of what they see and is dependent to a greater extent on their observational skills, training, and experience. This can lead to inconsistencies between petrographers and within the work of individual petrographers as they gain experience or encounter unfamiliar materials. But it also opens the way to adaptation and innovation in the way ceramic fabrics are examined.

Peacock (1977) likened ceramics fabrics to metamorphosed sedimentary rocks, but sedimentary petrography is aimed at coarse-grained rather than clay-based materials. Ceramic petrography is therefore better served by combining the methods of sedimentary petrography with soil micromorphology, while seeking to take account of features specific to ceramics, such as clay mixing and firing effects (Josephs 2005; Quinn 2013:71–116; Whitbread 1989, 1995: 365–396). One benefit of integrating soil micromorphology with ceramic petrography is evident in Kelly's analysis of argillaceous inclusions found in Hohokam decorated Red-on-buff ceramics of the Phoenix Basin, Arizona (this volume). The method of describing argillaceous inclusions in ceramics was developed in part because of difficulties in discriminating among grog, mudstone, clay pellets, and clay tempers

(Whitbread 1986). It was also because ceramic petrographers routinely appeared to ignore clay pellets as a potential source of information. Kelly was unable to establish a definitive origin for the argillaceous inclusions in red-on-buff ceramics using the descriptive method. She nevertheless exploited its real value by using the terminology as an exploratory tool with which to isolate and meticulously examine the critical properties of these inclusions. In so doing she was able to develop a range of technological models for the clay pellets/temper (whether they were naturally present or introduced through clay mixing) that can be tested in the future using alternative analytical techniques.

There are many ways of describing ceramic fabrics. It is of course essential that core properties are recorded so that results from different research projects can be used to build on previous analyses, as is well illustrated in Stovel et al.'s study (this volume), which integrates results from the two petrographic studies of San Pedro de Atacama ceramics by Cremonte and Echenique. Nevertheless, the diverse ways of recording and analyzing petrographic information presented in this volume (for example, by Heidke, Ownby, Stoltman, Kelly, Tonoike, Livingood, and Cordell) demonstrate the value of flexibility in the method. Its adaptation to different types of ceramic materials and regional contexts and to diverse research questions on provenance and technology encourages innovation and enhances integration with other techniques.

Technology: Petrographic Methods and Interpretations

Selecting and processing raw materials is fundamental to ceramic production and likewise to studies of past ceramic technologies. Potters evaluate several physical attributes when choosing the raw materials that they consider appropriate for particular types of ceramic. Some critical properties, such as plasticity, quantities and sorting of inclusions, and color, may be lost or significantly altered in the ceramic end product as a result of materials processing and firing. Consequently, locating and matching potential raw materials with ceramic fabrics is a complex process, but erroneous identifications can lead to unreliable interpretations of technological choices, procedures, and results.

Some regions appear to have an abundance of potential raw materials, but this raises questions as to whether these were used for ceramic production and if so how? Ancient Corinth, Greece, was a significant center of ceramic production throughout much of antiquity. Its regional geology, dominated by marl (calcareous clay) deposits, might appear to be rich in clay, but much of this material is unsuitable for ceramics without very careful clay selection and firing (Farnsworth et al. 1977; Whitbread 1995: 308–329). Likewise, loess forms extensive deposits in China and Europe, so its exploitation for ceramic production could easily be taken for granted. Stoltman (this volume) has shown that potters in widely separated locations appeared to draw similar conclusions as to its technological viability and limitations. Moreover, in not using loess for cookingware, they had to seek alternative materials in a landscape dominated by loess. In this respect, Stoltman shows how assessing the technological choices open to potters in selecting materials relies almost as much on finding out what they did not exploit as in identifying the materials that they used.

A comparable situation was encountered by Druc et al. (this volume), who found that despite extensive local volcanic deposits a third of the ceramic material that they studied at Kuntur Wasi, Peru, was tempered with intrusive igneous inclusions, with the nearest occurrence being 20–25 km from the site. A critical part of their investigation was recognizing that raw materials did not originate directly from the solid geology but rather were detrital sediments derived from various outcrops. Their chance discovery of a potential source (a soil with compacted, weathered intrusive material) only 8 km from the site shows that there is no substitute for fieldwork, an open mind, and a degree of serendipity in locating potential raw materials. Morrison and Horowitz (this volume) further demonstrate the importance of field prospection and replication experiments for developing insights into potters' decisions during materials selection and manipulation. Their search for ferruginous clays and calcite used in Middle Bronze Age cooking pots shows how distinctive fabrics in thin section can nevertheless be challenging to reconstruct through raw materials prospection. Methods of processing contribute to this problem by masking the presence and character of the original constituents as they appear in the field. In this case, the clays in Limenaria Cove are coarser than the ceramic fabrics, whereas mussel shells used for temper at Atchana may have been altered through heating before they were added as temper.

Both Druc et al. and Ford et al. (this volume) used ceramic petrography integrated with a range of other techniques to identify temporal changes in ceramic materials. Specific reasons behind these changes are unclear, but recognizing that a shift took place has to be the starting point from which strategies can be developed to look for connections with wider social events. Interpreting continuity of ceramic technology during periods of major social change (for example, Masucci et al.; Stovel et al.; and Tonoike, this volume) is especially challenging. In some cases, continued use of raw materials and processing methods may be attributable to a narrow range of lithologies, such as quartz-rich sediments, dominating the regional geology. It could equally result from day-to-day activities, such as ceramic production being largely unaffected by wider social events. Tonoike's pilot study at Tell Ziyadeh, northern Syria, found few significant differences in materials and manufacturing techniques between the culturally distinct Halaf and Ubaid periods. Certainly some sites, regions, or ceramic materials will not be productive for ceramic petrography owing to the absence of distinctive coarse inclusions, but in other cases important distinctions may be subtle and require different perspectives to isolate meaningful information. From a combination of ceramic petrology and NAA, Masucci et al. were able to conclude that black polished fineware, diagnostic of the Manteño phase of the Integration Period in coastal southern and central Ecuador, continued a long-established tradition in the use of ceramic materials. But enough compositional distinctions were isolated using the two techniques to indicate that products of different centers circulated within the region, possibly as a result of new ritual practices accompanying social change and settlement reorganization.

Hollenback's research (this volume) shows that a lack of change in ceramic materials can in itself constitute a major social statement during periods of severe disruption. She argues that the lack of pronounced changes in raw materials preparation following the 1782 Hidatsa epidemic in North Dakota may represent a concerted effort to maintain traditions in the face of an overwhelming event. A reduction in time taken to prepare ceramic materials after the 1837 outbreak, however, might have been a response to social stress but could in part be a reaction to the wider availability of Euroamerican goods. In other words, changes in the selection and preparation of ceramic materials may be the outcome of several compounded factors.

Programs of materials analysis are usually aimed at finding evidence of variation that may represent technological diversity or change. Hence results that show no significant differences between samples can be viewed negatively. Certainly this response is justified on some occasions, although negative results are inevitable at some point in research. The projects discussed above show, however, that significant differences might be subtle and require a particular combination of techniques and approaches in order to extract useful information. In other cases, continuity carries greater meaning than variation, but this can only be realized by integrating ceramic petrography with wider debates on cultural history and social organization.

Future Research Directions for Ceramic Petrography

As many of the chapters in this volume demonstrate, ceramic petrography can be adapted to new research questions and contexts. It continues to be the principal technique for mineralogical provenance determination, so the following discussion focuses on future developments in studying ceramic technology, processes of petrographic analysis, and integration with complementary analytical techniques.

Ceramic petrographers have long used inclusion compositions and sizes, fabric micromorphologies, and heterogeneities to characterize technological traits such as discriminating between temper (aplastic material intentionally added by the potter) and natural inclusions, identifying evidence of clay refining by sieving or levigation, noting potential clay mixing, and detecting traces of forming techniques. Nevertheless, interpretations can still rely on somewhat rudimentary explanations introduced when petrography was primarily aimed at provenance determination. Now the use of technological properties in defining social relationships creates a much greater need for justifying and testing these interpretations. For example, the presence of prominent angular inclusions in a ceramic fabric could be interpreted as evidence that they were crushed and added as temper. Such distinctive angular inclusions of chert dominate fabrics of Late Roman and Late Antique pottery in the Berbati Valley, Greece. Petrographic analysis of potential raw materials, however, has revealed the natural occurrence of similar inclusions in local Quaternary soils that chemically match the pottery (Whitbread et al. 2007; Whitbread 2011). In this case extrapolating any social interpretation based on assumed tempering practices would be incorrect. Degrees of uncertainty exist in all interpretations, of course, but developing approaches to critically evaluating technological evidence (for example, Miksa 2001) will enhance the future role of ceramic petrography in this area. Where it is feasible, prospecting for potential raw materials is an important part of this process. But no less important is the considerable potential for experimental ceramic petrography, using fabric properties generated under controlled conditions as a means of investigating their origins and technological significance.

One of the great strengths of ceramic petrography is the evidence that it provides on micromorphology. Digital image analysis therefore holds significant potential as a tool for supporting petrographic analysis (for example, Bouchain and Velde 2001), but its application is clearly challenging in both hand specimen and thin section (Livingood and Cordell, this volume; Livingood and Cordell 2009; Reedy 2006). Methods of image segmentation based on automated mineralogical and elemental maps (Knappett et al. 2011; Dal Sasso et al. 2014) offer petrography alternative routes for characterizing fabric micromorphologies, but they are restricted in the numbers of samples that can be analyzed and rely on sophisticated laboratory facilities and analytical expertise. In contrast, it is the ease with which large numbers of petrographic samples can now be recorded that enables digital images to be used in supporting routine visual assessments such as fabric sorting and comparison. The challenge therefore lies in developing approaches to digital image analysis that can enhance such processes with minimal impact on the time and cost of analysis.

Recent developments in the availability of instruments for elemental analysis that are portable or sustainable in smaller laboratories (for example, pXRF and μ-XRF) are introducing a shift toward greater integration of materials analysis within archaeological fieldwork, where petrography has traditionally been a key analytical method. There are clear advantages in enabling elemental analyses to be undertaken on ceramics not otherwise permitted for export to major laboratories, for potentially characterizing greater proportions of ceramic assemblages, and in evaluating likely raw materials during field collection. Numerous issues must be addressed in such developments, however, not least in establishing robust operating principles, maintaining data reliability, and clarifying the limitations of such instruments with respect to the continuing need for advanced laboratory facilities. Petrography needs to be developed as part of this process, recording visible properties not reflected in elemental analyses and adapted toward better integrating geological evidence during fieldwork.

Conclusions

The discussion above is a personal perspective on the range of case studies presented in this volume. Any collection of papers that focus on the application of a specific analytical technique runs the risk of being repetitive: same technique, different time and place. These chapters do indeed raise a number of overlapping themes. But more importantly they highlight the versatility of ceramic petrography as it is integrated into different research contexts, each with its own specific aims and constraints. Taken together they are a testament to strengths of ceramic petrography as an analytical technique that is open to integration, adaptation, and innovation.

Acknowledgments

I thank the editors for their invitation to contribute this concluding chapter to the volume.

References Cited

Beck, Curt W.
1987 Textbooks(?) of Archaeometry. *Journal of Field Archaeology* 14:487–490.
Bouchain, Isabella, and Bruce Velde
2001 Grain Distribution by Image Analysis of Thin Sections in some Gaulo-Roman Common Ware, St Marcel (Indre) France. In *Archaeology and Clays*, edited by Isabelle C. Druc, pp. 71–80. BAR International Series 942. British Archaeological Reports, Oxford.
Dal Sasso, Gregorio, Lara Maritan, Sandro Salvatori, Claudio Mazzoli, and Gilberto Artioli
2014 Discriminating Pottery Production by Image Analysis: A Case Study of Mesolithic and Neolithic Pottery from Al Khiday (Khartoum, Sudan). *Journal of Archaeological Science* 46:125–143.
Day, Peter M., Evangelia Kiriatzi, Alexandra Tsolakidou, and Vassilis Kilikoglou
1999 Group Therapy in Crete: A Comparison between Analyses by NAA and Thin Section Petrography of Early Minoan Pottery. *Journal of Archaeological Science* 26:1025–1036.
Dietrich, Ricard V., and Brian J. Skinner
1979 *Rocks and Rock Minerals*. John Wiley and Sons, New York.
Farnsworth, Marie, Isadore Perlman, and Frank Asaro
1977 Corinth and Corfu: A Neutron Activation Analysis of their Pottery. *American Journal of Archaeology* 81:455–468.
Gauss, Walter, and Evangelia Kiriatzi
2011 *Pottery Production and Supply at Bronze Age Kolonna, Aegina: An Integrated Archaeological and Scientific Study of a Ceramic Landscape*. Ägina-Kolonna. Forschungen und Ergebnisse, Band 5. Österreichischen Akademie der Wissenschaften, Vienna.
Gosselain, Olivier P.
1998 Social and Technical Identity in a Clay Crystal Ball. In *The Archaeology of Social Boundaries*, edited by Miriam T. Stark, pp. 78–106. Smithsonian Institution Press, Washington, D.C.
Jones, Richard E.
1986 *Greek and Cypriot Pottery: A Review of Scientific Studies*. Fitch Laboratory Occasional Paper 1. British School at Athens, London.
Josephs, Richard L.
2005 Short Contribution: Applying Micromorphological Terminology to Ceramic Petrography. *Geoarchaeology* 20:861–865.
Kilikoglou, Vassilis, Yannis Maniatis, and Apostolos P. Grimanis
1988 The Effect of Purification and Firing of Clays on Trace Element Provenance Studies. *Archaeometry* 30:37–46.
Knappett, Carl, Duncan Pirrie, Matthew R. Power, Irene Nikolakopoulou, Jillian Hilditch, and Gavyn K. Rollinson
2011 Mineralogical Analysis and Provenancing of Ancient Ceramics Using Automated SEM-EDS Analysis (QEMSCAN®): A Pilot Study on LB I Pottery from Akrotiri, Thera. *Journal of Archaeological Science* 38:219–232.
Livingood, Patrick C., and Ann S. Cordell
2009 Point/Counter Point: The Accuracy and Feasibility of Digital Image Techniques in the Analysis of Ceramic Thin Sections. *Journal of Archaeological Science* 36:867–872.
Loney, Helen L.
2000 Society and Technological Control: A Critical Review of Models of Technological Change in Ceramic Studies. *American Antiquity* 65:646–668.
Miksa, Elizabeth J.
2001 Criteria for Evaluating Multiple Components in Pottery Paste. In *Archaeology and Clays*, edited by Isabelle C. Druc, pp. 81–93. BAR International Series 942. British Archaeological Reports, Oxford.
Mommsen, Hans
2004 Short Note: Provenancing of Pottery: The Need for an Integrated Approach? *Archaeometry* 46:267–271.

Peacock, David P. S.
1977 Ceramics in Roman and Medieval Archaeology. In *Pottery and Early Commerce: Characterization and Trade in Roman and Later Ceramics*, edited by David P. S. Peacock, pp. 21–33. Academic Press, London.

Pfaffenberger, Bryan
1988 Fetishised Objects and Humanised Nature: Towards an Anthropology of Technology. *Man* 23:236–252.

Pollard, A. Mark, Cathy M. Batt, Ben Stern, and Suzanne M. M. Young
2007 *Analytical Chemistry in Archaeology*. Cambridge University Press, Cambridge.

Prehistoric Ceramics Research Group
2010 *The Study of Prehistoric Pottery: General Policies and Guidelines for Analysis and Publication*. 3rd ed. Occasional Papers Nos. 1 and 2. Prehistoric Ceramics Research Group, Oxford.

Quinn, Patrick S.
2013 *Ceramic Petrography: The Interpretation of Archaeological Pottery and Related Artefacts in Thin Section*. Archaeopress, London.

Quinn, Patrick S. (ed.)
2009 *Interpreting Silent Artefacts: Petrographic Approaches to Archaeological Ceramics*. Archaeopress, London.

Reedy, Chandra L.
2006 Review of Digital Image Analysis of Petrographic Thin Sections in Conservation Research. *Journal of the American Institute for Conservation* 45:127–146.

Shepard, Anna O.
1956 *Ceramics for the Archaeologist*. Carnegie Institution, Washington, D.C.

Sillar, Bill, and Michael S. Tite
2000 The Challenge of "Technological Choices" for Materials Science Approaches in Archaeology. *Archaeometry* 42(1):2–20.

Spataro, Michela, and Alexandra Villing
2009 Scientific Investigations of Pottery Grinding Bowl from the Archaic and Classical Eastern Mediterranean. *British Museum Technical Research Bulletin* 3:89–100.

Speakman, Robert J., and M. Steven Shackley
2013 Silo Science and Portable XRF in Archaeology: A Response to Frahm. *Journal of Archaeological Science* 40:1435–1443.

Stark, Miriam T. (ed.)
1998 *The Archaeology of Social Boundaries*. Smithsonian Institution Press, Washington, D.C.

Storey, Jeremy M. V.
1988 A Chemical Study of Clays and Roman Pottery from the Lower Nene Valley, Eastern England. *Journal of Archaeological Science* 15:35–50.

Thompson, Raymond H.
1991 Shepard, Kidder, and Carnegie. In *The Ceramic Legacy of Anna O. Shepard*, edited by Ronald L. Bishop and Frederick W. Lange, pp. 11–41. University Press of Colorado, Niwot.

Whitbread, Ian K.
1986 The Characterisation of Argillaceous Inclusions in Ceramic Thin Sections. *Archaeometry* 28(1):79–88.
1989 A Proposal for the Systematic Description of Thin Sections towards the Study of Ancient Ceramic Technology. In *Archaeometry: Proceedings of the 25th International Symposium*, edited by Yannis Maniatis, pp. 127–138. Elsevier, Amsterdam.
1995 *Greek Transport Amphorae: A Petrological and Archaeological Study*. Fitch Laboratory Occasional Paper 4. British School at Athens, Oxford.
2001 Ceramic Petrology, Clay Geochemistry and Ceramic Production: From Technology to the Mind of the Potter. In *Handbook of Archaeological Sciences*, edited by Don R. Brothwell and A. Mark Pollard, pp. 449–459. John Wiley and Sons, London.
2011 Petrographic Analysis of Ceramics from the Berbati Valley. In *Mastos in the Berbati Valley*, edited by Michael Lindblom and Berit Wells, pp. 143–176. Acta Instituti Atheniensis Regni Sueciae, Series 4, 54. Svenska Institutet i Athen, Stockholm.

Whitbread, Ian K., Matthew J. Ponting, and Berit Wells
2007 Temporal Patterns in Ceramic Production in the Berbati Valley, Greece. *Journal of Field Archaeology* 32:177–193.

Contributors

Clare Burke
University of Sheffield
Department of Archaeology
Sheffield, UK

Víctor Carlotto Caillaux
Universidad Nacional
San Antonio
Abad del Cusco, Peru

Connie Christensen
Ventura Community College
Geoscience Department
Ventura, CA

Pedro Navarro Colque
INGEMMET (Instituto Geológico Minero y Metalúrgico)
Lima, Peru

Ann Cordell
Florida Museum of Natural History
Gainesville, FL

Beatriz Cremonte
CONICET, Universidad Nacional de Jujuy
San Salvador de Jujuy, Argentina

Peter Day
University of Sheffield
Department of Archaeology
Sheffield, UK

Isabelle Druc
University Wisconsin
Department of Anthropology
Madison, WI

Ester Echenique
University of Arizona
School of Anthropology
Tucson, AZ

Anabel Ford
University of California
ISBER/MesoAmerican Research Center
Santa Barbara, CA

Michael Glascock
University of Missouri Research Reactor
Columbia, MO

Jim Heidke
Desert Archaeology, Inc.
Tucson, AZ

Kacy Hollenback
Southern Methodist University
Department of Anthropology
Dallas, TX

Mara Horowitz
Fordham University
Department of Sociology and Anthropology
Danbury, CT

Kinya Inokuchi
Saitama University
Faculty of Liberal Arts
Saitama, Japan

Sophia Kelly
U.S. Department of State
AAAS Science & Technology Policy Fellow
Washington, DC

Patrick Livingood
University of Oklahoma
Department of Anthropology
Norman, OK

Maria Masucci
Department of Anthropology
Drew University
Madison, NJ

Jerolyn Morrison
University of Leicester
School of Archaeology and Ancient History
Leicester, UK

Hector Neff
California State University, Long Beach
Department of Anthropology
Long Beach, CA

Mary Ownby
Desert Archaeology, Inc.
Tucson, AZ

Daniel Pullen
Florida State University
Department of Classics
Tallahassee, FL

Jeff Speakman
University of Georgia
Center for Applied Isotope Studies
Athens, GA

Frank Spera
University of California
Department of Earth Science & Earth Research Institute
Santa Barbara, CA

James Stoltman
University of Wisconsin
Department of Anthropology
Madison, WI

Emily Stovel
Instituto de Investigación Arqueológica y Museo Gustavo Le Paige
Universidad del Norte
San Pedro de Atacama, Chile

Yukiko Tonoike
Yale University
Department of Anthropology
New Haven, CT

Ian Whitbread
University of Leicester
School of Archaeology and Ancient History
Leicester, UK

Index